Microbes in Extreme Environments

Special Publications of the Society for General Microbiology

A complete list of books in this series appears at the end of the volume.

This book is based on a Symposium of the SGM held at Dundee, June 1984.

Microbes in Extreme Environments

Edited by

R. A Herbert and **G. A. Codd**

Department of Biological Sciences,
University of Dundee,
Dundee, UK

1986

Published for the

Society for General Microbiology

by

ACADEMIC PRESS

Harcourt Brace Jovanovich, Publishers

London   Orlando   San Diego   New York   Austin
Boston   Sydney   Tokyo   Toronto

ACADEMIC PRESS INC. (LONDON) LTD.
24–28 Oval Road
London

U.S. Edition published by
ACADEMIC PRESS INC.
Orlando, Florida 32887

British Library Cataloguing in Publication Data

Microbes in extreme environments.—(Special
  publications of the Society for General
  Microbiology; no. 17)
  1. Micro-organisms—Physiology
  2. Microbial ecology   3. Adaptation
  (Physiology)
  I. Herbert, R. A.   II. Codd, G. A.
  III. Society for General Microbiology
  IV. Series
  576'.15   QR97.A1
  ISBN 0–12–341460–1
  ISBN 0–12–341461–X (Pbk)

Typeset by Latimer Trend & Company Ltd,
and printed in Great Britain by
Galliard (Printers) Ltd, Great Yarmouth

Contributors

G. A. CODD   *Department of Biological Sciences, University of Dundee, Dundee DD1 4HN, UK*

C. S. DOW   *Department of Biological Sciences, University of Warwick, Coventry CV4 7AL, UK*

G. M. GADD   *Department of Biological Sciences, University of Dundee, Dundee DD1 4HN, UK*

H. VAN GEMERDEN   *Laboratorium voor Microbiologie, Kerklaan 30, Rijksuniversiteit Groningen, NN Haren (Gr.), The Netherlands*

W. D. GRANT   *Department of Microbiology, University of Leicester, Leicester LE1 7RH, UK*

R. A. HERBERT   *Department of Biological Sciences, University of Dundee, Dundee DD1 4HN, UK*

K. HORIKOSHI   *Department of Applied Microbiology, The RIKEN Institute, Wako-shi, Saitama 351, Japan*

P. MORGAN   *Department of Biological Sciences, University of Warwick, Coventry CV4 7AL, UK*

R. Y. MORITA   *Department of Microbiology, Oregon State University, Corvallis, Oregon 97731, USA*

M. J. MUNSTER   *Public Health Laboratory Service, Centre for Applied Microbiology and Research, Porton Down, Salisbury SP4 0JG, UK*

R. H. REED   *Department of Biological Sciences, University of Dundee, Dundee DD1 4HN, UK*

R. J. SHARP   *Public Health Laboratory Service, Centre for Applied Microbiology and Research, Porton Down, Salisbury SP4 0JG, UK*

B. J. TINDALL   *Institut für Mikrobiologie, Rheinische-Friedrich-Wilhelms-Universität, 5300 Bonn 1, FRG*

G. C. WHITELAM   *Department of Botany, University of Leicester, Leicester LE1 7RH, UK*

R. DE WIT   *Laboratorium voor Microbiologie, Kerklaan 30, Rijksuniversiteit Groningen, NN Haren (Gr.), The Netherlands*

Preface

Currently there is considerable interest, from both the academic and the biotechnological viewpoints, in the ecology and physiology of novel groups of microorganisms which are not only able to tolerate but able to grow at environmental extremes. An international conference organized under the auspices of the Scottish Branch of the Society for General Microbiology was held at the University of Dundee in June 1984. This book contains chapters on selected aspects of the subjects which were originally presented at this meeting. The subject of microbes in extreme environments is diverse and the book makes no attempt to cover the field exhaustively.

Rather, we have attempted to produce a volume which covers a selected number of topics in depth either where there have been recent developments in the subject or where the organisms have been used as the basis of new industrial processes. We have therefore included contributions on several environments (e.g. high sulphide and low nutrient concentrations) which have received somewhat less attention in previous reviews but which may be regarded as "extreme" in that they impose physiological limitations and stress and typically contain a specialized microflora. The inclusion of these chapters may encourage a broadening of the definition of what constitutes an extreme environment and may permit others to be recognized.

It is now widely acknowledged that microbes in extreme environments represent far more than an interesting collection of microbial curiosities which inhabit obscure ecological niches. These microorganisms not only represent a rich source of research material for future fundamental research into microbial adaptation but also contain valuable genetic resources, the potential applications of which are now gradually being appreciated. Microorganisms which until recently were only considered as microbiological curiosities may well in future years form the basis of new biotechnologically based industries.

We would like to thank Louise Burston for the cover layout and G. M. Gadd for the original design.

April 1986   R.A. Herbert
                                                                          G. A. Codd

Contents

1

The Ecology and Physiology
of Psychrophilic Microorganisms

R. A. HERBERT

*Department of Biological Sciences, The University,
Dundee DD1 4HN, UK*

Introduction

Temperature is one of the most important parameters regulating the activities of microorganisms in natural environments. The fundamental importance of temperature in regulating the rate of physicochemical reaction underlines its significance when considering the complexities of biological processes. Temperature influences the response of microorganisms either directly by its effect on growth rate, enzyme activity, cell composition and nutritional requirements or indirectly through its effect on the solubility of solute molecules, ion transport and diffusion, osmotic effects on membranes, surface tension, density and many colloidal properties of matter in aqueous systems (Oppenheimer, 1970). Considering the profound effects of temperature on the activities of microorganisms it is somewhat surprising that many microbiologists neglect the fact that most natural environments are at low ambient temperatures.

Water covers approximately 70% of the Earth's surface and the majority of this is marine. More than 90% by volume of the marine environment is at a temperature of 5 °C or less (Zobell, 1962). The range of temperature found in the ocean environment is between -2 °C in the polar regions and at great depth to $+40$ °C in the tropics (Svedrup *et al.*, 1942). Surface waters are subject to meteorological effects and vary in temperature with geographical location: however, below the thermocline, the water mass is almost uniformly cold. In addition to the marine environment, the polar regions, constituting some 14% of the Earth's surface, are at very low ambient temperature. The applied microbiologist equally has a considerable interest in the survival and growth of spoilage and potentially pathogenic microorganisms at low temperature since refrigerated storage is now a commonplace method of good preservation. It is convenient to subdivide low temperature environments into two major groups: those which are permanently cold such as deep ocean waters and those which are subject to periodic fluctuations in temperature. The latter category includes polar and tundra soils which may be subject to temperature extremes ranging from -88 to $+15\,°C$ (Weyant, 1966). In polar and tundra soils temperature fluctuations can occur over relatively short time periods. For example, Llano (1962) showed that at one location in the Horlick Mountains, Antarctica, the temperature of the rock surface rose from -15 to $+27.8\,°C$ within 3 hours owing to incident solar radiation. In this chapter the ecology and physiology of microorganisms capable of growth in permanently cold or seasonally cold environments will be discussed in relation to the mechanisms which they have evolved to compensate for the effects of low temperature on cell metabolism.

With so much of the Earth's surface and volume at low temperatures it is not surprising that so-called "cold-loving" microorganisms are ubiquitous. Forster (1887) is generally acknowledged as being the first person to report bacterial growth at $0\,°C$ with cultures of luminous bacteria isolated from marine fish. Within a year his work was substantiated by Fischer (1888) with further strains of luminous bacteria as well as freshly isolated strains of marine bacteria. Subsequently, Forster (1892) amongst others (Table 1) showed that bacteria able to grow at $0\,°C$ were widely distributed in nature by isolating them from such habitats as fresh and marine waters, the skin and intestinal contents of freshwater and marine fish, garden soil, ditch water and marine sediments. Later investigation (Table 1) showed that they were equally widely distributed in food products such as vegetables, meat, poultry and dairy products (e.g. milk, cream and butter). Bacteria able to grow at low temperatures have also been isolated from permanently cold environments such as deep ocean waters (Carey and Waksman, 1934; Harder and Veldkamp, 1971; Morita, 1975), polar soils (Straka and Stokes, 1960; Stanley and Rose, 1967; Vishniac and Hempfling, 1979a, b), Antarctic marine and fresh waters (Herbert and Bell, 1973; Tanner and Herbert, 1981) and cave systems (Gounot, 1969, 1973). It has also been demonstrated that microorganisms able to grow at $0\,°C$ can be readily isolated from less obviously cold environments such as estuaries and inshore coastal waters (Sieburth, 1967; Tajima et al., 1974). Whilst most authorities are agreed that microorganisms able to grow at low temperatures are widely distributed in nature, there is far less agreement on the terminology used to describe these organisms.

Traditionally, microorganisms have been classified on the basis of temperature optima for growth into thermophiles, mesophiles and psychrophiles according to their ability to grow at high, medium or low temperatures respectively. The first two groups can be readily differentiated by their

Table 1. *Early investigators reporting bacterial growth at temperatures of 0 °C or lower (modified from Morita, 1975)*

Investigator and year	Source material or organism
Forster (1887, 1892)	Fish, natural waters, foods, wastes, rubbish, soil, surface and intestines of fish
Fischer (1888)	Harbour water and soils, certain pathogenic bacteria
Conradi and Vogt (1901)	"*Proteus bacillus*"
Baur (1902)	*Bacterium lobatum*
Brandt (1902)	*Bacterium lobatum*
Schmidt-Nielsen (1902)	*Bacillus aquatile fluorescens, non-liquefacians, B. granulosum, B. paracoli gasoformans anandolicum, B. radiatum, B. tarde fluorescens, B. pestis* and *B. proteus fluorescens*
Müller (1903)	Sausage meat, fish, intestinal contents of fish, milk, vegetables, meal, garden soil
Feitel (1903)	Denitrifying bacteria from deep sea water
Tsiklinsky (1908)	Antarctic fish
Richardson (1908)	Non-proteolytic bacteria on frozen meat
Pennington (1908)	Milk
Conn (1910)	Frozen soils
Brown and Smith (1912)	Frozen soils
Reed and Reynolds (1916)	Milk

growth temperature optima. When defining the temperature characteristics of microorganisms able to grow at low temperatures the early investigators concentrated on the ability of the isolates to grow at or near 0 °C. Much confusion has therefore arisen over the naming of these cold-loving bacteria. The earliest use of the term psychrophile was by Schmidt-Nielsen (1902) who defined psychrophiles as those bacteria able not only to survive but also to grow at 0 °C. Almost immediately this definition was challenged by Müller (1903) who argued that in many cases the cold-loving bacteria had optimum growth temperatures between 20 and 30 °C and were therefore merely cold tolerant. This initiated almost a century of argument and counterargument for a variety of alternative definitions. For detailed discussions on the definition of psychrophiles the reader should consult reviews by Ingraham and Stokes (1959) and Morita (1966, 1975). During the last decade some semblance of agreement has been achieved amongst microbiologists in defining these microorganisms. The most widely accepted definition is that proposed by Morita (1975) who defined psychrophilic bacteria as those microorganisms which have an optimal growth temperature of 15 °C or lower, a maximal growth temperature of approximately 20 °C and a minimum growth temperature of 0 °C or lower. Microorganisms which grow at 0 to +5 °C and at maximum temperatures exceeding 25 °C will be considered to be psychrotrophic (cold tolerant) for the purposes of this review. An examination of the temperature characteristics of the isolates obtained by the early investigators shows that the majority were psychrotrophic microorganisms and the occurrence of truly psychrophilic bacteria was not unequivocally demonstrated until 1964 when Morita and Haight isolated *Vibrio marinus* MP-1 from oceanic waters off the Oregon coast. Subsequently, psychrophilic bacteria have been isolated from a range of natural habitats (Harder and Veldkamp, 1967; Stanley and Rose, 1967; Moiroud and Gounot, 1969; Herbert and Bhakoo, 1979).

Isolation Techniques for Psychrophilic Bacteria

The two principal reasons why early investigators were unsuccessful in isolating psychrophilic bacteria were probably the selection of suitable source material and experimental procedures for handling sample material and the isolates subsequently obtained. The past history of the source material is critical since it should never be exposed to lethal temperatures (about 20 °C) for long periods of time. Psychrophilic microorganisms are abnormally thermolabile and even exposure to moderate temperatures, e.g. room temperature, for extended periods of time is likely to prove lethal. When isolating psychrophiles from the natural environment every effort should therefore be made to ensure that the source material during its collection and subsequent transport is kept cool and that in all succeeding laboratory manipulations precooled pipettes, media, diluents, etc. are used (Morita, 1966; Brock and Rose, 1969; Wiebe and Hendricks, 1971). By observing these simple procedural precautions true psychrophilic micro-organisms can be readily isolated from natural environments.

Taxonomy of Psychrophilic Bacteria

One of the most notable features of psychrophilic bacteria is the predominance of Gram-negative species isolated and the relative rarity of Gram-positive organisms. The majority of psychrophilic bacteria isolated have been identified as belonging to the genera *Pseudomonas, Vibrio, Achromobacter, Flavobacterium* and *Cytophaga* (Morita and Haight, 1964; Herbert, 1981). The first psychrophilic bacterium to be well characterized was *Cytophaga psychrophila*, the aetiological agent for "cold-water disease" of salmon (Borg, 1960). Subsequently Eimjhellen isolated a psychrophilic bacterium from flounder eggs which was provisionally identified as a *Serratia* sp. (Hagen *et al.*, 1964). Subsequent taxonomic studies identified this bacterium as *Vibrio psychroerythrus* (Morita, 1975). Morita and Burton (1970) have isolated psychrophilic bacteria from the Arctic and North Pacific Oceans and Sieburth (1965, 1968) has isolated them from water off Narragansett Bay, Rhode Island. These isolates included *Achromobacter* spp., *Flavobacterium* spp. and *Pseudomonas* spp. Psychrophilic bacteria isolated from the North Sea include *Vibrio* spp., *Pseudomonas* spp. and *Spirillum* spp. (Harder and Veldkamp, 1967). However, an increasing number of workers have found that they were unable to classify the psychrophilic isolates they had obtained using existing monothetic classification schemes and so resorted to numerical taxonomy methods (Pfister and Burkholder, 1965; Kaneko *et al.*, 1979). Tanner and Herbert (1981) carried out a numerical taxonomic study of 144 psychrophilic bacteria isolated from Antarctic marine sediments. These workers reported that 55% of the isolates could be classified as *Vibrio/Aeromonas* spp., 22% as *Alcaligenes* spp., 4% as *Pseudomonas* spp. and 2% as *Flavobacterium* spp. The remaining isolates (17%) could not be readily identified with existing genera described in the eighth edition of *Bergy's Manual of Determinative Bacteriology* and may represent hitherto undescribed types of bacteria.

Whilst Gram-positive psychrophilic bacteria have been reported they are less frequently isolated than Gram-negative organisms. McLean *et al.* (1951) isolated a Gram-variable coccus, *Micrococcus cryophilus*; however, subsequent electron microscopy studies have demonstrated that this bacterium has a typical Gram-negative cell envelope (Russell, 1974). Sieburth (1967) isolated an *Arthrobacter* spp., from Narragansett Bay, Rhode Island, and a related bacterium, *Arthrobacter glacialis*, has been isolated from sediments below Arctic glaciers (Moiroud and Gounot, 1969). The phenomenon of obligately anaerobic growth appears to be relatively rare. Sinclair and Stokes (1965) successfully isolated *Clostridium* spp. which were capable of sporulating at 0 °C from soil, mud and sewage. Subsequently Liston *et al.* (1969) and then Finnes and Matches (1974) reported the presence of obligately psychrophilic *Clostridium* spp. from sediments in Puget Sound. Sixteen of these *Clostridium* isolates had growth temperature maxima below 15 °C. Psychrophilic *Clostridium* spp. not only are present in natural environments but also have been reported in food products such as raw milk (Bhadsavle *et al.*, 1972).

Herbert and Bell (1973, 1974) made an extensive survey of the waters and sediments of freshwater lakes and offshore coastal waters surrounding Signy Island in the maritime Antarctic for psychrophilic representatives of other bacterial groups. Table 2 shows the diversity of bacterial types that were isolated. The majority of the isolates obtained were psychrotrophic and only a relatively few psychrophilic types were isolated. It appears paradoxical that in permanently cold environments (-1 to $+5$ °C maximum) a greater proportion of the microflora are not psychrophilic.

The ability of microorganisms to grow at low temperature is not confined to prokaryotes, and psychrophilic representatives of yeasts, fungi and algae have also been reported. Psychrophilic yeasts identified as *Leucosporidium scottii*, *Torulopsis psychrophila* and *Cryptococcus vishniacii* have been isolated from Antarctic dry valley soils (Di Menna, 1966; Watson *et al.*, 1976;

Table 2. *Bacterial groups isolated from freshwater lakes and coastal offshore waters at Signy Island, South Orkney Islands, Antarctica (Herbert and Bhakoo, 1979)*

Bacterial group	Marine	Freshwater
Azotobacter sp.	+	+
Nitrosomonas sp.	−	+
Nitrobacter sp.	−	+
Denitrifying species	+	+
Proteolytic species	+	+
Purple non-sulphur species	+	+
Green sulphur species	+	+
Purple sulphur species	+	+
Sulphate reducers	+	+
Sulphur oxidizers	+	−
Cellulose decomposers	+	+

Vishniac and Hempfling, 1979a,b). Sinclair and Stokes (1965) determined the growth characteristics of a number of psychrophilic yeasts isolated from polar environments and showed that these organisms grow over the temperature range 0–20 °C with optimum growth temperatures of 15 °C. Pure cultures of snow algae have now been obtained and their temperature characteristics determined (Hohan, 1975). The optimum growth temperature for these isolates was 1 °C for *Chloromonas pichinchae*, 5 °C for *Rhaphidonema nivale* and 10 °C for *Cylindrocystis brebsonii* (Hohan, 1975). From this brief survey it is clearly evident that the ability to grow at low temperatures is widespread amongst microorganisms and is not solely confined to particular groups.

Ecological Significance of Psychrophilic Microorganisms in Natural Environments

Relatively few studies have been undertaken to determine the ecological significance of psychrophilic microorganisms in natural environments. Detailed seasonal studies have been made to assess the effect of temperature on the heterotrophic bacterial flora of Narragansett Bay, Rhode Island (Sieburth, 1967). During the winter months the water temperature reaches a minimum of − 2 °C and psychrophilic bacteria become the dominant component of the heterotrophic microflora. In contrast, during the summer months water temperatures reach + 23 °C and psychrotrophic and mesophilic bacteria become dominant. Laboratory studies on selected isolates collected throughout the year have substantiated the field data. A notable finding from this study was that there was no evidence for the enhancement or suppression of any individual taxonomic group as a consequence of these temperature changes. Tajima *et al.* (1974) have shown similar results for microbial populations present in offshore coastal waters in Hakodota Bay, Japan. Recently, King and Nedwell (1984) were able to demonstrate the development of distinct populations of nitrate-respiring bacteria in saltmarsh sediments in response to temperature. At low temperatures *Pseudomonas* spp. were the predominant nitrate-respiring bacteria present in the saltmarsh sediments whereas at 25 °C *Vibrio* spp. predominated. The principal conclusion that can be derived from these studies is that significant populations of psychrophilic bacteria develop in these environments at particular times of the year and that the selective influence is temperature. However, these studies do not demonstrate the ecological significance of psychrophilic bacteria in terms of microbial activity.

In a series of elegant experiments using chemostat-grown cultures Harder and Veldkamp (1971) demonstrated the effect of temperature upon the outcome of competition between defined mixed populations of psychrophilic and psychrotrophic bacteria. The psychrophile, a *Pseudomonas* sp. with an optimum growth temperature of 14 °C, outgrew the psychrotroph, a *Spirillum* sp., at all dilution rates employed at −2 °C, whilst at 16 °C the reverse occurred. At intermediate temperatures of 4 and 10 °C only the dilution rate influenced the outcome of the competition. These laboratory competition studies confirm that low temperatures exert a considerable selective influence and the bacterial populations which develop have growth characteristics appropriate to the imposed environmental conditions. It is therefore reasonable to assume that in environments which are permanently cold such as ocean waters psychrophilic microorganisms are principally responsible for mineralization processes. Tanner and Herbert (1981) have demonstrated that in the coastal waters surrounding Signy Island in the maritime Antarctic, which have a mean annual temperature of −1 °C, nutrient recycling is primarily due to the activities of psychrophilic bacteria.

Kinetics of Growth at Low Temperatures

Almost a century ago Arrhenius (1889) derived the following equation to describe the effects of temperature on the rates of chemical reactions:

$$K = A\,e^{-E/RT}$$

In this equation K is the reaction rate, R the gas constant, T the temperature in Kelvin, E the activation energy and A a constant. Some years later Arrhenius extended his equation to study the influence of temperature on the rates of biological processes and in this modified form he replaced E by the term μ which is referred to as the temperature characteristic of the process. By substituting bacterial growth rate for K in the equation the temperature characteristic for growth (μ) can be determined. It was postulated that the value could be used to determine whether a particular bacterium was a psychrophile or a mesophile since the former should have a lower activation energy and hence lower μ value (Ingraham, 1958, 1962). Subsequent workers have challenged Ingraham's data on the grounds that the calculations were erroneous (Jonato-Bassalik, 1963; Hanus and Morita, 1968). The latter workers precisely determined the values for a psychrophile, a psychrotroph and a mesophile and could show no difference. Similar results have been reported for yeasts (Shaw, 1967), Gram-positive bacteria (Brownlie, 1966) and *Vibrio* spp. (Herbert and Bhakoo, 1979). It is now generally accepted that μ values are of doubtful value in describing the growth characteristics of psychrophiles. Nevertheless the Arrhenius plot of the logarithm of the specific growth rate against the reciprocal of the absolute temperature is significant. Harder and Veldkamp (1971) showed that in the case of psychrophiles the slope of the Arrhenius plot is linear down to 0 °C, whilst psychrotrophs deviate from linearity at about 5 °C and mesophilic bacteria tend to deviate from a straight line at higher temperatures.

Effect of Temperature on Solute Uptake

Cell membranes are now recognized as complex heterogeneous systems whose properties are to a large degree determined by their composition and spatial organization as well as by external influences of which temperature is one of the most important. Most microorganisms alter their membrane lipid composition in response to changes in environmental temperature. By such mechanisms microorganisms can regulate the activity of solute transport systems and the function of essential membrane-bound enzymes. Since temperature can produce such fundamental changes in the structure and function of cell membranes it is not surprising that much interest has been shown in establishing the effects of low temperature on solute uptake processes. A number of workers have proposed that the minimum temperature for the growth of mesophiles is controlled by the low temperature inhibition of solute transport processes (Ingraham and Bailey, 1959; Morita and Buck, 1974). Experimental evidence to support this hypothesis came from comparative studies on the respiratory activity of a psychrotrophic *Candida* sp. and a mesophilic strain of *C. lipolytica* at different temperatures (Baxter and Gibbons, 1962). The psychrotroph was able to oxidize exogenous glucose at 0 °C whereas the mesophile showed no metabolic activity at temperatures less than 5 °C. However, the mesophile was able to metabolize endogenously at temperatures less than 5 °C, suggesting that the limiting factor was the ability to transport sugars into the cell at near-zero temperatures. Similar results were subsequently reported for mesophilic strains of *C. utilis* and *Arthrobacter* sp. (Rose and Evison, 1965).

In contrast, Baxter and Gibbons (1962) and Cirillo *et al.* (1963) showed that sugar transport was largely independent of temperature in psychrophilic *Candida* spp. Subsequent studies on a wide range of Gram-positive and Gram-negative bacteria have confirmed these findings (Wilkins, 1973). Wilkins demonstrated that the uptake of 2-deoxy glucose and L-leucine in Gram-positive bacteria was essentially unaffected by growth at low temperatures. Clinton (1968) not only studied the effect of temperature on lysine and glutamate uptake in a psychrophilic *Candida* sp. but also showed that the free amino pools in this yeast were largely unaffected by changes in growth temperature. In *M. cryophilus* the uptake of [^{14}C]lysine has been shown to be the same at 0 °C as at 20 °C (Russell, 1971) whereas Herbert and Bell (1977) reported that *Vibrio* AF-1, a psychrophilic bacterium isolated from an Antarctic lake, showed maximal rates of [^{14}C]glucose and [^{14}C]lactose uptake at 0 °C which decreased with increasing temperature.

These data indicate that psychrophiles and mesophiles differ in their ability to transport solute molecules across the cytoplasmic membrane at temperatures greater than 5 °C. The question that has to be addressed is how the transport processes differ in these two groups of microorganisms. Farrell and Rose (1967) postulated a number of mechanisms whereby solute uptake in mesophilic microorganisms could be inactivated at near-zero temperatures.

(1) Solute carrier proteins in the cytoplasmic membranes of psychrophiles were less susceptible to low temperature inactivation than their mesophilic counterparts.

(2) Solute carrier proteins in mesophilic microorganisms are not abnormally cold sensitive but, owing to changes in the spatial organization of the lipid bilayer, solute molecules are unable to combine with their respective carrier proteins.

(3) At low temperatures there is a shortage of energy to support active transport across the cytoplasmic membrane in mesophilic microorganisms.

To date there are no convincing data to show that individual carrier protein molecules are cold labile; neither has the lack of ATP for active uptake of solute molecules been demonstrated. As a result most studies have concentrated on determining the effects of low temperature on membrane lipid composition and the liquid-crystalline state of the lipid bilayer of the cell membrane.

Effect of Temperature on Membrane Structure and Composition

Since microorganisms can be grown easily under precisely controlled laboratory conditions they provide a convenient system for studying the effects of temperature on membrane structure and composition. During the growth of microorganisms under batch culture conditions a number of environmental parameters such as growth rate, nutrient concentration, pH and dissolved O$_2$ tension often change in an unpredictable manner which presents difficulties in the interpretation of the data obtained. In contrast, the use of chemostat cultures enables these variables to be maintained at constant values and this allows a more rigorous assessment to be made of the effects of temperature upon living cells.

Lipid Composition of Bacterial Membranes

Lipids are a primary component of all cell membranes and constitute some 40–70% of the total membrane dry weight (Quinn, 1976). As with higher organisms glycerol phosphatides represent the predominant structural lipid class present in bacterial cell membranes and consist of fatty acid esters of *sn*-glycerol-3-phosphate. The most commonly encountered phospholipid in bacteria is phosphatidylethanolamine and in Gram-negative species it is often the major component of the cell membrane. Phosphatidylglycerol and its derivative diphosphatidylglycerol (cardiolipin) are also frequently found in bacteria whereas phosphatidylserine is usually present in trace quantities since it is rapidly decarboxylated to yield phosphatidylethanolamine. Phosphatidylinositol and phosphatidylcholine whilst quantitatively important phospholipids in eukaryote cell membranes have seldom been reported as a component of bacterial membranes (Goldfine, 1972). Similarly, sterols such as cholesterol are rarely found in bacterial membrane systems, and their absence from prokaryote cell membranes has proved extremely useful in studying the effects of temperature on the transitions from gel to liquid-crystalline phase in lipid bilayers.

The fatty acids most commonly found in bacterial cells range in chain length from 10–20 carbon atoms with C_{16}–C_{18} acids usually predominating. The fatty acids present may be straight chain saturated, straight chain unsaturated, branched chain (*iso* or *anteiso*) or cyclopropane acids. Long chain polyunsaturated fatty acids are normally absent in bacteria. In contrast, cyclopropane fatty acids are commonly found in Gram-positive and Gram-negative bacteria but are rarely encountered in eukaryotes except in higher plants.

Effects of Temperature on Fatty Acid and Phospholipid Composition

Within the cell membrane, the phospholipid molecules are arranged in the form of a bilayer with the polar head groups at the intracellular and extracellular surfaces. These groups are thus able to interact with the aqueous phases on the inside and outside of the cell. The fatty acid acyl chains, in contrast, are stacked in a parallel fashion at right angles to the plane of the membrane, with the terminal methyl groups situated in the interior of the bilayer. It is well established that microorganisms like poikilotherms alter their membrane lipid composition in response to changes in environmental temperature (Marr and Ingraham, 1962; Kates and Hagen, 1964; Bhakoo and Herbert, 1979). These alterations in membrane lipid composition usually involve changes in the fatty acid moieties and less frequently the polar head groups (Russell, 1984). This is because the changes in fatty acid composition are much more effective in modifying membrane fluidity than changes in the head group are. The physiological significance of these observed changes in lipid composition with temperature is that optimum membrane fluidity and hence function are maintained. This process has been termed "homeoviscous adaptation" by Sinensky (1974) which implies that membrane fluidity is maintained more or less constant, i.e. the strategy is to maintain membrane function independent of temperature changes by modulating the viscosity of the membrane lipids. As a consequence numerous studies have been undertaken to determine the changes in fatty acid composition with temperature in psychrophilic, psychrotrophic and mesophilic microorganisms. However, the interpretation of some of these data is difficult because the microorganisms were grown in batch culture and as previously stated changes in growth conditions can greatly influence fatty acid composition.

When microorganisms are subjected to low temperatures (about 0 °C) a variety of changes in fatty acid composition may occur depending on the species involved. The most frequently observed response is an increase in the proportion of unsaturated fatty acids at the expense of saturated acids. For example, comparative studies using mesophilic and psychrophilic yeasts have clearly demonstrated that the psychrophiles synthesize elevated levels of unsaturated fatty acids, especially hexedecenoic (16:1) and octadecenoic (18:1), at low temperatures compared with their mesophilic counterparts (Kates and Baxter, 1962; Brown and Rose, 1969). In bacteria the response to a decrease in growth temperature is more complex and in addition to a change in the degree of unsaturation of the component fatty acids other alterations may occur, e.g. a shortening of acyl chain length or an increase in the proportion of branched chain fatty acids or a reduction in the proportion of cyclic fatty acids. All these changes produce lipids with a lower gel to liquid-crystalline transition temperature, thereby maintaining membrane mobility in response to the lowered growth temperature. Marr and Ingraham (1962) showed that *Escherichia coli* ML30, a typical mesophilic bacterium, responded to a decrease in growth temperature in the classical manner by synthesizing increased proportions of unsaturated fatty acids at the expense of saturated acids (Table 3). Similar changes in fatty acid composition have also been reported for psychrotrophic strains of *Ps. fluorescens* (Gill, 1975). Bhakoo and Herbert (1979) carried out a rigorous examination of the effects of temperature on the fatty acid composition of a number of psychrophilic *Vibrio* spp. isolated from coastal waters in the maritime Antarctic. The results are summarized in Table 3 and show that in *Vibrio* AM-1 the response to low growth temperatures was to synthesize increased quantities of

Table 3. Effect of growth temperature on the fatty acid composition (expressed as the percentage by weight of the total fatty acids) of some Gram-negative bacteria

Fatty acid [a]	Escherichia coli [b] for growth temperatures of			Pseudomonas T-6 [c] for growth temperatures of			Vibrio AM-1 [d] for growth temperatures of			Vibrio BM-2 [d] for growth temperatures of		
	10°C	20°C	30°C	0°C	8°C	20°C	0°C	8°C	15°C	0°C	8°C	15°C
9:1	—	—	—	—	—	—	—	—	—	11.4	6.6	5.4
12:0	—	—	—	1.5	2.3	2.1	—	—	0.8	0	6.4	9.0
12:1	—	—	—	—	—	—	8.2	10.7	1.0	13.7	1.0	—
13:0	—	—	—	2.2	2.6	—	—	—	—	—	—	—
13:1	—	—	—	4.4	8.1	9.9	—	—	—	—	—	—
14:0	3.9	4.1	3.8	5.1	6.1	9.9	1.2	1.7	21.8	—	1.0	1.2
14:1	—	—	—	4.3	1.7	—	5.5	—	—	20.6	7.5	12.0
2-OH 14:0	12.6	10.4	10.1	—	—	—	—	—	—	—	—	—
15:0	—	—	—	1.1	1.2	—	8.2	1.0	18.2	0.7	0.9	1.1
15:1	—	—	—	1.1	1.6	1.4	—	—	—	1.5	2.5	2.9
16:0	18.2	25.4	28.9	—	—	—	5.1	8.2	17.8	21.6	46.5	38.5
16:0Δ	1.3	1.5	3.4	—	—	—	—	—	—	—	—	—
16:1	26.0	24.4	23.3	56.6	54.6	59.0	38.6	39.6	—	20.3	17.7	22.0
17:0	—	—	—	—	—	—	8.6	11.1	13.5	0.4	0.4	0.4
17:0Δ	—	—	—	—	—	—	—	—	—	—	—	—
17:1	—	—	—	7.1	5.3	2.4	20.8	21.9	—	1.3	1.4	1.2
18:0	—	—	—	0.4	1.0	1.3	—	—	—	—	—	—
18:1	37.9	34.2	30.3	15.9	15.1	14.2	—	—	—	—	—	—

[a] In fatty acid designations, the first number indicates the number of carbon atoms and the second the number of double bonds; Δ indicates a cyclopropane ring.
[b] From Marr and Ingraham (1962).
[c] From Bhakoo and Herbert (1979).
[d] From Bhakoo and Herbert (1980).

hexadecenoic (16:1) and heptadecenoic acids (17:1). In contrast, *Vibrio* BM-2 synthesized increased proportions of shorter chain fatty acids as the growth temperature was lowered to 0 °C. This response is similar to that reported by Russell (1971) for *M*. cryophilus when grown at low temperatures.

Whilst many microorganisms are able to modify their fatty acid composition in response to growth at low temperatures it is not a universal requirement and several examples have been reported in the literature where little or no change in lipid profile occurs over a wide temperature range. For example, Hunter and Rose (1972) could demonstrate no significant differences in the fatty acid composition of *Saccharomyces cerevisiae* when grown at 15 and 30 °C. Similarly, Bhakoo and Herbert (1980) could demonstrate no significant differences in either the fatty acid or the phospholipid composition of four psychrotrophic pseudomonads isolated from Antarctic coastal waters. Data in Table 3 show that hexadecenoic acid (16:1) and octadecenoic (18:1) dominate the fatty acid profile of *Pseudomonas* T-6 and show no significant change over a 20 °C span of growth temperature. Similar acid profiles were observed for the remaining *Pseudomonas* spp. studied (Bhakoo and Herbert, 1980). These data indicate that there is no absolute requirement for changes in membrane lipid composition to compensate for microbial growth at low ambient temperatures. This has led several workers (Gill and Suisted, 1978; Silvius *et al.*, 1980) to question the need for prokaryotes to adapt their membrane lipids in response to change in growth temperature. Detailed studies carried out by Cronan and Gelman (1973) using fatty acid auxotrophs of *E. coli* have demonstrated that there are minimum requirements for both saturated and unsaturated fatty acids. At 37 °C the minimum requirement for unsaturated fatty acids is approximately 20% w/v and increases with decreasing growth temperature. Analysis of the data presented in Table 3 for the fatty acid profiles of *E. coli* ML30, the psychrophilic *Vibrio* spp. and the psychrotrophic *Pseudomonas* sp. shows that in all these bacteria the levels of unsaturated fatty acids are considerably in excess of the minima required for growth.

The pattern of change in membrane phospholipids in response to temperature is equally equivocal. Bhakoo and Herbert (1979) determined the effect of decreasing the growth temperature on the phospholipid composition of four psychrophilic *Vibrio* spp. and showed unequivocally that in these bacteria there was a significant increase in the total quantity of phospholipids synthesized at 0 °C. In contrast, these workers were unable to demonstrate any significant change in either phospholipid composition or total quantity of phospholipid synthesized in four psychrotrophic pseudomonads grown over the temperature range 0–20 °C (Bhakoo and Herbert, 1980). Brown and Minnikin (1973) were similarly unable to show any change in the phospholipid composition of four marine psychrotrophic pseudomonads grown at temperatures in the range 0–20 °C.

Effect of Temperature on Membrane Architecture and Fluidity

It has been well established that the classical lipid bilayer is a central structural feature of all biological membranes so far studied (Harrison and Lunt, 1975; McElhaney, 1976). In recent years detailed investigations have been undertaken to determine the effects of temperature on the molecular architecture of the plasma membrane. These studies have shown that the liquid-crystalline state is essential for correct membrane function such as solute transport (Overath et al., 1970), the assembly of transport proteins (Tsukagoshi and Fox, 1973) and the activity of membrane-bound enzymes (Kimelberg and Papahadjopoulos, 1972). As discussed earlier in this review prokaryote cell membranes contain little or no sterols and they are thus an ideal model system for investigating gel to liquid-crystalline phase transitions. In the absence of appreciable quantities of sterols prokaryote phospholipid layers can undergo a reversible thermotrophic gel to liquid-crystalline phase transition as a result of the cooperative melting of the hydrocarbon chains forming the interior of the bilayer (McElhaney, 1976). The unique feature of these transitions involving the selective melting of the phospholipid fatty acids is that they do not appear to involve any gross molecular rearrangement so that the overall bimolecular leaflet structure remains the same both above and below the phase transition temperature. The properties of the gel and the liquid-crystalline state, however, are significantly different. When the hydrocarbon chains are aligned at right angles to the plane of the membrane the lipid bilayer is essentially rigid and solute transport as a consequence is severely restricted. As the temperature is progressively increased the molecular motion of the hydrocarbon chains gradually increases until at a characteristic transition temperature the mobility of the hydrocarbon core abruptly increases, giving rise to the liquid-crystalline state. Under these conditions the cross-sectional area is reduced and the membrane is significantly thinner. Thus, the liquid-crystalline state is characterized by a fairly loosely packed, fluid and relatively permeable structure. McElhaney (1976) pertinently points out that in biological systems the transition from gel to liquid-crystalline state is not a sharp one owing primarily to the heterogeneity in the structure of the fatty acid side chains. Thus, within a given membrane system gel and liquid-crystalline states may exist simultaneously.

As has already been discussed the change from gel to liquid-crystalline state is due to the hydrocarbon chain's becoming partially melted and thus the temperature at which the transition occurs will be to a large degree governed by the nature of the fatty acids comprising the membrane lipid. Data presented in Table 4 show the melting points of individual fatty acids and the transition temperatures of diacylphosphatides commonly found in bacterial membranes. It is clearly apparent that the melting points of the saturated fatty acids decrease with decreasing chain length. More significantly, the introduction of a double bond into the longer chain fatty acids (16:1 and 18:1) profoundly decreases the melting point. Similarly, the introduction of a methyl group at the *anteiso* position results in fatty acids with significantly decreased melting points compared with their saturated counterparts whereas the insertion of a methyl group in the *iso* position has little effect (McElhaney, 1976). Whilst these data provide an index of the influence of carbon chain length, branching and degree of unsaturation on the melting point of individual authentic free fatty acids, they do not accurately reflect the situation in the lipid bilayer. Data presented in Table 4 show that the phase transition temperatures for pure diacylphosphatides in aqueous solution are considerably lower than those for free fatty acids.

Recent studies with model systems have established that the acyl positional distribution in phospholipids is also important in lipid fluidity, the *sn*-1 long, *sn*-2 short isomer having the lower melting temperature (Keogh and Davis, 1979; Davis *et al.*, 1981). McGibbon and Russell (1983) have shown that in *M. cryophilus* there is a preference for oleoyl acyl chains (18:1) in the *sn*-1 position of the two major phospholipids present, phosphatidylethanolamine and phosphatidylglycerol. The net result of the *sn*-1 long, *sn*-2 short isomeric configuration is to lower the lipid melting temperature and hence this may be an additional adaptation in this bacterium to ensure that the membrane

Table 4. *Melting points gel to liquid-crystalline transition temperatures of some mono-unsaturated and saturated fatty acids commonly found in the plasma membrane (modified from McElhaney, 1976)*

Fatty acid[a]	Melting Point (°C)	Transition temperature (°C) (diacylphosphatidylcholines)
14:0	54.4	23
16:0	62.9	41
18:0	70.1	60
16:1	≈ 0.0	ND[b]
18:1	13.4	−22
20:0	77.0	75

[a] In fatty acid designations, the first number indicates the number of carbon atoms and the second the number of double bonds.
[b] Not determined.

lipids remain in the liquid-crystalline phase when exposed to low temperatures. The principal conclusion that can be derived from these studies is that microorganisms have developed a number of strategies to maintain their membrane lipids fluid and functional at low growth temperatures.

Modification of Fatty Acid Composition in Response to Growth Temperatures

Although there is a considerable body of data to show the effects of temperature on lipid composition the mechanisms by which these changes are brought about are less well understood. A major difficulty is that these modification processes are regulated in a complex manner and at different levels. Clarke (1981) summarized the different levels at which regulation of membrane lipid modifications occurs as follows:

(1) regulation of transcription by *de novo* synthesis of desaturase enzymes;
(2) regulation at the level of enzyme activity by temperature-induced modifications of desaturase enzymes;
(3) regulation at the level of the membrane by modulating the activity and/or specificity of the acyl transferases;
(4) regulation by control of the relative balance of synthesis and desaturation of fatty acids available to the membrane.

A temperature-induced synthesis of desaturase has been reported in *Bacillus megaterium* (Fulco, 1970). In this bacterium the desaturase enzyme, which desaturates only at the Δ^5 position of acyl chains, is absent in cultures grown at 35 °C. Upon lowering the growth temperature to 20 °C the desaturase is synthesized. Addition of chloramphenicol to cell suspensions showed that enzyme synthesis was *de novo* (Fulco, 1970).

In *E. coli* the ratio of saturated to unsaturated fatty acids incorporated into membrane lipids is controlled by the relative activities of β-ketoacyl acyl carrier protein (ACP) synthase which exists in two forms known as synthase I and synthase II. The fatty acid synthetases in the bacterium produce both saturated and unsaturated acids owing to a branch in the biosynthetic pathway at β-OH-decanoyl-ACP. This C_{10} intermediate may give rise to either palmitic (16:0) or palmitoleic acid (16:1 Δ^9 *cis*). β-ketoacyl ACP synthase II is more efficient in elongating palmitoleate and relatively more temperature sensitive than synthase I. At low temperatures synthase II activity is enhanced relative to that of synthase I and a greater proportion of unsaturated acids are produced relative to the saturated counterparts (de Mendoza and Cronan, 1983). Fatty acid composition in *E. coli* is therefore mediated by temperature modulation of pre-existing enzymes. An alternative system is one in which bacteria manufacture unsaturated fatty acids by aerobic desaturation of existing fatty acids. Most authorities consider the substrate to be membrane lipid and this system provides a rapid initial modification of existing membrane in response to temperature change. The aerobic desaturation system is complex in its regulation and involves a temperature-sensitive modulator and enzyme turnover (Russell, 1984). The principal advantage of this system is that it allows short-term modification in fatty acid composition of existing membrane lipid in response to temperature changes. More detailed studies involving a wide range of microorganisms are now urgently required to determine the significance of these individual systems as mechanisms for controlling the fluidity of membrane lipids in response to lowered growth temperatures.

In conclusion the data presented in this review demonstrate that the liquid-crystalline state of membrane lipids is essential for the correct function of the plasma membrane. The mechanisms by which membrane fluidity is modified in response to temperature changes support the concept of an optimal state for membrane structure and function. Our present knowledge of the organization of cell membranes is still incomplete and must await the results of further studies in order to understand the subtle effects of changes in the fluidity of membrane lipids on the function of intrinsic membrane proteins and transport processes. However, it is self-evident that merely changing the physical state of membrane lipids will not transform a mesophile into a psychrophile.

References

Arrhenius, S. (1889). Cited by Farell, J. and Rose, A. H. (1967). Temperature effects on microorganisms. *In* "Thermobiology" (Ed. A. H. Rose), pp. 147–218. Academic Press, London and New York.

Baur, E. (1902). Uber zwei denitrifizierende Bakterien aus der Ostsee. *Wissenschaft Meeresuntersuchungen, Kiel* **6,** 9–22.

Baxter, R. M. and Gibbons, N. E. (1962). Observations on the physiology of psychrophilism in a yeast. *Canadian Journal of Microbiology* **8,** 511–517.

Bhadsavle, C. H., Shehata, T. E. and Collins, E. B. (1972). Isolation and identification of psychrophilic species of *Clostridium* from milk. *Applied Microbiology* **26,** 699–702.

Bhakoo, M. and Herbert, R. A. (1979). The effects of temperature on the fatty acid composition and phospholipid composition of four obligately psychrophilic *Vibrio* spp. *Archives of Microbiology* **121,** 121–127.

Bhakoo, M. and Herbert, R. A. (1980). Fatty acid and phospholipid composition of five psychrotrophic *Pseudomonas* spp. grown at different temperatures. *Archives of Microbiology* **126,** 51–55.

Borg, A. F. (1960). Studies on myxobacteria associated with diseases in salmonid fishes. *Wildlife Diseases* **8,** 1–85.

Brandt, K. (1902). Uber den Stoffwechsel im Meer. *Wissenschaft Meeresuntersuchungen, Kiel* **6**, 23–79.

Brock, T. D. and Rose, A. H. (1969). Psychrophiles and thermophiles. *In* "Methods of Microbiology" (Eds J. R. Norris and D. W. Ribbons), Vol. 3B, pp. 161–168. Academic Press, London and New York.

Brown, C. M. and Minnikin, E. E. (1973). The effect of growth temperature on the fatty acid composition of some psychrophilic marine pseudomonads. *Journal of General Microbiology* **75**, IX.

Brown, C. M. and Rose, A. H. (1969). Fatty acid composition of *Candida utilis* as affected by growth temperature and dissolved O_2 tension. *Journal of Bacteriology* **99**, 371–378.

Brown, P. E. and Smith, R. E. (1912). Bacterial activities in frozen soils. *Iowa Agricultural Experimental Station Research Bulletin* **4**, 155–184.

Brownlie, L. E. (1966). Effects of some environmental factors on psychrophilic *Microbacteria*. *Journal of Applied Bacteriology* **29**, 447–454.

Carey, C. L. and Waksman, S. A. (1934). The presence of nitrifying bacteria in deep seas. *Science* **79**, 349–350.

Cirillo, V. P., Wilkins, P. O. and Anton, J. (1963). Sugar transport in a psychrophilic yeast. *Journal of Bacteriology* **86**, 1259–1264.

Clarke, A. (1981). Effects of temperature on the lipid composition of *Tetrahymena*. *In* "Effects of Low Temperature on Biological Membranes" (Eds G. J. Morris and A. Clarke), pp. 55–82. Academic Press, London and New York.

Clinton, R. H. (1968). Effect of temperature on the uptake of glutamic acid and lysine in a psychrophilic yeast. *Antonie van Leeuwenhoek; Journal of Microbiology and Serology* **34**, 99–105.

Conn, H. J. (1910). Bacteria of frozen soil. *Centr. Bakterien Parasitenk, Abteilung* **2**, 337–340.

Conradi, H. and Vogt, H. (1901). Ein Betrag zur Aetiologie der Weilshen Krankheit. *Zeitschrift für Hygiene* **37**, 283–293.

Cronan, J. E. and Gelman, E. P. (1973). An estimate of the minimum amount of unsaturated fatty acid required for growth of *Escherichia coli*. *Journal of Biological Chemistry* **248**, 1188–1195.

Davis, P. J., Fleming, B. D., Coolbeas, K. P. and Keogh, K. M. W. (1981). Gel to liquid crystalline gel transition temperatures of water dispersions of two pairs of positional isomers of unsaturated mixed-acid phosphatidylcholines. *Biochemistry* **20**, 3633–3636.

Di Menna, M. E. (1966). Yeasts in Antarctic soils. *Antonie van Leeuwenhoek; Journal of Microbiology and Serology* **32**, 29–38.

Farrell, J. and Rose, A. H. (1967). Temperature effects on microorganisms. *In* "Thermobiology" (Ed. A. H. Rose), pp. 147–218. Academic Press, London and New York.

Feitel, R. (1903). Beiträge zur Kenntniss denitrifizierender Meeresbakterien. *Wissenschaft Meeresuntersuchungen, Kiel* **7**, 91–110.

Finnes, G. and Matches, J. R. (1974). Low temperature growing clostridia from marine sediments. *Canadian Journal of Microbiology* **20**, 1639–1645.

Fischer, B. (1888). Bakterien Wachstum bei 0 °C sowie über photographiren von kulturen leuchtender Bakterien in ihrem eigenen Lichte. *Zentralblatt fur Bakteriologie, Parasitenkunde, Infektionskrankheiten und Hygiene, Abteilung 1* **4**, 89–92.

Forster, J. (1887). Uber einige Eigenschaften leuchtender Bakterien. *Zentralblatt fur Bakteriologie, Parasitenkunde, Infektionskrankheiten und Hygiene, Abteilung 1* **2**, 337–340.

Forster, J. (1892). Uber die Entwichelung von Bakterien bei neideren Temperaturen. *Zentralblatt fur Bakteriologie, Parasitenkunde, Infektionskrankheiten und Hygiene, Abteilung 1* **12**, 431–436.

Fulco, A. J. (1970). Biosynthesis of unsaturated fatty acids in bacilli. *Journal of Biological Chemistry* **43**, 215–241.

Gill, C. O. (1975). Effect of growth temperature on the lipids of *Pseudomonas fluorescens*. *Journal of General Microbiology* **89**, 293–298.

Gill, C. O. and Suisted, J. R. (1978). The effects of temperature and growth rate on the proportion of unsaturated fatty acids in bacterial lipids. *Journal of General Microbiology* **104**, 31–36.

Goldfine, H. (1972). Comparative aspects of bacterial lipids. *In* "Advances in Microbiology Physiology" (Eds A. H. Rose and D. W. Tempest), Vol 8, pp. 1–58. Academic Press, London and New York.

Gounot, A. M. (1969). Contribution à l'étude des bactéries des grottes froides. *International Kongress Speläologie, Stuttgart* **4**, 1–6.

Gounot, A. M. (1973). Importance of temperature factor in the study of cold soils microbiology. *Bulletin Ecological Research Communications (Stockholm)* **17**, 172–173.

Hagen, P. O., Kushner, D. J. and Gibbons, N. E. (1964). Temperature induced death and lysis in a psychrophilic bacterium. *Canadian Journal of Microbiology* **10**, 813–823.

Hanus, F. J. and Morita, R. Y. (1968). Significance of the temperature characteristic of growth. *Journal of Bacteriology* **95**, 736–737.

Harder, W. and Veldkamp, H. (1967). A continuous culture study of an obligately psychrophilic *Pseudomonas* spp; *Archiv für Mikrobiologie* **59**, 123–130.

Harder, W. and Veldkamp, H. (1971). Competition of marine psychrophilic bacteria at low temperature. *Antonie van Leeuwenhoek; Journal of Microbiology and Serology* **37**, 51–63.

Harrison, R. and Lunt, G. C. (1975). "Biological Membranes". Blackie, Glasgow and London.

Herbert, R. A. (1981). Low temperature adaptation in bacteria. *In* "Effects of Low Temperatures on Biological Membranes" (Eds G. J. Morris and A. Clarke), pp. 41–53. Academic Press, London and New York.

Herbert, R. A. and Bell, C. R. (1973). Nutrient cycling in freshwater lakes on Signy Island, South Orkney Islands. *British Antarctic Survey Bulletin* **37**, 15–20.

Herbert, R. A. and Bell, C. R. (1974). Nutrient cycling in the Antarctic marine environment. *British Antarctic Survey Bulletin* **39**, 7–11.

Herbert, R. A. and Bell, C. R. (1977). Growth characteristics of an obligately psychrophilic *Vibrio* sp. *Archives of Microbiology* **113**, 215–220.

Herbert, R. A. and Bhakoo, M. (1979). Microbial growth at low temperatures. *In* "Growth in Cold Environments" (Ed. . A. D. Russell), Vol 13, pp. 1–16. Society of Applied Bacteriology, London.

Hoham, R. W. (1975). Optimum temperatures and temperature ranges for growth of snow algae. *Arctic Alpine Research* **7**, 13–24.

Hunter, K. and Rose, A. H. (1972). Lipid composition of *Saccharomyces cerevisiae* as influenced by growth temperature. *Biochimica et Biophysica Acta* **260**, 639–653.

Ingraham, J. L. (1958). Growth of psychrophilic bacteria. *Journal of Bacteriology* **76**, 75–80.

Ingraham, J. L. (1962). Temperature relationships. *In* "The Bacteria" (Eds I. C. Gunsalus and R. Y. Stanier), Vol. 3, pp. 265–296. Academic Press, London and New York.

Ingraham, J. L. and Bailey, G. F. (1959). Comparative effect of temperature on metabolism of mesophilic and psychrophilic bacteria. *Journal of Bacteriology* **77**, 609–613.

Ingraham, J. L. and Stokes, J. L. (1959). Psychrophilic bacteria. *Bacteriological Reviews* **23**, 97–108.

Janota-Bassalik, L. (1963). Growth of psychrophilic and mesophilic strains of peat bacteria. *Acta Microbiologica Polonica* **12**, 41–45.

Kaneko, T., Atlas, R. M. and Krichevsky, M. I. (1977). Bacterial diversity in the Beaufort Sea. *Nature, London* **270**, 596–599.

Kates, M. and Baxter, R. M. (1962). Lipid composition of mesophilic and psychrophilic yeasts (*Candida* spp.) as influenced by environmental temperatures. *Canadian Journal of Biochemistry and Physiology* **40**, 1213–1227.

Kates, M. and Hagen, P. O. (1964). Influence of temperature on fatty acid composition of psychrophilic and mesophilic *Serratia* spp. *Canadian Journal of Biochemistry* **42**, 481–488.

Keogh, K. M. W. and Davis, P. J. (1979). Gel to liquid crystalline phase transitions in water dispersions of saturated mixed-acid phosphatidylcholines. *Biochemistry* **18**, 1453–1459.

Kimelberg, H. J. and Papahadjopoulos, D. (1972). Phospholipid requirements for ($Na^+ + K^+$)-ATPase activity, head group specificity and fatty acid fluidity. *Biochimica et Biophysica Acta* **282**, 277–292.

King, D. and Nedwell, D. B. (1984). Changes in the nitrate reducing community of an anaerobic saltmarsh sediment in response to a seasonal selection by temperature. *Journal of General Microbiology* **130**, 2935–2941.

Liston, J., Holman, M. and Matches, J. (1969). Psychrophilic clostridia from marine sediments. *Bacteriological Proceedings* **35**, 40.

Llano, G. A. (1926). The terrestrial life of the Antarctic. *Scientific American* **207**, 212–230.

Marr, A. G. and Ingraham, J. L. (1962). Effect of temperature on the composition of fatty acids in *E. coli*. *Journal of Bacteriology* **84**, 1260–1267.

McElhaney, R. N. (1976). The biological significance of alterations in the fatty acid composition of microbial membrane lipids in response to changes in environmental temperature. *In* "Extreme Environments, Mechanisms of Microbial Adaptation" (Ed. M. R. Heinreich), pp. 255–281. Academic Press, London and New York.

McGibbon, L. and Russell, N. J. (1983). Fatty acid positional distribution in phospholipids of a psychrophilic bacterium during changes in growth temperature. *Current Microbiology* **9**, 241–244.

McLean, R. A., Sulzbacter, W. L. and Mudd, S. (1951). *Micrococcus cryophilus sp. nov.*, a large coccus especially suitable for cytological study. *Journal of Bacteriology* **62**, 723–728.

de Mendoza, D. and Cronan, J. E. (1983). Thermal regulation of membrane lipid fluidity by bacteria. *Trends in Biochemical Sciences* **8**, 49–52.

Moiroud, A. and Gounot, A. M. (1969). Sur une bactéria psychrophile obligatoire isolée de limins glaciaires. *Comptes Rendus de l'Academie des Sciences, Series D* **269**, 2150–2152.

Morita, R. Y. (1966). Marine psychrophilic bacteria. *Oceanography and Marine Biology Annual Reviews* **4**, 105–121.

Morita, R. Y. (1975). Psychrophilic bacteria. *Bacteriological Reviews* **39**, 146–167.

Morita, R. Y. and Buck, G. E. (1974). Low temperature inhibition of substrate uptake. *In* "Effects of the Ocean Environment on Microbial Activities" (Eds

R. R. Colwell and R. Y. Morita), pp. 124–129. University Park Press, Baltimore, Maryland.

Morita, R. Y. and Burton, S. D. (1970). Occurrence, significance and metabolism of obligate psychrophiles in marine waters. In "Organic Matter in Natural Waters" (Ed. D. Wood), pp. 275–285. Institute of Marine Science, Fairbanks, Alaska.

Morita, R. Y. and Haight, R. D. (1964). Temperature effects on the growth of an obligately psychrophilic marine bacterium. Limnology and Oceanography 9, 103–106.

Müller, M. (1903). Uber das Wachstum und die Lebenstatigkeit von Bakterien sowie den Ablauf fermentativer Progresse bei niederer Temperatur unter spezieller Berucksichtung der fleisches. Archives für Hygiene 47, 127–193.

Oppenheimer, C. H. (1970). Temperature. In "Microbial Ecology" (Ed. O. Kinne), Vol. 1, pp. 347–361. Wiley, New York.

Overath, P., Schairer, H. V. and Stoffel, W. (1970). Correlation of in vitro and in vivo phase transitions of membrane lipids in E. coli. Proceedings of the National Academy of Sciences of the United States of America 67, 606–612.

Pennington, M. E. (1908). Bacterial growth and chemical changes in milk kept at low temperatures. Journal of Biological Chemistry 4, 353–393.

Pfister, R. M. and Burkholder, P. R. (1965). Numerical taxonomy of some bacteria isolated from Antarctic and tropical seawaters. Journal of Bacteriology 90, 863–872.

Quinn, P. J. (1976). "Molecular Biology of Cell Membranes". Macmillan, London.

Reed, H. S. and Reynolds, R. R. (1916). Some effects of temperature upon the growth and activity of bacteria in milk. Virginia Agricultural Experimental Station Research Bulletin 10, 1–30.

Richardson, W. D. (1908). The cold storage of beef and poultry. In "Premier Congress International du Froid", Vol. 2, pp. 261–316.

Rose, A. H. and Evison, L. M. (1965). Studies on the biochemical basis of the minimum temperature for growth of certain psychrophilic and mesophilic microorganisms. Journal of General Microbiology 38, 131–141.

Russell, N. J. (1971). Alteration in fatty acid chain length in Micrococcus cryophilus growth at different temperatures. Biochimica et Biophysica Acta 231, 254–256.

Russell, N. J. (1974). The lipid composition of the psychrophilic bacterium Micrococcus cryophilis. Journal of General Microbiology 80, 217–225.

Russell, N. J. (1984). Mechanisms of thermal adaptation in bacteria: blueprints for survival. Trends in Biochemical Sciences 9, 108–112.

Schmidt-Nielsen, S. (1902). Uber einige psychrophile Mikroorganismen und ihr Vorkommen. Zentralblatt für Bakteriologie, Parasitenkunde, Infektionskrankheiten und Hygiene, Abteilung II 9, 145–147.

Shaw, M. K. (1967). Effect of abrupt temperature shifts on the growth of mesophilic and psychrophilic yeasts. Journal of Bacteriology 93, 1332–1336.

Sieburth, J. McN. (1965). Microbiology of the Antarctic. In "Biogeography and Ecology in Antarctica" (Eds P. V. Oye and J. V. Mieghem). W. Junk Publishers, The Hague.

Sieburth, J. McN. (1967). Seasonal selection of estuarine bacteria by water temperature. Journal of Experimental Marine Biology and Ecology 1, 98–121.

Sieburth, J. McN. (1968). Observations on planktonic bacteria in Narragansett Bay, Rhode Island; a résumé. Bulletin of the Misaki Marine Biology Institute, Kyoto University 12, 49–64.

Silvius, J. R., Maku, N. and McElhaney, R. N. (1980). In "Membrane Fluidity" (Eds M. Kates and A. Kuksis), pp. 213–222. Humana, New Jersey.

Sinclair, N. A. and Stokes, J. L. (1965). Isolation of obligately anaerobic psychrophilic bacteria. *Journal of Bacteriology* **87**, 562–565.

Sinensky, M. (1974). Homeoviscous adaptation—a homeostatic response that regulates the viscosity of membrane lipids in *E. coli*. *Proceedings of the National Academy of Sciences of the United States of America* **71**, 522–525.

Stanley, S. O. and Rose, A. H. (1967). Bacteria and yeasts from lakes on Deception Island. *Proceedings of the Royal Society of London, Series B: Biological Sciences* **252**, 199–207.

Straka, R. P. and Stokes, J. L. (1960). Psychrophilic bacteria from Antarctica. *Journal of Bacteriology* **80**, 622–625.

Svedrup, H. V., Johnson, M. W. and Fleming, R. H. (1942). "The Oceans". Prentice-Hall, Englewood Cliffs, New Jersey.

Tajima, K., Daiku, K., Ezura, Y., Kimura, T. and Sakai, M. (1974). Procedure for the isolation of psychrophilic marine bacteria. *In* "Effect of Ocean Environment on Microbial Activities" (Eds R. R. Colwell and R. Y. Morita), pp. 375–386. University Park Press, Baltimore, Maryland.

Tanner, A. C. and Herbert, R. A. (1981). A numerical taxonomic study of Gram-negative bacteria isolated from the Antarctic marine environment. *In* "Deuxième Colloque de Microbiologie Marine, Marseille, Vol. 13, pp. 31–38. CNEXO, Marseilles.

Tsiklinsky, M. (1908). La flore microbienne dans les regions du Pôle Sud. Expedition antarctique francaise 1903–1905. Masson et Cie, Paris.

Tsukagoshi, N. and Fox, C. D. (1973). Abortive assembly of the lactose transport in *E. coli*. *Biochemistry* **12**, 2816–2822.

Vishniac, H. S. and Hempfling, W. P. (1979a). Evidence of an indigenous microbiota (yeast) in the dry valleys of Antarctica. *Journal of General Microbiology* **112**, 301–314.

Vishniac, H. S. and Hempfling, W. P. (1979b). *Cryptococcus vishiacii sp. nov.*, an Antarctic yeast. *International Journal of Systematic Bacteriology* **29**, 153–158.

Watson, K., Arthur, H. and Shipton, W. A. (1976). *Leucosporidium* yeasts: obligate psychrophiles which alter membrane lipid and cytochrome composition with temperature. *Journal of General Microbiology* **97**, 11–18.

Weyant, W. S. (1966). The Antarctic climate. *In* "Antarctic Soils and Soil Forming Processes" (Ed. J. C. F. Tedrowe), pp. 47–49. Geography Union Publishing, Washington, District of Columbia.

Wiebe, W. J. and Hendricks, C. W. (1974). Distribution of heterotrophic bacteria in a transect of the Antarctic Ocean. *In* "Effect of the Ocean Environments on Microbial Activity" (Eds R. R. Colwell and R. Y. Morita), pp. 524–525. University Park Press, Baltimore, Maryland.

Wilkins, P. O. (1973). Psychrotrophic Gram-positive bacteria: temperature effects on growth and solute uptake. *Canadian Journal of Microbiology* **19**, 909–915.

Zobell, C. E. (1942). "Marine Microbiology." Chronica Botanica, Waltham, Massachusetts.

2

The Alkaline Saline Environment

W. D. GRANT AND B. J. TINDALL

Department of Microbiology, University of Leicester, Leicester LE1 7RH, UK and
Institut für Mikrobiologie, Rheinische-Friedrich-Wilhelms-Universität, 5300 Bonn 1, FRG

Introduction

Alkalinity in the terrestrial and aquatic environment may be generated by a number of factors, either naturally occurring or brought about by man's activities. Commercial processes ranging from cement manufacture to the preparation of indigo dye give rise to environments with high pH. In the case of environments produced by cement manufacture (or casting) and electroplating, the pH is usually in excess of 12, and this factor in conjunction with the presence of high concentrations of toxic ions (both anions and cations) seems to inhibit the establishment of an actively growing microbial population (Grant, unpublished results). Rather less alkaline industrial effluents are generated by the lye treatment of potatoes, rayon manufacture and traditional methods of indigo production. These wastes are capable of supporting the growth of a wide variety of organisms, although certain bacteria may predominate in any one particular environment (Gee *et al.*, 1980; Horikoshi and Akiba, 1982; Collins *et al.*, 1983).

Naturally occurring alkaline environments may arise by biological activity, such as ammonification, sulphate reduction and oxygenic photosynthesis (Abd-el-Malek and Rizk, 1963a,b,c; Brewer and Goldman, 1976; Langworthy, 1978). In these instances the alkalinity may be localized and relatively short lived, or even diurnal (as in the case of photosynthetically generated alkalinity).

The most stable, and most significant, naturally occurring alkaline environments are caused by a combination of geological, geographical and climatic conditions. Alkaline lakes and deserts so produced are geographically widely distributed (Table 1). These environments are characterized by

Table 1. *World-wide distribution of soda lakes and soda deserts (Te-Pang, 1890)*

Africa	Egypt: Wadi Natrun East Africa: Magadi, Lake Natron, Lake Bogoria, etc. Libya: Fezzan Central Africa: Lake Chad

The Americas	USA
	    California: Owens Lake (Inyo County), Searles Lake, Trona
	      (San Bernardino County), Borax Lake (Lake County)
	    Nevada: Ragtown Soda Lakes (near Carson Sink)
	    Wyoming: Union Pacific Lakes (near Green River and along
	      Union Pacific Railroad)
	    Oregon: Albert Lake
	Mexico: Lake Texcoco
	Venezuela: Lagunilla Valley
	Chile: Antofagasta

Asia	Siberia: Chita and Lake Baikal region, Barnaul, Slavgorod,
	    Kulunda Steppe region
	Armenia: Araxes Plain Lakes
	India: Lake Looner, etc.
	China:
	    Outer Mongolia: various "nors"
	    Sui-Yuan: Cha-Han-Nor, Na-Lin-Nor
	    Heilungkiang: Hailar, Tsitsihar
	    Kirin: Fu-U-Hsein, Taboos-Nor
	    Liao-Ning: Tao-Nan Hsien (Polishan Lake, Tafusu Lake, etc.)
	    Jehol: various soda lakes
	    Chahar: Cheng-Lang-Chi
	    Shansi: U-Tsu-Hsien
	    Shensi: Shen-Hsia-Hsien
	    Kansu: Ning-Hsia-Hsien
	    Tibet: alkali deserts

Europe	Russia: Caspian Sea region Hungary: Szegedin district

the presence of large amounts of sodium carbonate (or complexes of this salt), formed by evaporative concentration. In the course of the formation of alkalinity in this way, other salts (particularly NaCl) also concentrate, giving rise to alkaline saline environments. Thus naturally occurring alkalinity is usually associated with salinity. Examples of such environments that have been studied both geochemically and biologically include the alkaline saline (soda) lakes of East Africa (Beadle, 1932; Rich, 1932, 1933; Jenkin, 1932, 1936; Baker, 1958; Talling and Talling, 1965; Eugster, 1970; Talling *et al.*, 1973; Hecky and Kilham, 1973; Melack and Kilham, 1974; Melack, 1978;

Grant *et al.*, 1979; Tindall, 1980), and the Wadi Natrun in Egypt (Jannasch, 1957; Abd-el-Malek and Rizk, 1963a, b, c; Imhoff *et al.*, 1978, 1979). More restricted geochemical or biological investigations have been made on a variety of soda lakes in western America (Eugster and Hardie, 1978; Tew, 1966, 1980; Irgens, 1983), Hungary (Uherkovich, 1970) and the Kulunda Steppe region of the Soviet Union (Isachenko, 1951).

The formation of alkaline saline brines is generally characterized by a combination of three factors:

(a) the presence of geological conditions which favour the formation of alkaline drainage waters;
(b) suitable topography which restricts surface outflow from the drainage basin (often a closed basin);
(c) climatic conditions conducive to evaporative concentration.

The mechanisms contributing to the formation of alkalinity have been reviewed by a number of researchers (Cole, 1968; Eugster, 1970; Hardie and Eugster, 1970; Eugster and Hardie, 1978). The major influencing factor is the geology, since the ionic composition of the drainage water is determined by leaching of the surrounding rocks. However, the presence of large amounts of sodium carbonate-bearing rocks, as seen near Oldoinya Lengai in Tanzania, does not appear to be the sole factor in determining whether or not a lake will be alkaline (Eugster, 1970), since in most cases the carbonate is derived from solution of atmospheric or respired CO_2. One of the most significant factors governing the evolution of saline brines is the relative concentration of divalent cations (Mg^{2+} and Ca^{2+}) to anions (CO_3^{2-}, SO_4^{2-} and SiO_2). Even in situations of low evaporative concentration, groundwater becomes rapidly saturated with respect to $CaCO_3$, resulting in the deposition of calcite (often with coprecipitation of $MgCO_3$). It is this initial step, apparently universal to all drainage systems, which seems to determine whether or not alkalinity will be generated (Eugster and Hardie, 1978). When carbonate is present in concentrations greater than that of Ca^{2+} plus Mg^{2+}, then these cations are removed from solution, and carbonate may further concentrate to give rise to an alkaline soda brine. Geological conditions which favour this are alkali-rich (K^+ and Na^+ predominate) rather than calc-alkali-containing (Ca^{2+} and Mg^{2+} predominate) strata. In other situations where the concentrations of Ca^{2+} and Mg^{2+} exceed that of CO_3^{2-}, or where they are equimolar, the carbonate is removed from solution, and the genesis of alkaline brines is inhibited. In these situations the brines which result may be enriched in Ca^{2+} and Mg^{2+}, as in the first case, or depleted of these cations, as in the second case. Further modifications of the brines occur if significant amounts of SiO_2 or SO_4^{2-} are present, causing the precipitation of sepiolite or gypsum respectively. Fig. 1 illustrates in a much simplified form the three major

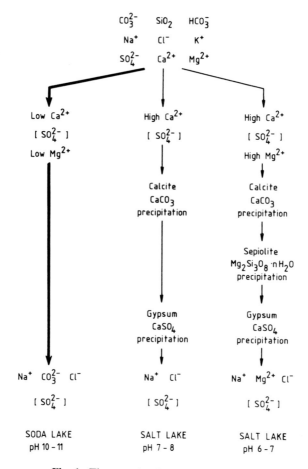

Fig. 1. The genesis of highly saline brines.

pathways. The Dead Sea is a typical example of a more acid saline lake enriched with cations, whereas the Great Salt Lake is a typical example of a brine depleted of neutral divalent cations.

The Kenyan–Tanzanian Rift Valley contains a number of alkaline soda lakes (Fig. 2), whose development is a consequence of the factors described previously. Geologically, alkaline trachyte lavas predominate and, owing to the topography of the graben, form a number of distinct, isolated drainage basins (Baker, 1958; Thompson and Dodson, 1963; McCall, 1967). In the majority of cases these are closed basins, although in some cases there may be subterranean outflow through established fault lines (Thompson and Dod-

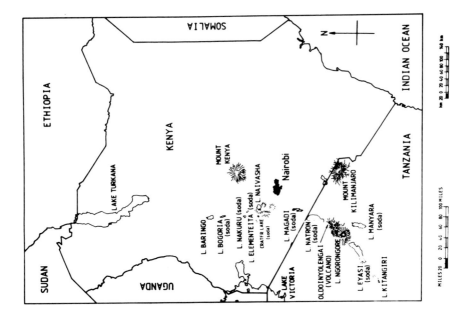

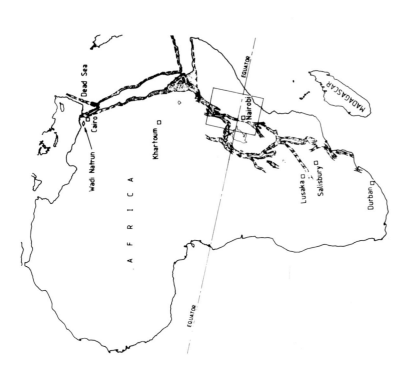

Fig. 2. The lakes of the East African Rift Valley.

son, 1963; McCall, 1967). Owing to variations in the climatic conditions, the total salinity of these lakes varies according to the geographical location, and also according to the seasonal weather conditions. The salinity of the alkaline saline lakes in the Kenyan Rift Valley ranges from 5% w/v, in the case of the more northerly lakes (Lakes Bogoria, Nakuru and Elmentieta), to saturation (30% or greater) in parts of Lake Magadi (Baker, 1958; Talling and Talling, 1965; Tindall, 1980). Chemical analyses of the lakes made in two different years are shown in Table 2, and they indicate that Na^+, Cl^- and CO_3^{2-} are the major ions responsible for the salinity of the lakes. The relative concentrations of Cl^- and CO_3^{2-} vary according to the stage of development of the brine. In lakes of lower salinity the concentration of CO_3^{2-} usually exceeds that of Cl^-, but in brines of higher salinity Cl^- exceeds the CO_3^{2-} concentration. Analyses of a greater number of lakes at earlier periods are to be found in the papers of Talling and Talling (1965) and Melack and Kilham (1974). Lake Magadi has been the subject of most study because of vast trona (natron) deposits ($NaHCO_3 \cdot Na_2CO_3 \cdot 2H_2O$—the stable crystalline product of Na_2CO_3, $NaHCO_3$ and CO_2 equilibrium), which are harvested and kilned to produce anhydrous Na_2CO_3 (soda ash) used in glass manufacture. There is also a common salt (NaCl) harvesting operation at the site with a series of solar evaporation ponds (salterns) where NaCl preferentially precipitates during the cooler night. The results of analyses on the salterns at Lake Magadi are also included in Table 2.

In contrast, the lakes of the Wadi Natrun, a localized depression outside the Rift Valley containing a dozen small lakes, do not show such a variation in chemical composition (Imhoff et al., 1979). These lakes most closely resemble Lake Magadi in composition, although the concentration of sulphate is higher than that found in the Kenyan lakes (Talling and Talling, 1965). The genesis of the brines in the Wadi Natrun has not been extensively studied, although it must be presumed that similar parameters prevail as in other soda lakes. Studies carried out in the laboratory, and in the field, suggest that sulphate reduction in the surrounding area may also significantly contribute to the formation of alkalinity (Abd-el-Malek and Rizk, 1963a, b, c).

The soda lakes of the Kulunda Steppe, studied by Isachenko (1951), were not analysed with respect to their water chemistry, although the pH of the three Tanatar Lakes is recorded. More detailed chemical analyses of the soda lakes in western America are available from a variety of sources (see Eugster and Hardie (1978) for a review). Because of the geographical distribution of these lakes there is a large degree of variation in their water chemistry, although there are no significant features not observed in either the lakes of the Kenyan Rift Valley or the Wadi Natrun.

Other sources of alkaline aquatic environments include natural springs.

Table 2. Chemical analyses of Kenyan lake waters (Tindall, 1980)

	Lake Bogoria	Lake Elmenteita	Lake Magadi	Lake Nakuru	Naivasha Crater Lake	Salterns
Analyses made in 1976						
Na^+ (mM)	494	1725	4600	410	154	—
K^+ (mM)	4.64	31.8	42.0	7.8	5.8	—
Mg^{2+} (mM)	0.002	0.001	0.001	0.005	0.038	—
Ca^{2+} (mM)	0.061	0.006	0.006	0.018	0.039	—
Cl^- (mM)	37.5	200	640	32.5	11.0	—
CO_3^{2-} (mM)	75	445	1950	60	28	—
Total alkalinity (mM)	382	1075	2380	291	137	—
pH (measured)	10.1	10.4	10.5	10.15	10.0	—
pH (calculated)	9.76	10.22	11.03	9.79	9.78	—
Analyses made in 1978						
Na^+ (mM)	652	300	2000	400	—	10 200
K^+ (mM)	5.0	4.5	16.0	3.1	—	215
Mg^{2+} (mM)	0.002	0.001	b.l.d	0.001	—	b.l.d
Ca^{2+} (mM)	0.045	0.165	0.0175	0.040	—	b.l.d
Cl^- (mM)	85	67	400	33	—	4500
CO_3^{2-} (mM)	220	54	582	57	—	1654
Total alkalinity (mM)	546	133	1304	123	—	2233
pH (measured)	10.4	10.5	10.35	10.45	—	10.8
pH (calculated)	10.2	10.21	10.28	10.31	—	10.8
PO_4^{3-} (mM)	0.004	0.05	0.008	0.06	—	0.10
NH_4^+ (mM)	0.23	0.042	0.42	0.042	—	0.25
Total nitrogen (mM)	0.25	0.25	2.2	0.20	—	0.80

b.l.d., ion concentrations below the level of detection; —, not done.

In some cases these are also thermal, and may or may not be associated with alkaline lakes. These springs are generally of low salinity and have not usually been found to have pH values in excess of 9.0. The origin of the alkalinity is similar to that of alkaline lakes. However, other springs are known where the genesis of the alkalinity is not understood but is not due to sodium carbonate (Souza and Deal, 1977).

The Ecology of the Saline Alkaline Environment

Until comparatively recently, studies on the biota of alkaline saline environments were restricted to the macroflora and macrofauna, microbiological aspects with the exception of algae and cyanobacteria being largely neglected. Microorganisms inhabiting alkaline environments may be divided into alkaliphiles and alkali-tolerant types. A general working definition is that an alkaliphile is an organism with an obligate requirement for alkaline growth conditions (i.e. above pH 8.0), while an alkali-tolerant organism is able to grow at alkaline pH but grows optimally at lower pH.

In common with other extreme environments, in alkaline environments species diversity decreases as the salinity and/or alkalinity increases. Thus, the less saline soda lakes of the Kenyan Rift Valley contain a wide variety of invertebrates, algae and cyanobacteria (Beadle, 1932; Jenkin, 1932, 1936; Rich, 1932, 1933; Lind, 1968; Hecky and Kilham, 1973). The phytoplankton is dominated by cyanobacteria, although diatoms of the genera *Nitzchia, Navicula* and *Cyclotella* are often present in considerable numbers, particularly in less concentrated and less alkaline conditions (Hecky and Kilham, 1973). Zooplankton is characteristically the rotifer *Brachtonus plicatilis* and the copepod *Paradiaptomus africanus*. In some of the lakes *Artemia* spp. and the cichlid fish *Tilapia grahami* are to be found. The highly saline and alkaline regions of Lake Magadi and the lakes of the Wadi Natrun, in contrast, do not appear to contain organisms ranking higher than protozoa (Imhoff *et al.*, 1979; Grant and Tindall, unpublished results).

One of the most striking features of naturally occurring alkaline lakes is the predominance of microorganisms, particularly phototrophs, as permanent or seasonal blooms (Melack, 1978; Imhoff *et al.*, 1978, 1979; Tindall, 1980). This is reflected in the high primary productivity commonly encountered in these lakes (Talling *et al.*, 1973; Melack and Kilham, 1974). In the less saline lakes of the Kenyan Rift Valley the primary productivity is due to the presence of dense populations of cyanobacteria, while in the highly saline lakes of the Wadi Natrun, or Lake Magadi, anoxygenic phototrophic bacteria as well as cyanobacteria may be bloom forming (Jannasch, 1957;

Imhoff *et al.*, 1978, 1979; Tindall, 1980; Grant and Tindall, unpublished results).

Investigations of the cyanobacteria present in the soda lakes indicate that in the less saline lakes the predominant species are the filamentous *Anabaenopsis arnoldii* (Fig. 3), *Spirulina platensis* (Fig. 4), or a related species *Spirulina maxima*, and certain unicellular species which may be *Chroococcus* sp., *Synechococcus* sp. or *Synechocystis* sp. (Fig. 5). Although detailed studies have yet to be carried out on the ecology of the cyanobacteria of these lakes, it is clear from casual observations that they are subject to seasonal changes, probably dependent on the salinity of the lake (Brown, 1959; Tindall, 1980). The influence of the cyanobacteria on the ecology of the lakes is dramatically illustrated by the changes in the flamingo population of the lakes. The flamingoes are filter feeders, concentrating their food by a sieve-like mechanism in the beak. The lesser flamingoes in particular are able to feed on the large filamentous cyanobacteria, such as *Spirulina* sp. or *Anabaenopsis* sp., and these birds may be seen to migrate preferentially to those lakes which contain these cyanobacteria. Such is the primary productivity of the lakes that at Lake Nakuru (an area of 11.5 square miles) numbers of up to 1 million birds have been counted, consuming on average 180 tons day^{-1} (Brown, 1959). It appears that the disappearance of the filamentous forms from the lakes is not due to overgrazing but is a result of changes in the water

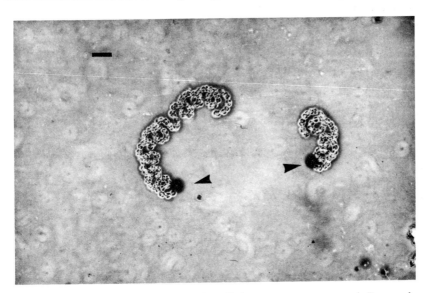

Fig. 3. *A. arnoldii* sample from Lake Magadi; heterocysts are arrowed. Bar marker 10 μm. Phase contrast.

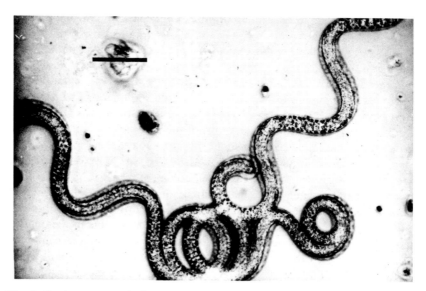

Fig. 4. *S. platensis* sample from Lake Bogoria. Bar marker 20 μm. Phase contrast.

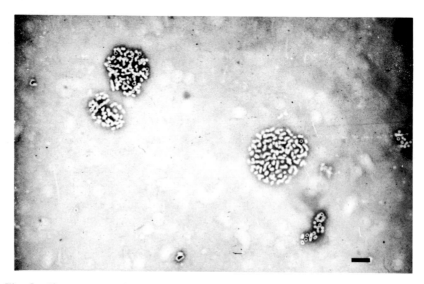

Fig. 5. *Chroococcus* sp.? sample from Lake Elmentieta. Bar marker 10 μm. Phase contrast.

chemistry of the lakes brought about by evaporative concentration or rainfall. Vareschi (1978, 1982) has recorded variations in flamingo populations in East Africa over a decade in relation to cyanobacterial blooms.

Apart from serving as a good source for flamingoes, members of the genus *Spirulina* have also served as a human food source in various parts of the world. Owing to the presence of gas vacuoles, the large coiled filaments float to the surface and may collect as a visible scum on the water, making collection and drying of the organisms relatively easy. Studies on the nutritive value of *Spirulina* spp. have shown that, apart from having a balanced amino acid composition suitable for use as a human food supplement, they contain comparatively little nucleic acid and cell wall material and have no apparent toxicity (Ciferri, 1983). This food supplement historically used by the Aztecs (tecuitlatl) and the natives around Lake Chad (dihé) is now receiving closer study as a potential prokaryote protein source. At present pilot-scale work is being carried out on the production of *Spirulina* spp. as a human and animal food supplement, although the current market is largely as a health food (Ciferri, 1983).

Despite the economic interest shown in members of the genus *Spirulina* there does not appear to be agreement as to the taxonomic position of this genus. In many of the older works, such as the studies on the soda lakes by Jenkin (1932, 1936) and Rich (1932, 1933), the generic name *Arthrospira* was used for some species currently bearing the generic epithet *Spirulina* (Ciferri, 1983). More detailed studies are required to resolve the taxonomy of this genus fully, since in some cases members of the genus have been misclassified into the genus *Microcoleus*, while other workers are of the opinion that *Spirulina* spp. are no more than spiral forms of the genus *Oscillatoria* (Ciferri, 1983). Previous studies have shown that *S. platensis* and *S. maxima* may be separated from other members of this genus as they appear to be typical alkaliphiles, growing optimally between pH 8.0 and 11.0 and being inhibited by high levels of divalent cations, such as Mg^{2+}.

Anoxygenic phototrophic bacteria are also capable of forming visible blooms in soda lakes. In early studies, based on morphological data, Isachenko (1951) and Jannasch (1957) observed visible accumulations of anoxygenic phototrophs, which they classified as members of the genera *Chromatium* and *Thiospirillum*. In recent years a number of anoxygenic phototrophs have been isolated from soda lakes, including members of the genera *Chromatium, Rhodopseudomonas, Lamprocystis, Thiocapsa* and *Ectothiorhodospira* (Tew, 1966, 1980; Imhoff *et al.*, 1978, 1979; Grant *et al.*, 1979; Tindall, 1980; Irgens, 1983; Tindall and Trüper, 1984). Physiological studies on the majority of strains isolated indicate that pH optima greater than 8.5 are exceptional. Thus many of these strains are only alkali tolerant and are not characteristic of the soda lakes (Tindall, unpublished results).

The most notable exceptions are members of the genus *Ectothiorhodospira*. Under appropriate conditions these organisms may form visible blooms in alkaline lakes of differing salinities. In the less saline lakes of the Kenyan Rift Valley cyanobacterial blooms predominate throughout the year, but in exceptional circumstances visible blooms of the less halophilic members of the genus *Ectothiorhodospira* may be found (Tindall, 1980). In the more saline lakes of the Wadi Natrun and at Lake Magadi halophilic members of this genus are also apparent (Imhoff *et al.*, 1979). At Lake Magadi the central part of the lake consists of a thick deposit of trona some hundreds of metres deep. Digging down through the upper layers reveals a clear (Fig. 6) stratification. The uppermost region is orange–pink (with some bleaching at the surface) owing to the presence of archaebacterial halophiles (see later). Below this are several distinct green- or purple–red-coloured horizons. The layer immediately underlying the surface is pigmented green owing to the presence of cyanobacteria (Tindall, 1980; Grant and Tindall, unpublished results). The next horizon is purple–red and contains a number of different species of the genus *Ectothiorhodospira* (Tindall, 1980). The presence of a second green layer below this is suggestive of the green-coloured members of the genus, i.e. *E. halochloris* and *E. abdelmalekii* (Imhoff and Trüper, 1977, 1981), or members of the family Chlorobiaceae, although no evidence was found for the presence of these organisms (Tindall, 1980). The green-

Fig. 6. Section of trona crust, Lake Magadi.

coloured *E. halochloris* and *E. abdelmalekii* have to date only been isolated from the Wadi Natrun lakes (Imhoff and Trüper, 1977, 1981). The remaining visible layer comprised trona impregnated with black alkaline (pH 11.0) odorous, sediment-laden water. Although sulphide production by sulphate-reducing bacteria has been postulated in Lake Magadi, and in the lakes of the Wadi Natrun, no such organisms have so far been isolated (Grant and Tindall, unpublished results).

Members of the genus *Ectothiorhodospira* are found in a variety of environments, including brackish water, marine water and highly saline lakes (Trüper, 1968, 1970; Raymond and Sistrom, 1967; Imhoff *et al.*, 1978, 1979, 1981; Trüper and Imhoff, 1981), although the pH optimum of all known species is in the alkaline region. Probably the first members of this group to be isolated were the strains designated "pseudomonas-type" purple sulphur bacteria (van Niel, 1931) and "*E. mobile*" (Pelsh, 1937). Both van Niel (1931) and Pelsh (1937) recognized the physiological distinction between their isolates and other members of the family Chromatiaceae (at that time "Thiorhodaceae"). van Niel was able to demonstrate the alkaline pH optima of these strains (van Niel, 1931), as did Chesnokov and Shaposhnikov (1936) for "*E. mobile*" when grown on sulphide-containing media. The ability of the strains to oxidize sulphide to sulphate, with the extracellular deposition of sulphur as an intermediate, was also noted as a characteristic of the organisms (van Niel, 1931; Chesnokov and Shaposhnikov, 1936; Pelsh, 1937), and led Pelsh (1937) to propose the creation of a new family, the "Ectothiorhodaceae", to accommodate them.

Much of the debate regarding the taxonomic position of these organisms subsided with the loss of the type species. It was not until Trüper (1968) re-isolated *E. mobilis* that the debate was reopened, although the existence of an autotrophic "*Rhodopseudomonas* sp.", later called *E. shaposhnikovii*, had been known for some time (Kondrat'eva, 1956; Moshentseva and Krondrat'eva, 1962; Uspenskaya and Kondrat'eva, 1962; Kondrat'eva and Malofeeva, 1964). The existence of an extremely halophilic member of this genus, *E. halophila*, was reported by Raymond and Sistrom (1967, 1969). On the basis of the ability of all these organisms to utilize sulphide, Trüper (1968) suggested that they be placed in the family Chromatiaceae, since sulphide utilization by members of the "purple non-sulphur" bacteria was at that time not known. However, Pfennig (1977) argued that on Molisch's (1907) original criteria these organisms should be grouped with the family Rhodospirillaceae, since the cells did not store sulphur intracellularly. The ability of members of the genus *Ectothiorhodospira* to deposit sulphur outside the cell does not distinguish this genus from certain members of the Rhodospirillaceae and, although *E. mobilis* and *E. shaposhnikovii* are capable of further oxidation of the sulphur, no evidence has been found that the extremely halophilic members of this genus photooxidize sulphur (Tindall, 1980; Then, 1984; Then and Trüper, 1984). More detailed investigations of the fine structure and chemical composition of the cells (Remsen *et al.*, 1968; Cherni *et al.*, 1969; Holt *et al.*, 1971; Oyewole and Holt, 1976; Grant *et al.*, 1979; Tindall, 1980) indicated a unique intracellular membrane structure (Fig. 7), which is easily distinguished from other similar cytoplasmic intrusions found in phototrophs (Tindall, 1980; Tindall and Grant, in press). Such features, taken together with a polar lipid composition unique among the phototrophs (Kenyon, 1978; Tindall, 1980), led Tindall (1980) to suggest that a family, the Ectothiorhodospiraceae (equivalent to Pelsh's "Ectothiorhodaceae"), be created to accommodate these organisms. Following the isolation of new species, *E. vacuolata, E. halochloris* and *E. abdelmalekii* (Imhoff and Trüper, 1977, 1981; Tindall, 1980; Imhoff *et al.*, 1981), further studies on the lipids, quinones, cytochromes and 16S rRNA cataloging (Collins *et al.*, 1981; Asselineau and Trüper, 1982; Imhoff, 1982, 1984a; Imhoff *et al.*, 1982; Meyer, 1984; Stackebrandt *et al.*, 1984) have confirmed this view and have led to the creation of the family *Ectothiorhodospiraceae* (Imhoff, 1984b). Current data appear to indicate a clear division of the halophilic species. *E. halophila, E. abdelmalekii* and *E. halochloris* from the moderately halophilic species *E. vacuolata, E. mobilis* and *E. shaposhnikovii*, perhaps as two distinct genera (Stackebrandt *et al.*, 1984).

In the highly saline alkaline lakes of the Wadi Natrun, and at Lake

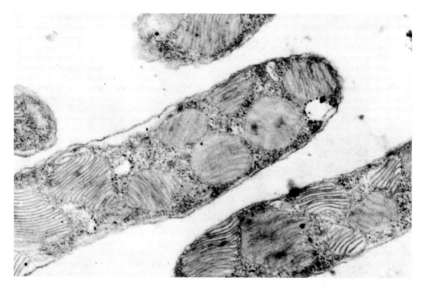

Fig. 7. *E. mobilis* isolated from Lake Bogoria. Bar marker 1 μm. Thin section.

Magadi, the brines are usually coloured by the presence of characteristic halophilic archaebacteria as well as members of the genus *Ectothiorhodospira* (Imhoff *et al.*, 1978, 1979; Grant and Tindall, 1980; Tindall, 1980; Soliman and Trüper, 1982). While the coexistence of supposedly aerobic, halophilic archaebacteria and anaerobic anoxygenic phototrophs can be demonstrated microscopically, simple laboratory experiments illustrate environmental selectivity. If a sample of saltern liquor is incubated in the medium of Imhoff and Trüper (1977) anaerobically in the light, then members of the genus *Ectothiorhodospira* are enriched, while samples inoculated into the medium of Tindall *et al.* (1980) and incubated aerobically, with shaking, in the light enrich for halophilic archaebacteria. However, when the medium of Tindall *et al.* (1980) is inoculated and left static in the light, without exclusion of air, then a coculture of archaebacterial halophiles and halophilic members of the genus *Ectothiorhodospira* develops.

These differences can be seen in the absorption spectra (Fig. 8) and in the colour of cultures. Cultures containing only halobacteria are typically orange–red, while those enriched with halophilic *Ectothiorhodospira* spp. are violet–red. The mixed culture has a characteristic claret colour, which is

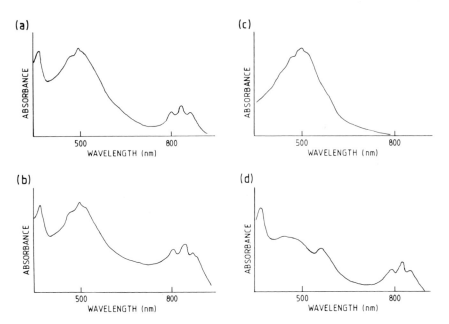

Fig. 8. Absorption spectra of saltern liquor and enrichment cultures: (a) saltern liquor; (b) enrichment culture for haloalkaliphilic archaebacteria; 1 week shaken or unshaken; (c) enrichment culture for haloalkaliphilic archaebacteria; 2 weeks shaken; (d) enrichment culture for haloalkaliphilic archaebacteria; 2 weeks unshaken.

almost identical with that of natural brine samples (Tindall, 1980; Grant and Tindall, unpublished results), indicating the importance of both organisms in these environments. Anaerobic anoxygenic phototrophs are able to grow in these shallow waters because Na_2CO_3–NaCl-saturated brines are essentially devoid of dissolved oxygen (Grant and Tindall, unpublished results).

Archaebacterial halophiles, both coccoid (halococci) and rod-shaped (halobacteria) forms, are known to be characteristically associated with salt lakes and salterns in tropical areas of the world, causing striking red pigmentation of the liquor similar to that seen at Lake Magadi (Kushner, 1978), although the majority of salterns operate at neutral pH. It is now clear that extremely alkaline salt ponds and salt lakes contain large numbers of archaebacterial halophiles quite distinct from the halobacteria and halococci found in neutral ponds and salt lakes. Initial isolations of halophilic archaebacteria from Lake Magadi used media normally employed for the cultivation of members of the genus *Halobacterium* but adjusted to pH 9.5 (Tindall, 1980). Poor growth in such media could be attributed to the toxic effects of Mg^{2+} (Tindall, 1980; Tindall *et al.*, 1980; Soliman and Trüper, 1982). Thus, these halophilic archaebacteria differed in two fundamental ways from normal halobacteria and halococci, in that they showed alkaline pH optima and, instead of requiring high concentrations of Mg^{2+} for growth, were inhibited by its presence. Initial studies indicated that these haloalkaliphiles were represented only by rod-shaped organisms (Tindall, 1980), although a single isolate of a non-motile coccus is now known (Ross, 1982). Tindall *et al.* (1984) have recently classified these haloalkaliphilic archaebacteria into two new genera *Natronococcus* and *Natronobacterium* for the coccoid and rod-shaped isolates respectively.

Natronococci and natronobacteria are unusual in that they have membranes based on a C_{20},C_{25} diether core lipid as well as the universal C_{20},C_{20} archaebacterial core lipid (Ross *et al.*, 1981; De Rosa *et al.*, 1982; Tindall *et al.*, 1984). Some isolates also contain a third C_{25},C_{25} diether core lipid (Fig. 9) (De Rosa *et al.*, 1983). Examination of the polar lipids of the haloalkaliphilic strains indicated that they had an unusual polar lipid pattern (Tindall, 1980; Tindall *et al.*, 1984). Extensions of this work have led to the discovery of a number of other distinct polar lipid patterns among halophilic archaebacteria, suggesting a number of other major taxa (Ross, 1982). Ross and Grant (1985) have recently shown that nucleic acid hybridization studies further support the existence of several new taxa among the archaebacterial halophiles in addition to *Natronococcus* and *Natronobacterium*.

Haloalkaliphlic archaebacteria appear to possess similar carotenoid pigments to other halophilic archaebacteria, these probably serving a photoprotective function as well as a role in the cell membrane (Bayley and Morton, 1978; Tindall, 1980). However, haloalkaliphilic archaebacteria do not appear

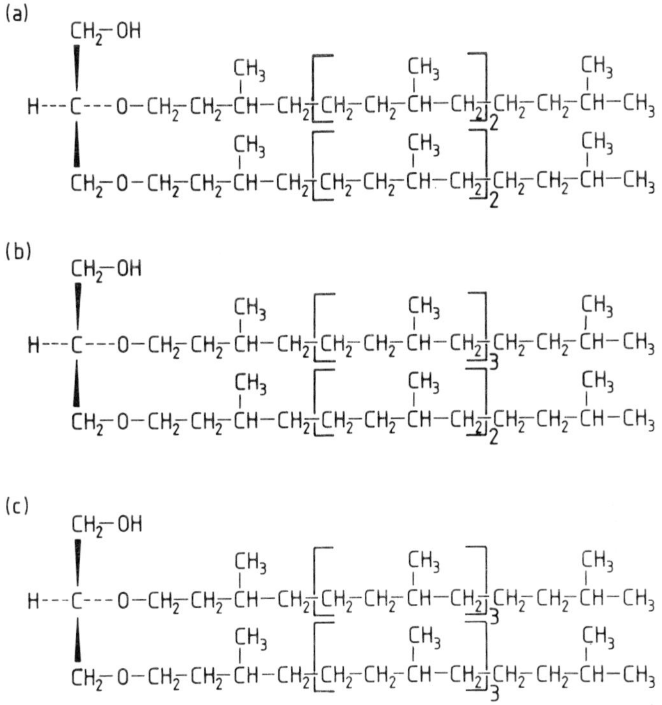

(a)

(b)

(c)

Fig. 9. Isopranoid diether core lipids from haloalkaliphilic archaebacteria: (a) 2,3-di-O-phytanyl-sn-glycerol (C_{20}, C_{20}); (b) 2-O-sesterterpanyl-3-O-phytanyl-sn-glycerol (C_{20}, C_{25}); (c) 2,3-di-O-sesterterpanyl-sn-glycerol (C_{25}, C_{25}).

to contain a bacteriorhodopsin proton pump, although they do contain small amounts of pigments which are known to be photoactive and retinal based (Stoeckenius, personal communication).

Other organisms present in visible accumulations in alkaline environments are the cyanobacteria–*Chloroflexus* mats characteristically associated with alkaline hot springs throughout the world (Brock, 1978; Tindall, 1980). These organisms do not appear to be subjected to very high pH for long periods of time, and are probably not extremely alkaliphilic. However, the cyanobacteria–*Chloroflexus* associations (Fig. 10) found in the hot springs on the shores of Lake Bogoria seem to be tolerant of seasonal submersion in the alkaline waters at times of high rainfall and flooding of the lake.

A variety of heterotrophic organisms have also been isolated from a number of alkaline environments. In the studies of Horikoshi and coworkers, reviewed by Horikoshi and Akiba (1982), the majority of these isolates are

Fig. 10. *Chloroflexus*-cyanobacterial mat at the west shore of Lake Bogoria (68 °C, pH 9.0).

from alkaline, but not saline, environments. The vast majority of heterotrophic organisms which have been studied are aerobic, spore-forming, Gram-positive, motile rods, and have consequently been placed in the genus *Bacillus*. Alkaliphilic members of this genus include *B. alcalophilus, B. alcalphilus* subsp. *halodurans, B. pasteurii, B. firmus, B. lentus, B. rotans, B. sphaericus* and *B. pantothenticus*. A number of other species are apparently capable of adapting to growth in alkaline media (Gordon, 1981). A large number of alkaliphilic *Bacillus* spp. (Horikoshi and Akiba, 1982) do not appear to have been formally classified, while the studies of Gordon *et al.* (1977) indicate that the *B. lentus–B. firmus* group represents a microbial spectrum. Comparative studies on *B. alcalophilus, B. alcalophilus* subsp. *halodurans* and a haloalkaliphilic *Bacillus* sp. from the Wadi Natrun indicated that the latter strain inhibited greater salt tolerance, and was probably taxonomically distinct from the other two species (Weißer, 1983). Comparative studies on the lipids of some alkaliphilic strains of *Bacillus* indicate specific differences within the group and also suggest that they are taxonomically distinct from other, non-alkaliphilic *Bacillus* sp. (Tindall and Grant, unpublished results). At present the taxonomy of the genus *Bacillus* appears to be in a state of flux (Berkeley and Goodfellow, 1981), and the exact taxonomic status of the alkaliphilic members of this genus remains to be determined.

An unusual "coryneform"-like bacterium which has been isolated from the waste of a potato lye treatment factory (Gee *et al.*, 1980) has been recently placed in a new genus represented by the species *Exiguobacterium aurantiacum* (Collins *et al.*, 1983). Careful comparison of this organism with the data given by Horikoshi and Akiba (1982) for *Corynebacterium* sp. strain 93–1 suggests that their organism is probably also a member of this genus.

A number of other bacteria from alkaline environments have also been described, and include members of the genera *Rhizobium* (Allen and Holding, 1974), *Thiobacillus* (Rupela and Tauro, 1973a, b), and *Flavobacterium* (Souza and Deal, 1977). Perhaps the most interesting organism reported from alkaline environments is the so-called living fossil "*Kakabekia umbellata*" (Siegel and Giumarro, 1966). However, there are apparently no extant cultures, and attempts to isolate this organism from the same source as Siegel and Giumarro (1966) did not prove successful (Grant and Tindall, unpublished results).

In a number of different bacterial taxa there are examples of species which are alkali tolerant. The alkali tolerance of the enteric bacteria is well known, and isolation of members of the enteric streptococci (some of which now belong to the genus *Enterococcus*) may be achieved at alkaline pH. Other organisms capable of growing at alkaline pH include members of the genera *Pseudomonas, Sarcina, Micrococcus, Proteus* and *Bacillus*, although they are not obligate alkaliphiles (Horikoshi and Akiba, 1982).

Properties of Alkaliphiles

A general property of alkaliphiles is their ability to grow optimally at pH values above 8.0. Investigations of a number of different taxa have indicated that it is common for alkaliphiles to exhibit more than one pH optimum (Gee *et al.*, 1980; Tindall, 1980; Weißer, 1983). For heterotrophic organisms of the genera *Exiguobacterium* and *Bacillus* there appears to be no explanation for the occurrence of two distinct optima (Gee *et al.*, 1980; Weißer, 1983). In these cases the lower maximum is between pH 8.0 and 9.0, while the upper maximum is between pH 9.5 and 10.0. Closer examination of the pH optima in members of the genus *Ectothiorhodospira* (Chesnokov and Shaposhnikov, 1936; Tindall, 1980) indicated that the pH optimum may be related to the redox potential of the photosynthetic electron acceptor. Thus, different sulphur compounds may give different pH optima, ranging from pH 7.0 to 8.5, while organic compounds give more alkaline optima (pH 9.0–10.0). In such cases the organism may appear to be both an alkaliphile and alkali-tolerant, depending on the cultural conditions. The description of these organisms as alkaliphiles in such instances requires qualification.

A simple chemical constraint on alkaliphiles is the scarcity of Mg^{2+} and Ca^{2+} in the environment. In typical naturally occurring alkaline environments, such as the soda lakes, the presence of high concentrations of CO_3^{2-} results in the immediate precipitation of these alkali earth metals as insoluble carbonates. Thus, it is not surprising that not only do alkaliphiles survive in media with low concentrations of these divalent cations but also higher levels are toxic (Grant and Tindall, 1980; Tindall, 1980). The most striking example of this is in the genera *Natronobacterium* and *Natronococcus* (Tindall, 1980; Tindall *et al.*, 1980, 1984; Ross, 1982; Soliman and Trüper, 1982). These organisms are, phylogenetically, members of the extremely halophilic archaebacteria (Ross, 1982; Tindall *et al.*, 1984; Ross and Grant, 1985), which are usually characterized by their high salt optima and requirement for high levels of Mg^{2+} (0.2–1.0 M). However, the optimal Mg^{2+} concentration for the haloalkaliphilic strains is 2 mM or less (Tindall *et al.*, 1980; Soliman and Trüper, 1982). While it is known that the ribosomes of certain *Halobacterium* species require elevated Mg^{2+} concentrations, whether this also applies to the haloalkaliphiles and if so how this is achieved in the haloalkaliphiles is not known.

A feature of certain alkaliphiles is their absolute requirement for Na^+ (Horikoshi and Akiba, 1982). In some strains capable of growing at both alkaline and neutral pH a requirement for NaCl is observed at lower pH which is not apparent at high pH, presumably because of the abundance of Na_2CO_3. Studies on a number of organisms have indicated that Na^+ is an important symporter in alkaliphiles (Koyama *et al.*, 1976; Krulwich *et al.*, 1979; Karazonov and Ivanosky, 1980; Garcia *et al.*, 1983; Krulwich and Guffanti, 1983), presumably because of the scarcity of protons and the difficulties in establishing a normal proton gradient (internal alkaline) at alkaline pH.

Studies on the internal pH of a number of alkaliphiles favour the concept that it does not exceed pH 9.0–9.5 (Guffanti *et al.*, 1978; Horikoshi and Akiba, 1982; Krulwich and Guffanti, 1983). In some strains when the pH exceeds 9.5 the pH gradient is reversed, becoming internal acid (Guffanti *et al.*, 1978). Under such conditions the ΔpH is reversed, but the Δψ increases with the increasing pH to such an extent that the transmembrane potential changes from −84 mV at pH 9.0 to −152 mV at pH 11.5. In contrast, the ΔpH values change from +36 mV at pH 10.0 to +151 mV at pH 11.5. Presumably growth ceases at pH 11.5 because no net transmembrane protonmotive force exists (Guffanti *et al.*, 1978), i.e. the transmembrane electrical and proton gradients are equal and opposite.

The alkaliphile *B. pasteurii* is unusual in that it has an obligate requirement for ammonia which can be met by a combination of high pH and NH_4^+ ions (NH_4^+ being in equilibrium with ammonia at high pH). However, when the cells are disrupted enzyme activity does not require these two factors (Wiley and Stokes, 1962, 1963). Measurement of the internal pH suggests that it is near neutral, and the principal role of the ammonia is substrate transport across the cell membrane. No such effects have been observed in other alkaliphiles studied to date.

Investigations of the chemical composition of the cells of alkaliphiles have not often sought to explain the alkaliphilic nature of the bacteria. However, investigations of both proteins and lipids lead to some interesting observations. Considering the polar lipid composition of the alkaliphiles studied to date, phylogenetically diverse organisms show similar properties (Tindall, 1980; Asselineau and Trüper, 1982; Imhoff et al., 1982; Nishihara et al., 1982; Ross, 1982; Tindall et al., 1984; Tindall, unpublished results). In all alkaliphiles so far studied the polar lipids are composed exclusively of phospholipids, and they do not possess glycolipids. In members of the genus Bacillus (Nishihara et al., 1982; Tindall, unpublished results) the major polar lipids present are phosphatidylglycerol (PG), phosphatidylethanolamine (PE), diphosphatidyglycerol (DPG) and bis(monoacyl)glycerol phosphate (BMP). An interesting difference has been observed between the less halotolerant and more halotolerant strains of alkaliphilic Bacillus spp. (Tindall, unpublished results). In the more halotolerant strain WN13 (Weißer, 1983) and similar organisms from Owens Lake, California (Tindall, unpublished results), the levels of PE are very low, while in all other strains there are significant amounts of this lipid. A correlation between these results and the menaquinone composition of the two groups has also been observed (Tindall, unpublished results). The polar lipid pattern of the taxonomically unrelated E. aurantiacum (Collins et al., 1983) also indicates significant quantities of DPG and PE but small amounts of PG and a phospholipid with an R_f value similar to that of BMP.

The polar lipids of the species E. mobilis, E. vacuolata and E. shaposhnikovii contain, in addition to DPG, PG, PE and a lipid apparently similar to BMP, i.e. the quaternary nitrogen-containing lipid phosphatidylcholine (Tindall, 1980; Asselineau and Trüper, 1982; Imhoff et al., 1982; Tindall, unpublished results). Lyso derivatives of these lipids may also be present. In the extremely halophilic members of this genus the levels of PE are significantly reduced, and PE is usually present in only trace amounts (Asselineau and Trüper, 1982; Imhoff et al., 1982; Tindall, unpublished results).

The haloalkaliphilic archaebacteria, currently comprising the two genera Natronobacterium and Natronococcus (Tindall et al., 1984), show differences in their polar lipids from other halobacteria. Like the alkaliphilic eubacteria they contain only phospholipids, largely the diether derivatives of PG and phosphatidyglycerophosphate (Tindall, 1980; Ross, 1982; Tindall *et al.*, 1984; Ross and Grant, 1985). The occurrence of the core diether lipids 2,3-di-*O*-phytanyl-*sn*-glycerol and 2-*O*-sesterterpanyl-3-*O*-phytanyl-*sn*-glycerol (Ross *et al.*, 1981; De Rosa *et al.*, 1982) is common to all alkaliphilic archaebacteria, although the presence of these two diethers is not confined to haloalkaliphiles (Ross *et al.*, 1981; Ross, 1982; Tindall *et al.*, 1984), indicating that alkalinity is not the principal reason for the synthesis of the different diethers but has a phylogenetic basis (Ross, 1982).

Proteins of alkaliphiles have not been extensively studied to date, although it is clear that there may be specific changes in protein structure in response to alkalinity. In one study, the response of an alkali-tolerant *Bacillus* sp. to growth at pH 7.5 and 10.2, it was shown that there were quantitative differences in the cell membrane protein composition (Koyama *et al.*, 1983). Studies on an alkaliphilic *Bacillus* sp. (Koyama and Nosoh, 1976; Koyama *et al.*, 1976) grown at pH 10.0 and 8.2 also indicated that there were specific differences in the protein composition, which were reflected in the greater negative charge of the proteins at pH 10.0. Similarly, analysis of the cell envelope of a large number of alkaliphilic *Bacillus* spp. (Horikoshi and Akiba, 1982) indicated that in some strains there were large amounts of negatively charged compounds. This may be one of the reasons why the cell membrane requires alkaline pH for the transport of charged substances, while intracellular enzymes function optimally at neutral pH (Ohta *et al.*, 1975). Studies on the protein and amino acid composition of the haloalkaliphilic archaebacteria also indicated specific differences between neutrophilic halobacteria and the haloalkaliphilic strains (Tindall, 1980). However, the interpretation of the data is complicated, because the differences may also be due to the phylogenetic relationships of the organisms. At present no generalization can be made as to the influence of alkalinity on the proteins of alkaliphilic organisms.

Apart from the environmental stress of alkalinity, organisms living in alkaline saline environments are also subjected to osmotic stress. The alternatives for an organism subjected to increasing salinity are to permit the entry of ions or to develop mechanisms for excluding ions. Active mechanisms for the exclusion of Na^+ include the substitution of this ion by other ions or the use of organic molecules to generate an intracellular osmotic potential. Active exclusion of Na^+ and its replacement with K^+ is typical of the halophilic archaebacteria. This mechanism results in the development of large transmembrane K^+ and Na^+ gradients. The internal ion concentration of a typical halobacterium such as *Halobacterium salinarium* is approximately 4.5 M K^+ and 1.3 M Na^+ (Bayley and Morton, 1978), while the external concentrations of these ions are 0.03 M and 4.0 M respectively. Studies on osmoregulation in alkaliphilic, halophilic or halotolerant eubacteria are scarce and are restricted to *Ectothiorhodospira* spp. (Galinski and Trüper, 1982, 1984), *Bacillus* sp. (Weißer, 1983) and cyanobacteria of the genera *Synechococcus, Synechocystis* and possibly *Spirulina* (Mackay *et al.*, 1984). Comparison with other data from halophilic or halotolerant eubacteria suggests that no exceptional osmoregulatory mechanism exists in alkaliphiles.

Typical organic osmoregulatory compounds (compatible solutes) may be divided into three groups:

(a) sugars, e.g. sucrose, glucose, trehalose and fructose;
(b) polyols, e.g. glycerol and glycosylglycerol;
(c) nitrogenous compounds, e.g. glutamine betaine and glycine betaine.

Organisms containing sugars as the sole compatible solute are less relatively salt tolerant, while those capable of synthesizing large amounts of the quaternary nitrogen compounds glycine betaine and glutamine betaine are tolerant of high salinity (Mackay *et al.*, 1984). Organisms producing glycosylglycerol appear to occupy an intermediate position. In some organisms producing betaine derivatives, other compatible solutes such as sucrose or trehalose are also present (Galinski and Trüper, 1984; Mackay *et al.*, 1984). In the extremely halophilic members of the genus *Ectothiorhodospira* a novel amino acid has been found with possible osmoregulatory function (Galinski, personal communication), although it may not be unique to this genus. The quantitative studies on *Ectothiorhodospira* spp. indicate that apart from an internal Na^+ concentration of 1.0 M there is a maximum glycine betaine concentration of 2.5 M, while the maximum concentrations of trehalose and the novel compound both reach 0.5 M depending on the growth conditions (Galinski and Trüper, 1984; Galinski, personal communication). Studies on enzymes of *Ectothiorhodospira* spp. (Tabita and MacFadden, 1976; Galinski, 1980; Meyer, 1984), and *Bacillus* sp. (Weißer, 1983) indicate that they are in general not halophilic, confirming the findings of lower internal salinity (Galinski, 1980). An exception appears to be the cytochrome c-551 isolated from *E. halochloris* and *E. abdelmalekii* by Then (1984). Surprisingly in some cases where the enzymes were not halophilic betaine was not found to act as an efficient protective solute (Meyer, 1984). The qualitative data of Mackay *et al.* (1984) include data for alkaliphilic members of the genera *Synechococcus* and *Synechocystis*, as well as for halophilic *Spirulina* sp., and quite clearly show that these organisms fit into the same compatible solute groups as other similar non-alkaliphilic strains, the halotolerance being directly related to the nature of the compatible solute (Reed *et al.*, 1984; Mackay *et al.*, 1984). Thus, salt tolerance in eubacteria and cyanobacteria is related to the nature of the compatible solute (salt tolerance increasing in the order sugars $\langle$ polyols $\langle$ betaine derivatives), and the compatible solute appears to be unrelated to the pH optimum of the strain.

Summary

Present microbiological studies on the alkaline and saline environment are rather limited, but they clearly indicate that this environment is a fruitful source of a phylogenetically diverse range of organisms. The range of organisms isolated from these lakes is restricted to a small number of genera to date. However, it is clear from microscopic observations, and enrichments made for a variety of physiological groups (Imhoff et al., 1979; Grant and Tindall, unpublished results) that the physiological, biochemical and taxonomic range of organisms living in the alkaline environment is far from limited. Perhaps with improvement in enrichment methods, and a more detailed microbiological survey of these environments, a new range of interesting and possibly useful organisms will be found to rival some of the commercially useful alkaliphilic Bacillus spp. described by Horikoshi and colleagues (Horikoshi and Akiba, 1982) (see Chapter 10 by Horikoshi, this volume). That organisms such as Spirulina sp. are now receiving attention as a commercial source of single-cell protein is a further encouraging sign that these highly productive environments will in the future provide some useful function to industry, rather than remaining uninviting, harsh environments of purely geological and geochemical interest.

Clearly the study of organisms from such environments with respect to their unusual requirements not only helps us to understand how they can survive in an apparently hostile environment but also adds to our appreciation of biological diversity and limitation. It can only be hoped that the work surveyed here will be seen as the first steps in the study of these lakes and will provide a useful insight into possible future research.

Acknowledgements

We would like to thank E. A. Galinski (University of Bonn) for helpful discussions concerning compatible solutes and for making available unpublished data. We also appreciate the constructive criticisms from Professor H. G. Trüper during the preparation of the manuscript. B. J. Tindall wishes to acknowledge the financial support of a grant from the Deutsche Forschungsgemeinschaft to Professor H. G. Trüper.

References

Abd-el-Malek, Y. and Rizk, S. G. (1963a). Bacterial sulphate reduction and the development of alkalinity. I. Experiments with synthetic media. *Journal of Applied Bacteriology* **26**, 7–13.

Abd-el-Malek, Y. and Rizk, S. G. (1963b). Bacterial sulphate reduction and the development of alkalinity. II. Laboratory experiments with soils. *Journal of Applied Bacteriology* **26**, 14–19.

Abd-el-Malek, Y. and Rizk, S. G. (1963c). Bacterial sulphate reduction and the development of alkalinity. III. Experiments under natural conditions. *Journal of Applied Bacteriology* **26**, 20–26.

Allen, O. N. and Holding, A. J. (1974). Gram-negative aerobic rods and cocci. Family III. *Rhizobiaceae. In* "Bergey's Manual of Determinative Bacteriology" (Eds R. E. Buchanan and N. E. Gibbons), 8th edn., pp. 264–267. Williams and Wilkins, Baltimore, Maryland.

Asselineau, J. and Trüper, H. G. (1982). Lipid composition of six species of the phototrophic bacterial genus *Ectothiorhodospira. Biochimica et Biophysica Acta* **712**, 111–116.

Baker, B. H. (1958). Geology of the Magadi area. *Geological Survey of Kenya*, Report No. 42.

Bayley, S. T. and Morton, R. A. (1978). Recent developments in the molecular biology of extremely halophilic bacteria. *Critical Reviews in Microbiology* **6**, 151–205.

Beadle, L. C. (1932). Scientific results of the Cambridge Expedition to the East African Lakes, 1930. 4. The waters of some East African lakes in relation to their flora and fauna. *Journal of the Linnean Society of London, Zoology* **38**, 157–211.

Berkeley, R. C. W. and Goodfellow, M. (Eds) (1981). "The Aerobic Endospore-forming Bacteria". Academic Press, London.

Brewer, P. G. and Goldman, J. C. (1976). Alkalinity changes generated by phytoplankton growth. *Limnology and Oceanography* **21**, 108–117.

Brock, T. D. (1978). "Thermophilic Microorganisms and Life at High Temperatures". Springer, New York and Heidelberg.

Brown, L. (1959). "The Mystery of the Flamingoes". Hamlyn, London.

Cherni, N. E., Solov'eva, J. V., Fedorova, V. D. and Kondrat'eva, E. N. (1969). The ultrastructure of two species of purple sulphur bacteria. *Microbiologiya* **38**, 479–484.

Chesnokov, V. A. and Shaposhnikov, D. I. (1936). The influence of pH on the development of purple sulphur bacteria. *Biokhimiya* **1**, 63–74.

Ciferri, O. (1983). *Spirulina,* the edible microorganisms. *Microbiological Reviews* **47**, 551–578.

Cole, G. A. (1968). *In* "Desert Biology" (Ed. G. W. Brown), Vol. 1, pp. 423–486. Academic Press, New York.

Collins, M. D., Ross, H. M. N., Tindall, B. J. and Grant, W. D. (1981). The distribution of isoprenoid quinones in halophilic bacteria. *Journal of Applied Bacteriology* **50**, 559–565.

Collins, M. D., Lund, B. M., Farrow, J. A. E. and Schliefer, K. H. (1983). Chemotaxonomic study of an alkalophilic bacterium *Exiguobacterium aurantiacum* gen. nov., sp. nov. *Journal of General Microbiology* **129**, 2037–2042.

De Rosa, M., Gambacorta, A., Nicolaus, B., Ross, H. N. M., Grant, W. D. and

Bu'lock, J. D. (1982). An asymmetric archaebacterial diether lipid from alkaliphilic halophiles. *Journal of General Microbiology* **128**, 343–348.

De Rosa, M., Gambacorta, A., Nicolaus, B. and Grant, W. D. (1983). A C_{25},C_{25} diether core lipid from archaebacterial haloalkaliphiles. *Journal of General Microbiology* **129**, 2333–2337.

Eugster, H. P. (1970). Chemistry and origins of the brines of Lake Magadi. *Mineralogical Society of America, Special Publication* **3**, 215–235.

Eugster, H. P. and Hardie, L. A. (1978). *In* "Lakes: Chemistry, Geology and Physics" (Ed. A. Lerman), pp. 237–293. Springer, New York.

Galinski, E. A. (1980). Untersuchungen zur Osmoregulation bei *Ectothiorhodospira halochloris*. Diplom Arbeit. Universitat Bonn.

Galinski, E. A. and Trüper, H. G. (1982). Betaine, a compatible solute in the extremely halophilic phototrophic bacterium *Ectothiorhodospira halochloris*. *FEMS Microbiology Letters* **13**, 357–360.

Galinski, E. A. and Trüper, H. G. (1984). Salt-stress strategies of haloalkaliphilic phototrophs. *Systematic and Applied Microbiology* **5**, 277.

Garcia, M. L., Guffanti, A. A. and Krulwich, T. A. (1983). Characterisation of the Na^+/H^+ antiporter of alkalophilic bacilli *in vivo*: $\Delta\psi$-dependent $^{22}Na^+$ efflux from whole cells. *Journal of Bacteriology* **156**, 1151–1157.

Gee, J. M., Lund, B. M., Metcalf, G. and Peel, J. L. (1980). Properties of a new group of alkalophilic bacteria. *Journal of General Microbiology* **117**, 9–17.

Gordon, R. E. (1981). *In* "The Aerobic Endospore-forming-bacteria" (Eds R. C. W. Berkeley and M. Goodfellow), pp. 1–16. Academic Press, London and New York.

Gordon, R. E., Hyde, W. C. and Pang, H.-N. (1977). *Bacillus firmus–Bacillus lentus*: a series or one species. *International Journal of Systematic Bacteriology* **27**, 256–262.

Grant, W. D. and Tindall, B. J. (1980). *In* "Microbial Growth and Survival in Extremes of Environment" (Eds G. W. Gould and J. E. L. Corry), pp. 27–36. Academic Press, London and New York.

Grant, W. D., Mills, A. A. and Schofield, A. K. (1979). An alkalophilic species of *Ectothiorhodospira* from a Kenyan soda lake. *Journal of General Microbiology* **110**, 137–142.

Guffanti, A. A., Susman, P., Blanco, R. and Krulwich, T. A. (1978). The protonmotive force and γ-aminoisobutyric acid transport in an obligately alkalophilic bacterium. *Journal of Biological Chemistry* **253**, 708–715.

Hardie, L. A. and Eugster, H. P. (1970). The evolution of closed-basin brines. *Mineralogical Society of America, Special Publication* **3**, 273–290.

Hecky, R. E. and Kilham, P. (1973). Diatoms in alkaline saline lakes; ecology and geochemical implications. *Limnology and Oceanography* **18**, 53–71.

Holt, S. G., Trüper, H. G. and Tackacs, B. J. (1971). Fine structure of *Ectothiorhodospira mobilis* strain 8113 thylakoids: chemical fixation and freeze-etching studies. *Archiv für Mikrobiologie* **62**, 111–128.

Horikoshi, K. and Akiba, T. (1982). "Alkalophilic Microorganisms". Springer, Berlin, Heidelberg and New York.

Imhoff, J. F. (1982). Taxonomic and phylogenetic implications of lipid and quinone compositions in phototrophic microorganisms. *In* "Biochemistry and Metabolism of Plant Lipids" (Eds J. F. G. M. Wintermanns and P. J. C. Kruper), pp. 541–544. Elsevier Biomedical, Amsterdam.

Imhoff, J. F. (1984a). Quinones of phototrophic bacteria. *FEMS Microbiology Letters* **25**, 85–89.

Imhoff, J. F. (1984b). Reassignment of the genus *Ectothiorhodospira* Pelsh 1936 to a new family, Ectothiorhodospiraceae fam. nov., and emended description of the Chromatiaceae Bavendamm 1924. *International Journal of Systematic Bacteriology* **34**, 338–339.

Imhoff, J. F. and Trüper, H. G. (1977). *Ectothiorhodospira halochloris* sp. nov., a new extremely halophilic phototrophic bacterium containing bacteriochlorophyll b. *Archives of Microbiology* **114**, 115–121.

Imhoff, J. F. and Trüper, H. G. (1981). *Ectothiorhodospira abdelmalekii* sp. nov., a new halophilic and alkaliphilic phototrophic bacterium. *Zentralblatt für Bakteriologie, Mikrobiologie und Hygiene I. Abteilung Originale C* **2**, 228–234.

Imhoff, J. F., Hashwa, F. and Trüper, H. G. (1978). Isolation of extremely halophilic phototrophic bacteria from the alkaline Wadi Natrun, Egypt. *Archiv für Hydrobiologie* **84**, 381–388.

Imhoff, J. F., Sahl, H. G., Soliman, G. S. H. and Trüper, H. G. (1979). The Wadi Natrun: chemical composition and microbial mass developments in alkaline brines of eutrophic desert lakes. *Geomicrobiology Journal* **1**, 219–234.

Imhoff, J. F., Tindall, B. J., Grant, W. D. and Trüper, H. G. (1981). *Ectothiorhodospira vacuolata* sp. nov., a new phototrophic bacterium from soda lakes. *Archives of Microbiology* **130**, 238–242.

Imhoff, J. F., Kushner, D. J., Kushwaha, S. C. and Kates, M. (1982). Polar lipids in phototrophic bacteria of the Rhodospirillaceae and Chromatiaceae families. *Journal of Bacteriology* **150**, 1192–1201.

Irgens, R. L. (1983). Thioacetamide as a source of hydrogen sulphide for colony growth of purple sulphur bacteria. *Current Microbiology* **8**, 183–186.

Isachenko, V. L. (1951). Chlorous, sulphate and soda lakes of the Kulunda Steppe and the biogenic processes in them. *In* "Selected Works", Vol. II, pp. 143–162. Academy of Sciences of the Union of Soviet Socialist Republics, Moscow.

Jannash, H. W. (1957). Die Bakteriella rotfarbung der Salzseen des Wadi Natrun. *Archiv für Hydrobiologie* **53**, 425–433.

Jenkin, P. M. (1932). Reports on the Percy Sladen expedition to some Rift Valley lakes in Kenya in 1929. I. Introductory account of the biological survey of five freshwater and alkaline lakes. *Annuals and Magazine of Natural History, Series X* **9**, 533–553.

Jenkin, P. M. (1936). Reports on the Percy Sladen expedition to some Rift Valley lakes in Kenya in 1929. VII. Summary of the ecological results with special reference to the alkaline lakes. *Annuals and Magazine of Natural History, Series X* **18**, 133–181.

Karazanov, V. V. and Ivanovsky, R. N. (1980). Sodium-dependent succinate uptake in purple bacterium *Ectothiordospira shapshnikovii*. *Biochimica et Biophysica Acta* **598**, 91–99.

Kenyon, C. N. (1978). Complex lipids and fatty acids in photosynthetic bacteria. *In* "The Photosynthetic Bacteria" (Eds R. K. Clayton and W. R. Sistrom), pp. 281–316. Plenum, New York.

Kondrat'eva, E. N. (1956). The utilisation of organic compounds by purple bacteria in the presence of light. *Mikrobiologiya* **25**, 393–400.

Kondrat'eva, E. N. and Malofeeva, V. (1964). On the study of the purple sulphur bacteria. *Mikrobiologiya* **33**, 758–762.

Koyama, N. and Nosoh, Y. (1976). The effect of the pH of culture medium on the alkalophilicity of a species of *Bacillus*. *Archives of Microbiology* **109**, 105–108.

Koyama, N., Kiyomiya, A. and Nosoh, Y. (1976). Na$^+$-dependant uptake of amino acids by an alkalophilic *Bacillus*. *FEBS Letters* **72**, 77–78.

Koyama, N., Takinishi, H. and Nosoh, Y. (1983). A possible relation of membrane proteins to the alkalostability of a facultatively alkalophilic *Bacillus*. *FEMS Microbiology Letters* **16**, 213–216.

Krulwich, T. A. and Guffanti, A. A. (1983). Physiology of acidophilic and alkalophilic bacteria. *Advances in Microbial Physiology* **24**, 173–213.

Krulwich, T. A., Mandel, K. G., Borstein, R. F. and Guffanti, A. A. (1979). A non-alkalophilic mutant of *Bacillus alcalophilus* lacks the Na$^+$/H$^+$ antiporter. *Biochemical and Biophysical Research Communications* **91**, 58–62.

Kushner, D. J. (1978). Life in high salt and solute concentrations: halophilic bacteria. *In* "Microbial Life in Extreme Environments" (Ed. D. J. Kushner), pp. 318–368. Academic Press, London and New York.

Langworthy, T. A. (1978). Microbial life in extreme pH values. *In* "Microbial Life in Extreme Environments" (Ed. D. J. Kushner), pp. 279–317. Academic Press, London and New York.

Lind, E. M. (1968). Notes on the distribution of phytoplankton of some Kenyan waters. *British Phycological Bulletin* **3**, 481–493.

Mackay, M. A., Norton, R. S. and Borowitzka, L. J. (1984). Organic osmoregulatory solutes in cyanobacteria. *Journal of General Microbiology* **130**, 2177–2191.

McCall, G. J. H. (1967). Geology of the Nakuru-Thompson's Falls—Lake Hannington area. *Geological Survey of Kenya*. Report No 78.

Melack, J. M. (1978). Morphometric, physical, and chemical features of the volcanic crater lakes of Western Uganda. *Archives of Hydrobiology* **84**, 430–453.

Melack, J. M. and Kilham, P. (1974). Photosynthetic rates of phytoplankton in East African Lakes. *Limnology and Oceanography* **19**, 743–755.

Meyer, B. (1984). Die Kopplungsfaktor-ATPase aus dem haloalkaliphilen Bakterium *Ectothiorhodospira halochloris*: Anreicherung und Charakterisierung. Diplom Arbeit, Universitat Bonn.

Molisch, E. (1907). "Die Purpurbakterien nach neuen Untersuchungen", pp. 1–95. Fischer, Jena.

Moshentseva, L. and Kondrat'eva, E. N. (1962). Production of chlorophylls by purple and green bacteria during photoautotrophic and photoeterotrophic growth. *Mikrobiologiya* **31**, 199–202.

van Niel, C. B. (1931). On the morphology and physiology of the purple and green sulphur bacteria. *Archiv für Mikrobiologie* **3**, 1–112.

Nishihara, M., Morri, H. and Koga, Y. (1982). Bis(monoacylglycero)phosphate in alkalophilic bacteria. *Journal of Biochemistry* **92**, 1469–1479.

Ohta, K., Kiyomiya, A., Koyama, N. and Nosch, Y. (1975). The basis of the alkalophilic property of a species of *Bacillus*. *Journal of General Microbiology* **86**, 259–266.

Oyewole, S. H. and Holt, S. C. (1976). Structure and composition of intracytoplasmic membranes of *Ectothiorhodospira mobilis*. *Archives of Microbiology* **107**, 167–173.

Pelsh, A. D. (1937). Photosynthetic sulphur bacteria of the eastern reservoir of Lake Sakskoe. *Mikrobiologiya* **6**, 1090–1100.

Pfennig, N. (1977). Phototrophic green and purple bacteria: a comparative systematic survey. *Annual Review of Microbiology* **31**, 275–290.

Raymond, J. C. and Sistrom, W. R. (1967). The isolation and preliminary characterisation of a halophilic photosynthetic bacterium. *Archiv für Mikrobiologie* **59**, 255–268.

Raymond, J. C. and Sistrom, W. R. (1969). *Ectothiorhodospira halophila*: a new species of the genus *Ectothiorhodospira*. *Archiv für Mikrobiologie* **69**, 121–126.

Reed, R. H., Richardson, D. L., Warr, S. C. R. and Stewart, W. D. (1984). Carbohydrate accumulation and osmotic stress in cyanobacteria. *Journal of General Microbiology* **130**, 1–4.

Remsen, C. C., Watson, S. W., Waterbury, J. B. and Trüper, H. G. (1968). The film structure of *Ectothiorhodospira mobilis* Pelsh. *Journal of Bacteriology* **95**, 2374–2392.

Rich, F. (1932). Phytoplankton from the Rift Valley lakes in Kenya. *Annuals and Magazine of Natural History, Series X* **10**, 233–262.

Rich, F. (1933). Scientific results of the Cambridge expedition to the East African lakes, 1930–1931. VII. The algae. *Journal of the Linnean Society of London, Zoology* **38**, 249–275.

Ross, H. N. M. (1982). The halophilic archaebacteria. Ph.D. Thesis, University of Leicester.

Ross, H. N. M. and Grant, W. D. (1985). Nucleic acid studies of halophilic archaebacteria. *Journal of General Microbiology* **131**, 165–173.

Ross, H. N. M., Collins, M. D., Tindall, B. J. and Grant, W. D. (1981). A rapid procedure for the detection of archaebacterial lipids in halophilic bacteria. *Journal of General Microbiology* **123**, 75–80.

Rupela, O. P. and Tauro, P. (1973a). Isolation and characterisation of *Thiobacillus* from alkali soils. *Soil Biology and Biochemistry* **5**, 891–897.

Rupela, O. P. and Tauro, P. (1973b). Utilisation of *Thiobacillus novellus* to reclaim alkali soils. *Soil Biology and Biochemistry* **5**, 899–901.

Siegel, S. M. and Giumarro, C. (1966). On the culture of a microorganism similar to the Pre-Cambrian microfossil, *Kakebekia umbellata* Barghoorn, in ammonia-rich atmospheres. *Proceedings of the National Academy of Sciences of the United States of America* **55**, 349–353.

Soliman, G. S. H. and Trüper, H. G. (1982). *Halobacterium pharaonis* sp. nov., a new, extremely haloalkaliphilic archaebacterium with low magnesium requirement. *Zentralblatt für Bakteriologie, Mikrobiologie und Hygiene I. Abteilung Originale C* **3**, 318–329.

Souza, K. A. and Deal, P. H. (1977). Characterisation of a novel extremely alkalophilic bacterium. *Journal of General Microbiology* **101**, 103–109.

Stackebrandt, E., Fowler, V. J., Schubert, W. and Imhoff, J. F. (1984). Towards a phylogeny of phototrophic purple sulphur bacteria the genus *Ectothiorhodospira*. *Archives of Microbiology* **137**, 366–370.

Tabita, F. R. and McFadden, B. A. (1976). Molecular and catalytic properties of ribulose-1, 5-biphosphate carboxylase from the photosynthetic extreme halophile, *Ectothiorhodospira halophila*. *Journal of Bacteriology* **126**, 1271–1277.

Talling, J. F. and Talling, I. B. (1965). The chemical composition of African lake waters. *Internationale Revue der gesamten Hydrobiologie und Hydrographie* **59**, 421–463.

Talling, J. F., Wood, R. B., Prosser, M. V. and Baxter, R. M. (1973). The upper limit of photosynthetic productivity by phytoplankton: evidence from Ethiopian soda lakes. *Freshwater Biology* **3**, 53–76.

Te-Pang, H. (1890). "Manufacture of Soda, with Special Reference to the Ammonia

Process: a Practical Treatise", American Chemical Society Monograph No 65. Hafner, New York.

Tew, R. W. (1966). Photosynthetic halophiles from Owens Lake. NASA Report No CR-361.

Tew, R. W. (1980). Halotolerant *Ectothiorhodospira*: survival in mirabilite-experiments with a model of chemical stratification by hydrate deposition in saline lakes. *Geomicrobiology Journal* **2**, 13–20.

Then, J. (1984). Beitrage zur Sulfidoxiation durch *Ectothiorhodospira abdelmalekii* und *Ectothiorhodospira halochloris*. Doktor Arbeit, Universitat Bonn.

Then, J. and Trüper, H. G. (1984). Utilisation of sulfide and elemental sulfur by *Ectothiorhodospira halochloris*. *Archives of Microbiology* **139**, 295–298.

Thompson, A. O. and Dodson, R. G. (1963). Geology of the Naivasna area. *Geological Survey of Kenya*, Report No 55.

Tindall, B. J. (1980). Phototrophic bacteria from Kenyan soda lakes. Ph.D. Thesis, Leicester University.

Tindall, B. J. and Grant, W. D. (1986). *In* "Anaerobic Bacteria in Habitats Other than Man" (Eds E. M. Barnes and G. Mead). Society for Applied Bacteriology Symposium Series Number 13, Blackwell Scientific Publications, London, 115–155.

Tindall, B. J. and Trüper, H. G. (1984). Phototrophic bacteria from alkaline, saline lakes. *Systematic and Applied Microbiology* **5**, 276.

Tindall, B. J., Mills, A. A. and Grant, W. D. (1980). An alkalophilic red halophilic bacterium with a low magnesium requirement from a Kenyan soda lake. *Journal of General Microbiology* **116**, 257–260.

Tindall, B. J., Ross, H. N. M. and Grant, W. D. (1984). *Natronobacterium* gen. nov. and *Natronococcus* gen. nov., two new genera of haloalkaliphilic archaebacteria. *Systematic and Applied Microbiology* **5**, 41–57.

Trüper, H. G. (1968). *Ectothiorhodospira mobilis* Pelsh, a photosynthetic sulphur bacterium depositing sulphur outside the cell. *Journal of Bacteriology* **95**, 1910–1920.

Trüper, H. G. (1970). Culture and isolation of phototrophic sulphur bacteria from the marine environment. *Helgoländer Wissenschaftliche Meeresuntersuchungen* **20**, 6–16.

Trüper, H. G. and Imhoff, J. F. (1981). The genus *Ectothiorhodospira*. *In* "The Prokaryotes" (Eds M. P. Starr, H. Stolp, H. G. Trüper, A. Balows and H. G. Schlegel), Vol. I, pp. 274–278. Springer, London, Heidelberg and New York.

Uherkovich, G. (1970). Contributions to the knowledge of the algae of the alkaline waters of Hungary. 3: The phytoseston of the alkaline ponds near Kunfeherto. *Acta Botanica Academiae Scientiarum Hungaricae* **16**, 405–426.

Uspenskaya, V. E. and Kondrat'eva, E. N. (1962). Relationships between photoautotrophic bacteria and vitamin synthesis in these organisms. *Mikrobiologiya* **31**, 396–400.

Vareschi, E. (1978). The ecology of Lake Nakuru (Kenya). I. Abundance and feeding of lesser flamingo. *Oecologia* **32**, 11–35.

Vareschi, E. (1982). The ecology of Lake Nakuru (Kenya). III. Abiotic factors and primary production. *Oecologia* **32**, 81–101.

Weißer, J. (1983). Ein neuer haloalkaliphiler *Bacillus* aus Egyptischen Sodaseen: Untersuchungen zur Osmoregulation. Diplom Arbeit, Universitat Bonn.

Wiley, W. R. and Stokes, J. L. (1962). Requirements of an alkaline pH and ammonia for substrate oxidation by *Bacillus pasteurii*. *Journal of Bacteriology* **84**, 730–734.

Wiley, W. R. and Stokes, J. L. (1963). Effect of pH and ammonia ions on the permeability of *Bacillus pasteurii*. *Journal of Bacteriology* **86**, 1152–1156.

3

Halotolerant and Halophilic Microbes

R. H. REED

Department of Biological Sciences, University of Dundee, Dundee DD1 4HN, UK

Introduction

Microrganisms of one kind or another can survive and grow over a wide range of salt concentration. In aquatic environments the conditions range from fresh waters (containing less than 0.05% w/v dissolved salts), through sea waters (with total salinities of 3.2–3.8% w/v) to saturated salt solutions (up to 30% w/v, and above). In fresh waters, the lack of inorganic salts and micronutrients may be a limiting factor to microbial growth. Furthermore, microbes in freshwater habitats, where the total extracellular solute concentration is lower than that of the intracellular fluid, will be exposed to a dilution stress. Thus, in the absence of any other forces, water entry due to the osmotic imbalance between cell and medium would lead to swelling and, eventually, to osmotic lysis. However, this is prevented in most cases by the presence of a rigid or semi-rigid cell wall, which allows a positive hydrostatic (turgor) pressure to develop and thus restricts water entry and cell enlargement. Wall-less, freshwater microbes must counteract the osmotically driven uptake of water by using contractile vacuoles to concentrate and then to remove water from the cell interior (Aaronson and Behrens, 1974; Raven, 1976). Dilution stress will be greatest in pure water free of added salts and thus represents the opposite form of water stress to that encountered under highly saline conditions.

In contrast, microorganisms growing in sea waters are faced with a combination of salt and osmotic stresses. Since there are no mechanisms available that might enable a microbial cell to maintain a dilute interior (see Brown, 1964a, 1976), marine microbes must therefore counterbalance the salt present in their external medium by the intracellular accumulation of solutes to an equivalent osmotic level. However, marine environments are not generally regarded as extreme, owing to the abundance of oceanic and marine habitats (with approximately 70% of the surface of the Earth covered by sea waters). Rather, it is the hypersaline habitats and non-ionic environments of high osmotic strength which are considered to be extreme. These hypersaline habitats are the domain of the halotolerant and halophilic microbes and this chapter considers some of the problems encountered by microorganisms which inhabit such environments, characterized by a combination of elevated ionic and osmotic strengths. Non-ionic environments of high osmolality (e.g. nectaries and preserved foodstuffs) will not be considered, although some physical features (and microbial responses) are shared by environments of high salinity and high non-ionic osmotic strength (see Brown, 1976, 1978). Similarly, desiccation stress will not be considered in detail. Desiccation stress can be viewed as a more extreme form of osmotic and (lack of) water stress, where water availability may be severely limited and where many of the modifications of microbes are designed either (i) to reduce the rate of water loss (stress avoidance) or (ii) to enable a rapid recovery upon rehydration (stress tolerance). The former category is not relevant to microorganisms in aquatic environments; because of the presence of liquid water in the extracellular medium, coupled with the high hydraulic conductivity (water permeability) of the plasma membrane, aquatic microbes are always in osmotic equilibrium with their environment. The latter category encompasses those modifications which enable microorganisms to *tolerate* (survive) rather than to *grow* in extreme conditions and, as such, they fall outside the scope of this chapter.

Terminology

The adjectives "halophilic" and "halotolerant" have been widely used in the literature, with a range of definitions and categories (see, for example, Kushner, 1978; Brown, 1983). For the purposes of this chapter, adequate definitions of halophilic and halotolerant microorganisms can be obtained by using the marine ("non-extreme") environment as a datum. Thus halophiles, or salt-loving microbes, should have out-of-the-ordinary requirements for salt which separate them from other microorganisms; any organism that has a minimum requirement for salt (usually, but not exclusively, as NaCl) in excess of the level found in sea water (i.e. more than 0.5 M NaCl, or thereabouts) can therefore be regarded as a halophile. In contrast, the term halotolerant implies no specific requirement for salt, although it suggests an upper salt limit for growth which is rather higher than that of the average microbe. By using sea water as a reference, a reasonable definition can be obtained by suggesting that any organism capable of sustained growth at a salinity which is double that of sea water (i.e. 1 M NaCl or greater), but with no absolute requirement for salt in amounts greater than 0.5 M NaCl or its equivalent, may be regarded as halotolerant. These two rather broad categories define the scope of this chapter; further subdivision (e.g. into moderately halotolerant and extremely halotolerant categories) becomes more difficult owing to the complexities of microbial responses to salt and osmotic stresses and the effects of other environmental parameters (e.g. temperature, divalent cation concentration and pH) upon salt tolerances and salt requirements, as measured in laboratory culture.

Units of Measurement

Several parameters and units can be used to describe the osmotic strength of solutions, in relation to either increasing solute levels or decreasing amounts of (solvent) water. All are interrelated and can usually be interconverted by simple mathematical calculations and approximations (see Wyn Jones and Gorham, 1983).

In research studies using microalgae, the parameter most frequently employed has been ψ_w, the water potential (a term derived from the chemical potential of water; see Dainty (1976). The water potential ψ_w^c of a walled microbial cell can be regarded as the sum of two major components, as follows:

$$\psi_w^c = \psi_p^c + \psi_\pi^c \tag{1}$$

where ψ_p^c and ψ_π^c represent hydrostatic (turgor) pressure and osmotic potential respectively. At equilibrium, the intracellular water potential will be equal to that of the surrounding medium (ψ_w^m), which is determined by its osmotic potential ψ_π^m. Thus

$$\psi_w^m = \psi_\pi^m = \psi_w^c = \psi_p^c + \psi_\pi^c \tag{2}$$

In general, cell turgor has a positive value while ψ_π^c is negative (with respect to pure water) (Dainty, 1976). Thus each component of ψ_w^c tends to act in opposition; increasing ψ_p^c will lead to a more positive value for ψ_w^c, causing water to move outwards (down a gradient of water potential) to restore equilibrium, while the intracellular osmotic potential produces an inwardly directed force with a tendency for water to enter the cell when the internal osmotic strength is increased (and ψ_π^c becomes more negative). Water potential and its component parts are usually expressed in units of pressure (megapascals, where $1\,\text{MPa} = 10\,\text{bar} = 9.87\,\text{atm}$, at 25 °C). Several studies have been carried out to quantify ψ_p^c and ψ_π^c in microalgal cells maintained in a range of saline media, to determine the effects of variations in ψ_w^m upon the individual components of ψ_w^c (see below, p. 66).

Microbial physiologists, working with bacterial cells, have tended to express the osmotic strength of fluids in terms of water activity a_w, a parameter related directly to water potential by the following equation:

$$\psi_w = \frac{RT \ln a_w}{\bar{V}_w} + \psi_p \tag{3}$$

where $\bar{V}_w$ is the partial molar volume of water while R and T have their usual meanings as the universal gas constant and absolute temperature respectively. Since the water activity of the bathing medium will contain no hydrostatic pressure term (ψ_p), the a_w value of a walled microbial cell will usually be lower (smaller) than the corresponding a_w value for the surrounding fluid (see Horowitz, 1979). In a pure, ideal solvent, $a_w = 1$ and the presence of solutes serves to reduce a_w. Thus a 1 osmolal solution (that is, 1 osmol of solute per kilogram of solvent water) has a water activity of 0.982; this is the approximate a_w value for sea water, with a total salt level equivalent to 0.5 M NaCl (≈ 1 osmol kg^{-1}).

Further details of the derivation and usage of water relations parameters are given in several texts, including Nobel (1974, 1983), Brown (1976), Borowitzka (1981) and Wyn Jones and Gorham (1983).

Microbial Diversity in Hypersaline Environments

Table 1 lists some of the principal microbes present in hypersaline habitats, together with their (approximate) upper and lower salt tolerances. This is not intended as a comprehensive list of all known halotolerant and halophilic

Table 1. *Halotolerant and halophilic microbes: representative examples*

Microorganism[a]	Growth range (M NaCl)[b]	Category
Aphanocapsa halophytica (*Synechococcus*)	>0.5–4+	Halophilic
Aphanothece halophytica (*Synechococcus*)	<0.5–4	Halotolerant
Asteromonas gracilis	<0.5–4.5	Halotolerant
Dunaliella salina	0.1–5	Halotolerant
Dunaliella viridis	1–4+	Halophilic
Ectothiorhodospira halochloris	1.5–4	Halophilic
Halobacterium spp.	2–5+	Halophilic
Halococcus spp.	2–5+	Halophilic
Synechocystis DUN52	0.3–3+	Halotolerant
Synechocystis PCC6714	0–1	Halotolerant

[a] See references in text for details.
[b] Or its approximate equivalent.

microbes, although it includes microorganisms that are regarded generally as characteristic of extremely saline environments. Several of the microorganisms listed in Table 1 have been studied in detail by microbial ecologists and physiologists; consequently the physiological basis for halotolerance and halophilism is most clearly understood in these organisms.

Hypersaline habitats, in common with other so-called extreme environments, typically show a reduced microbial species diversity (see Brock, 1979). The major microbial groups represented in the most extreme saline environments (approaching salt saturation) are bacteria (e.g. *Halobacterium, Halococcus* and *Ectothiorhodospira*), cyanobacteria (*Synechococcus* (≡ *Aphanothece* ≡ *Aphanocapsa* ≡ *Coccochloris*) and/or *Synechocystis* and *Dactylococcopsis*) and eukaryotic photosynthetic flagellates belonging to the genus *Dunaliella* or to related genera e.g. *Chlamydomonas* (Yamamoto and Okamoto, 1967) and *Asteromonas* (Ben Amotz and Grunwald, 1981).

Cells of *Halobacterium* (an elongated, Gram-negative, rod-shaped bacterium) and *Halococcus* (a thick-walled, Gram-positive coccus) share several major features, including a heterotrophic (aerobic) mode of growth, an *absolute* requirement for salt (NaCl) at 2 M or greater for viability and sustained growth, pink or red pigmentation, lack of peptidoglycan in the cell envelope/cell wall and a preference for amino acids as sources of fixed carbon for energy metabolism. Some species of *Halobacterium*, notably *Halobacterium halobium*, also have a capacity for photophosphorylation (via the pigment bacteriorhodopsin) under anaerobic or microaerophilic conditions. While the significance of this reaction has yet to be established for cells of *Halobacterium* growing in natural hypersaline habitats, it has been well characterized in the laboratory (see Caplan and Ginzburg, 1978; Stoeckenius *et al.*, 1979).

Ectothiorhodospira halochloris is an anaerobic, halophilic, phototrophic bacterium with a requirement for NaCl at more than 1.5 M and a salinity optimum for growth of approximately 3 M. *E. halochloris* grows as spiral-shaped rods that are approximately 0.5 μm wide and up to 15 μm long (Imhoff and Trüper, 1977). Cells are flagellate, with a bipolar arrangement of flagella, and contain stacked intracytoplasmic membranes; these intracytoplasmic membranes are sites of photosynthetic pigment localization. *E. halochloris* is slightly thermophilic, with a temperature optimum for growth of 35–50 °C. Cells grow best under anaerobic conditions with H_2S as an electron donor and CO_2 as a source of fixed carbon, although photoheterotrophic growth can be sustained by a range of amino acids. Further details are given by Imhoff and Trüper (1977).

Although cyanobacteria have been frequently regarded as minor constituents of the microbial flora in hypersaline habitats (Brown, 1976, 1983), it is clear that salt-tolerant forms occur in a wide range of extreme saline environments, with isolates obtained from the Dead Sea (Volcani, 1944), the Great Salt Lake, USA (Brock, 1976), salt ponds in Puerto Rico (Kao *et al.*, 1973), solar evaporation ponds in California, USA (Yopp *et al.*, 1977), Solar Lake, Sinai (Cohen *et al.*, 1977; Walsby *et al.*, 1983), hypersaline ponds at Yallahs, Jamaica (Golubic, 1980), and the Al-Khiran area of Kuwait (Mohammad *et al.*, 1983). Most of these cyanobacterial isolates are unicellular, and they can be grouped within the form genera *Synechococcus* or *Synechocystis* (Rippka *et al.*, 1979). Filamentous isolates have also been obtained from hypersaline habitats, including *Oscillatoria limnetica*, a cyanobacterium capable of facultative anoxygenic photosynthesis (Cohen *et al.*, 1975), and *Phormidium* spp. (Golubic, 1980).

Recently, Walsby *et al.* (1983) have described a novel gas-vacuolate unicellular cyanobacterium, *Dactylococcopsis salina*, from Solar Lake, Sinai. The cells of this cyanobacterium have a characteristic morphology, growing as slender, fusiform rods up to 80 μm long, with tapering ends and granular cytoplasm. *D. salina* has an absolute requirement for salt, showing no growth at salinities below those of sea water, and can therefore be regarded as truly halophilic. A species of *Dactylococcopsis* has also been reported from hypersaline ponds at Yallahs, Jamaica (Golubic, 1980).

A number of cyanobacterial isolates from hypersaline habitats have been variously assigned to *Aphanothece halophytica* (Yopp *et al.*, 1978), *Aphanocapsa halophytica* (Brock, 1976) or *Coccochloris elabans* (Kao *et al.*, 1973), although there are no unequivocal morphological features that separate these isolates. The recent classification scheme of Rippka *et al.* (1979) has recognized only five form genera for all unicellular cyanobacteria which reproduce by binary fission. Of these, the genus *Synechococcus* includes all organisms that divide in one plane, having no sheath surrounding the cell; this definition encompasses *Aphanothece halophytica, Aphanocapsa halophytica* and *C. elabens*. However, each of these isolates has its own characteristic salinity minimum, optimum and maximum for growth, with some strains showing halotolerant features while others are truly halophilic, and it is possible that these strains represent separate species within the same genus (see Golubic, 1980). The other major genus of unicellular cyanobacteria recorded from hypersaline environments is *Synechocystis* (Mohammad *et al.*, 1983). Isolates of *Synechocystis* are characterized by their lack of a sheath surrounding the cell, together with cell division in two or three successive planes at right angles to one another. However, when cell separation occurs soon after cell division, the planes of successive divisions may not be easily determined (Rippka *et al.*, 1979) and Mohammad *et al.* (1983) have noted that their isolate of *Synechocystis* sp. showed cell division in more than one plane only at salinities below 7% w/v sea salt; above this salt concentration, the cells appeared to be very similar to *Synechococcus*. Adequate classification of these unicellular forms must await further characterization of their biochemistry, physiology and cellular composition, since morphological features are limited.

The genus *Dunaliella* includes a variety of ill-defined species of biflagellate unicellular green algae, lacking a rigid polysaccharide cell wall. Within present knowledge *Dunaliella* is the most euryhaline photosynthetic eukaryote, often present as the dominant primary producer in many hypersaline lakes. Individual species of *Dunaliella* demonstrate a capacity to grow over a wide range of salt concentration (in contrast with *Halobacterium* and *Halococcus* which are more exacting in their salt requirements and are less able to tolerate changes in external salinity); the genus is well represented in freshwater, marine and hypersaline habitats. Some species from hypersaline environments (e.g. *D. salina* (see Brown, 1978), and *D. bardawil* (Ben Amotz and Avron, 1981) grow in media ranging from less than 0.5 to 5 M NaCl and can thus be regarded as halotolerant, while other strains grow only in hypersaline media (e.g. *D. viridis*, which requires in excess of 1 M NaCl for sustained growth (Borowitzka and Brown, 1974)) and can be regarded as halophilic (Table 1).

In concluding this section, it is also worth noting that natural hypersaline waters are not always Na^+ dominated; for example, the Dead Sea contains greater amounts of Mg^{2+} than Na^+. Other examples are given by Kushner (1978) and Brock (1979).

Living with Salt and Water Stress: the Principles of Halotolerance and Halophilism

All microorganisms growing in saline waters are faced with similar problems owing to the elevated osmotic and ionic strengths of their bathing medium. However, their methods of dealing with these extreme stresses are not necessarily identical. It is possible to divide the (genetic) adaptations shown by halophilic and halotolerant microbes into three rather broad categories, as follows.

(1) *Insulation*. This category encompasses those adaptations which reduce the impact of the external environment upon the cell interior by preventing (or limiting) the entry of salt; such features must, of necessity, reside in the cell wall and/or plasma membrane.

(2) *Protection*. Any feature that protects intracellular function and metabolic activity from the inhibitory effects of (toxic) Na^+ can be assigned to this category, since NaCl is regarded generally as being toxic to most of the processes that occur within living cells at concentrations similar to those

found in hypersaline habitats (see Brown, 1976; Wyn Jones and Gorham, 1983).

(3) *Modification*. This category includes those aspects of cellular metabolism which show optimal function in high-salt conditions. Such processes show modification when compared with similar processes in salt-sensitive microbial cells.

These three categories should only be regarded as rather general and they are not necessarily mutually exclusive. Thus, many halotolerant and halophilic microbes show a combination of adaptations, as described below. The categories can also be envisaged as "lines of defence" against salt stress, with insulation acting as the first defensive barrier and modification providing a "last resort" measure. Further details are given below for each category.

Insulation

High concentrations of Na^+ are accepted generally as being toxic to most biological systems. Thus the majority of bacteria, fungi and algae possess membranes with a lower permeability to (toxic) Na^+ than to K^+ and, in addition, active Na^+ extrusion mechanisms to lower the intracellular Na^+ content below the (equilibrium) value predicted (by the Nernst equation) from the interior-negative membrane potential (see Harold, 1977; Raven, 1980; Jennings, 1983). Marine and freshwater microbes thus contain substantially less Na^+ than K^+ when grown in saline media (Christian and Waltho, 1961, 1962; Kushner, 1978).

Published values for the intracellular cation levels in *Halobacterium* show a wider range; for example, cells grown in media containing 3.3–5.1 M NaCl contain 0.36–3.6 M Na^+ (see Brown, 1983). While some of the variation in cell Na^+ may be due to differences between exponentially growing and stationary-phase cells (Brown and Sturtevant, 1980), the remaining variation is likely to be due to technical difficulties associated with the contamination of samples of cells by extracellular Na^+. The experimental techniques employed to separate cells of *Halobacterium* from their bathing media involve either filtration or centrifugation and extracellular Na^+ is accounted for either by use of an appropriate correction factor (using an impermeable marker compound, such as ^{14}C-labelled dextran (Ginzburg, 1969)) or by rinsing in an isotonic medium free of NaCl. Because of the technical problems of accurately assessing the extent of (extracellular) Na^+ contamination in these cell samples, it is likely that some of the published values for cell Na^+ in *Halobacterium* are overestimates. However, certain general comments can still be applied:

(1) Despite the high $Na^+:K^+$ ratios in their growth medium, cells of *Halobacterium* invariably contain more K^+ than Na^+. Thus the $Na^+:K^+$ ratio of actively growing cells is generally 0.1–0.2, rising in stationary-phase cells to 0.7 or more (Brown, 1983).

(2) The cells show efficient active Na^+ extrusion, with Na^+ efflux from the cell linked to proton entry (Lanyi, 1978, 1979) by an "antiporter" system which functions with an apparent stoichiometry of $2H^+(in):1Na^+(out)$.

(3) The plasma membrane of *Halobacterium* has a low passive Na^+ permeability. This feature is essential if intracellular Na^+ is to be maintained at a substantially lower value than in the medium and Bayley and Morton (1979) have suggested that the unique composition of the cell membrane of *Halobacterium* may reduce its permeability to ions.

Insulation of the cell from Na^+ entry is clearly a feature of the osmotic physiology of *Halobacterium*. However, while cell Na^+ levels are lower than external Na^+ levels by an order of magnitude or so, cell K^+ concentrations are generally higher than extracellular K^+ concentrations, with accumulation ratios (intracellular:extracellular) as high as 1000:1 (Ginzburg *et al.*, 1971). The significance of high intracellular monovalent cation levels to other aspects of the physiology of *Halobacterium* is discussed below (p. 73).

In *Dunaliella* there is considerable variation in the published values for cell Na^+, with some research studies suggesting that internal Na^+ concentrations are rather low in cells grown in high-salt media (e.g. Borowitzka and Brown, 1974; Ginzburg, 1981) and others suggesting a two-compartment model with an inner, well-insulated compartment of low Na^+ concentration and an outer "leaky" compartment that contains Na^+ at levels similar to those in the bathing medium (e.g. Ginzburg, 1978; Ehrenfeld and Cousin, 1982, 1984; Zmiri *et al.*, 1984). However, there are considerable technical difficulties associated with the measurement of Na^+ concentration in a so-called "leaky" outer compartment of a fragile wall-less organism suspended in an Na^+-rich medium and some of the controversy regarding cation levels in *Dunaliella* may be due to differences in methodology and techniques. Many of the studies that appear to support the two-compartment model rely upon the use of two different methods to measure ion levels in the so-called "inner" and "outer" compartments, and they may therefore be measuring variations due to differences in technique rather than providing clear evidence of a multicompartment system (for Na^+) in cells of *Dunaliella*. Unequivocal proof of either the one-compartment or the two-compartment model must await further experimental refinement and methodological comparisons. However, it is clear that these experiments show evidence in support of the hypothesis that all, or part, of the cell interior is insulated from the surrounding NaCl, with low internal Na^+ levels and a low passive Na^+

permeability. Furthermore, the presence of other intracellular (organic) solutes within the cell also supports the notion that intracellular cation levels must be low, since *Dunaliella* cells are incapable of generating large hydrostatic (turgor) pressures owing to their lack of a rigid cell wall and to the presence of flagella (Raven, 1980).

Low intracellular Na^+ levels have also been reported for several marine, halotolerant and halophilic cyanobacterial cells. Thus, cells of the unicellular marine isolate *Agmenellum quadruplicatum* (*Synechococcus* sp.) contain Na^+ at approximately 0.03 M (calculated from the data of Batterton and Van Baalen (1971) using unpublished values for the cell volume-to-dry weight 64 ratio of *A. quadruplicatum*) when grown in a medium containing 1.03 M NaCl, while the euryhaline halotolerant unicell *Synechocystis* DUN52 contains less than 0.30 M Na^+ when grown in 1.87 M NaCl and 0.16 M Na^+ when grown in 0.47 M NaCl (Reed *et al.*, 1984a); low intracellular Na^+ levels have also been reported for a halophilic isolate of *Apanothece halophytica* (*Synechococcus*) by Miller *et al.* (1976). Freshwater cyanobacteria also maintain low internal Na^+ levels when grown in saline media (Batterton and Van Baalen, 1971; Richardson *et al.*, 1983) and it appears that low intracellular Na^+ may be characteristic of salt-sensitive and salt-tolerant strains (see Warr *et al.*, 1984).

Intracellular accumulation of K^+ occurs in halotolerant cyanobacteria and in *Dunaliella*, although not to the same extent as in *Halobacterium*. Most reports have suggested that intracellular K^+ in halotolerant cyanobacteria and *Dunaliella* is maintained at approximately 0.2–0.3 M for cells which have been fully adapted in low-salt and high-salt media (see Ginzburg, 1981; Reed *et al.*, 1984a). One exception to this rather general observation comes from the research studies of Miller *et al.* (1976) who have shown that K^+ is accumulated to levels in excess of 1.0 M in a halophilic isolate of *Aphanothece halophytica* (*Synechococcus*). In contrast with other cyanobacteria, which accumulate organic osmotica rather than inorganic ions to achieve osmotic balance with their surrounding medium (see Table 2), this isolate appears to use K^+ as a major intracellular solute and merits further investigation.

It is clear from the above discussion that all halotolerant and halophilic microbes show some degree of insulation of the cell interior from external Na^+, although the extent of insulation varies between organisms (and, seemingly, between research laboratories) and is also subject to variation during the cell cycle. Since there is no known mechanism that would enable the inside of the cell to remain in a dilute state with respect to its surrounding fluid, alternative compounds must be used to generate intracellular osmotic potential and maintain osmotic equilibrium (equation (2)). For walled cells there is a further requirement for an excess of intracellular solutes to provide the turgor pressure that is needed for cell expansion growth (Taiz, 1984). In

Table 2. *Organic solutes in salt-stressed phototrophic microbes: representative examples*

Microorganism	Principal organic solute(s)	Reference
Anabaena cylindrica	Sucrose	Reed *et al.* (1984b)
Anabaena CA	Sucrose	Reed *et al.* (1984b)
Asteromonas gracilis	Glycerol	Ben-Amotz and Grunwald (1981)
Calothrix PCC 7426	Glutamate betaine and disaccharides	Mackay *et al.* (1984)
Chlorella emersonii	Sucrose and proline	Setter and Greenway (1979)
Cyclotella cryptica	Proline	Liu and Hellebust (1976)
Dunaliella spp.	Glycerol	Borowitzka and Brown (1974)
Ectothiorhodospira halochloris	Glycine betaine	Galinski and Trüper (1982)
Lyngbya spp.	Trehalose	Reed *et al.* (1984b)
Pavlova sp.	Cyclohexanetetrol	Craigie (1969)
Poterioochromonas sp.	Galactosylglycerol	Kauss (1978)
Stichococcus spp.	Sorbitol and proline	Brown and Hellebust (1980)
Synechococcus N100	Glucosylglycerol	Borowitzka *et al.* (1980)
Synechocystis PCC6714	Glucosylglycerol	Reed *et al.* (1985b)
Synechocystis DUN52	Glycine betaine	Reed *et al.* (1984a)
Tetraselmis sp.	Mannitol	Kirst (1977)

Halobacterium and *Halococcus* this requirement for additional solutes is most probably achieved by using K^+, as discussed above. Further evidence in support of this hypothesis comes from the observation that intracellular K^+ concentration is markedly affected by variations in extracellular osmolality (see Brown, 1976), as predicted for a solute involved in osmotic adjustment. However, photoautotrophic organisms generally accumulate organic solutes rather than inorganic ions to achieve osmotic balance (Hellebust, 1976). In *Dunaliella* glycerol is synthesized when cells are grown in saline media (Craigie and McLachlan, 1964; Wegmann, 1971; Ben-Amotz and Avron, 1973). Glycerol is also synthesized in the halotolerant alga *Asteromonas* (Ben-Amotz and Grunwald, 1981) and in halotolerant isolates of *Chlamydomonas* (see Hellebust, 1976). In four cyanobacterial isolates from hypersaline habitats, the quaternary ammonium compound glycine betaine has been identified as the principal intracellular solute (Mohammad *et al.*, 1983; Reed *et al.*, 1984a) and Mackay *et al.* (1984) have recently shown that the novel quaternary ammonium compound glutamate betaine is a major organic solute in two N_2-fixing halotolerant filamentous cyanobacteria belonging to the genus *Calothrix*. Glycine betaine has also been identified as an organic osmoticum in cells of *E. halochloris* (Galinski and Trüper, 1982) and in other bacterial cells (Imhoff and Rodriguez-Valera, 1984).

The technique that has been most widely used to investigate intracellular solute accumulation in halotolerant and halophilic microbes is natural abundance ^{13}C nuclear magnetic resonance spectroscopy. This technique is particularly useful since it allows the principal low molecular weight organic compounds present in samples of living cells and/or cell extracts to be identified by non-destructive means (Borowitzka *et al.*, 1980). Once the major organic compounds have been identified in this manner, subsequent assays can then be carried out using more rapid analytical procedures (e.g. periodide precipitation techniques to measure intracellular levels of quaternary ammonium compounds) (Reed *et al.*, 1984a).

In all the microorganisms described above, the level of the major intracellular solute has been shown to be sensitive to changes in external osmolality, increasing in response to "upshock" (an increase in external salt concentration) and decreasing upon "downshock" (a reduction in external salinity). This osmotic adjustment process has been well characterized for many halotolerant and halophilic microbes and is also common to many freshwater and marine organisms (although intracellular solute levels are generally lower in the latter groups). The term *osmoregulation* has been applied to these changes in intracellular solute concentration, and to the processes responsible for their adjustment (see Hellebust, 1976; Zimmermann, 1978). However, there is a case for replacing this rather misleading term with more appropriate terminology, e.g. *turgor/volume regulation* or *osmotic adjustment* (see Cram, 1976; Reed, 1984a), in recognition of the fact that it is cell turgor and/or volume (rather than intracellular osmolality or osmotic potential) which is restored by these changes in internal solute levels, following a change in external osmolality. Thus osmoregulation should strictly be used to refer to the *maintenance* of intracellular osmotic potential in the face of external perturbation (see, for example, Bisson and Bartholomew, 1984). Usage of osmo-terminology is further discussed by Reed (1984a).

The studies of A. D. Brown and coworkers (see Brown, 1976) have shown that polyols and other organic molecules are not inhibitory to enzyme activity when assayed *in vitro* at physiologically relevant concentrations, in contrast with NaCl. The term "compatible solute" has been used to describe all compounds showing this feature of compatibility with physiological function. This was first demonstrated for polyols in sugar-tolerant yeasts and for glycerol in *Dunaliella* (Brown and Simpson, 1972; Borowitzka and Brown, 1974). Thus glucose-6-phosphate dehydrogenase from *Dunaliella* was not inhibited by the action of glycerol at up to 4 M, in contrast with the inhibitory effects of salt at substantially lower concentrations (0.4 M) (Borowitzka and Brown, 1974). These findings have now been extended to cover a wider range of organisms, including salt-tolerant higher plants (halophytes) which accumulate glycine betaine (Storey and Wyn Jones, 1977) or proline (Stewart and Lee, 1974; Stewart and Hanson, 1980) when subjected to salt stress, and a diverse range of marine animals which accumulate various amino acids and quaternary ammonium compounds in response to osmotic stress (see Yancey *et al.*, 1982). In all cases where enzymes have been isolated from organisms growing in saline media, NaCl has proved to be more inhibitory than the dominant intracellular organic solute (Yancey *et al.*, 1982; Wyn Jones and Gorham, 1983).

Our research studies have recently confirmed that glycine betaine acts as a compatible solute in cells of the halotolerant cyanobacterium *Synechocystis* DUN52 (Warr *et al.*, 1984). Low concentrations of NaCl (up to 0.75 M) stimulated the enzyme glutamine synthetase *in vitro* by up to 30%, with inhibition above 0.75 M NaCl; maximum inhibition of glutamine synthetase activity occurred at 2.0 M NaCl. In contrast, glycine betaine concentrations up to 1.8–2.0 M resulted in no loss of enzyme activity (see Fig. 1b). These findings can be compared with the effects of the disaccharide sucrose, which is accumulated by some freshwater and marine cyanobacterial strains (Reed *et al.*, 1984b). Sucrose was found to be inhibitory to glutamine synthetase extracted from the marine N_2-fixing filamentous cyanobacterium *N. harveyana* at concentrations in excess of 0.3 M, with a 60% reduction in activity at 2.0 M and somewhat similar findings for NaCl (Fig. 1a). These results suggest that the degree of halotolerance of a particular cyanobacterial isolate may depend, in part at least, on the effects of compounds that are accumu-

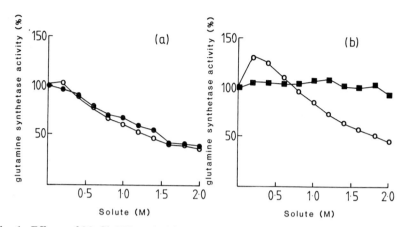

Fig. 1. Effects of NaCl (○) and either glycine betaine (■) or sucrose (●) on the activity of glutamine synthetase from (a) the marine cyanobacterium *Nodularia harveyana* and (b) the halotolerant cyanobacterium *Synechocystis* DUN52 (adapted from Warr *et al.* (1984)).

lated as internal osmotica. It thus appears that glycine betaine may be tolerated at higher intracellular levels than sucrose owing to its increased compatibility with enzyme activity and metabolic function (Warr *et al.*, 1984). The increased halotolerance of betaine-accumulating cyanobacteria (Mackay *et al.*, 1984; Reed *et al.*, 1984a) may also be a reflection of this phenomenon.

As discussed above, the principal intracellular solute in *Halobacterium* and *Halococcus* is K(Cl), rather than an organic osmoticum. This sets these organisms apart from salt-tolerant eukaryotes (Yancey *et al.*, 1982) and cyanobacteria (Reed *et al.*, 1984a,b) and requires the enzymes in these organisms to function in an environment of high ionic strength, with KCl acting as a compatible solute. Studies using isocitrate dehydrogenase from *H. salinarium* (Aitken *et al.*, 1970; Aitken and Brown, 1972) have shown that KCl is a much less effective inhibitor of enzyme activity than NaCl when assayed *in vitro* at physiologically relevant concentrations. In addition, enzyme activity was found to be relatively insensitive to changes in KCl concentration, as would be expected for a solute involved in osmotic adjustment, while NaCl produced a stimulation at low concentration and a rapid decline in activity at supra-optimal concentrations. A similar pattern of NaCl stimulation and inhibition is shown in Fig. 1b for the halotolerant cyanobacterium *Synechocystis* DUN 52. The role of KCl as a compatible solute is discussed in some detail by Brown (1978, 1983).

With the exception of *Halobacterium* and *Halococcus* organic solutes are employed generally as intracellular solutes and a diverse range of carbohydrates, amino acids, quaternary ammonium and tertiary sulphonium compounds and related molecules are employed by salt-stressed organisms to effect osmotic balance. Current knowledge of the range of intracellular organic solutes is summarized by Ben-Amotz and Avron (1983) for halotolerant algae, Yancey *et al.* (1982) for animals, Measures (1975) for amino acid solutes in bacteria, Reed *et al.* (1984a,b) and Mackay *et al.* (1984) for cyanobacteria and Wyn Jones and Gorham (1983) for higher plants. Representative examples of well-characterized systems in phototrophic microorganisms are given in Table 2 while Fig. 2 gives the structures of some organic osmotica. In photosynthetic organisms, these solutes are consistently and rapidly labelled with ^{14}C upon incubation in $NaH^{14}CO_3$ or $^{14}CO_2$. The patterns of incorporation of radiocarbon into low molecular weight organic solutes can thus be employed as a taxonomic marker, since the types of accumulated solutes are specific to different groups of photosynthetic organisms (Kremer, 1980). For example, this approach has recently been used to confirm that the acidothermophilic unicell *Cyanidium caldarium* is a red alga, as it accumulates the heteroside galactosylglycerol (floridoside) in response to salt stress (Reed, 1983) in common with other members of the Rhodophyta (Reed *et al.*, 1980; Reed, 1985).

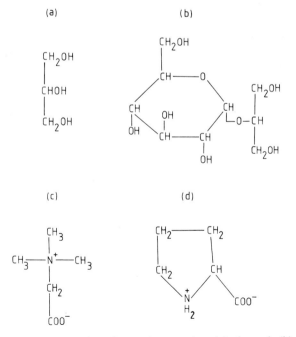

Fig. 2. Representative examples of organic osmotica: (a) glycerol; (b) glucosylglycerol; (c) glycine betaine; (d) proline.

One of the most intriguing aspects of osmotic response concerns the sensory and control mechanisms of osmotic adjustment. It is far from clear how cells sense a change in external water status, increasing their intracellular solute levels to restore cell turgor (walled cells) or volume (wall-less cells). A number of studies have shown that osmotic adjustment occurs in response to changes in external water status *per se*, rather than to specific ionic effects (see Cram, 1976; Hastings and Gutknecht, 1976; Borowitzka *et al.*, 1977; Reed *et al.*, 1980). Gutknecht and coworkers, using giant-celled algae, have proposed that an interaction between the plasma membrane and cell wall may lead to deformation of the plasma membrane and, consequently, to changes in the activity of membrane-associated enzymes and transport systems (see Bisson and Gutknecht, 1980). Zimmermann and coworkers have formulated a more quantitative model of a turgor-sensing system, based on electromechanical principles (see Zimmermann, 1978). The electromechanical model postulates that pressure-induced changes in the dimensions of the plasma membrane are transformed into changes in the electrical field distribution within the membrane, and that osmotic adjustment is linked to (i) changes in plasma membrane thickness by compression or stretching and/ or (ii) the plasma membrane electrical potential. Most of the experiments that have been designed to test the above models have used giant-celled algae, e.g. *Valonia* or *Chara* (Hastings and Gutknecht, 1974, 1976; Zimmermann *et al.*, 1976, 1977). However, the electromechanical model is applicable to all walled and well-less cells, including halotolerant and halophilic microbes.

Protection

Having established that all halotolerant and halophilic microorganisms discriminate to some extent against external NaCl, thus maintaining lower intracellular Na^+ concentrations by a combination of low plasma membrane Na^+ permeability and active Na^+ extrusion, we must also recognize that the internal Na^+ levels of salt-tolerant organisms are generally rather higher than those of their freshwater or marine counterparts. To illustrate this, halotolerant cyanobacteria contain Na^+ at 0.10–0.30 M over the range 0.23– 1.87 M NaCl (Reed *et al.*, 1984a) while freshwater and marine forms contain less than 0.05 M (Batterton and Van Baalen, 1971; Dewar and Barber, 1973; Paschinger, 1977; Richardson *et al.*, 1983; Warr *et al.*, 1984). Microorganisms growing at high external NaCl concentrations may thus be faced with a combination of a lowered (more negative) intracellular water potential and an increased internal Na^+ content (if insulation is less than 100% effective).

The "compatible solute" theory has been extended in recent years to accommodate these aspects of salt and osmotic stress. Schobert (1977) has suggested that the water-like hydroxyl groups of polyols (e.g., glycerol and mannitol (Fig. 2)) can replace water molecules and thus maintain hydrophobically enforced water structure within the cytoplasm under conditions of lowered cellular water potential. The hydrophobic methyl groups of quaternary ammonium compounds and the imino acid proline might also function in a similar manner (see Fig. 2), and Schobert (1977, 1980) has proposed the term "water structure regulation" to describe the protective effects of organic solutes during water stress (see also Schobert and Marsh, 1982). No similar proposals have been made for K^+ in *Halobacterium* and, since K^+ is known to affect water structure to a lesser extent than Na^+ does, it is unlikely that similar effects would be possible (Brown, 1976).

In the case of organic solutes, there is also evidence that protection against NaCl inactivation may occur when high levels of these solutes are used *in vitro*. Thus Pollard and Wyn Jones (1979) have shown that glycine betaine can partially protect malate dehydrogenase (from *Hordeum vulgare*) against NaCl inhibition. Somewhat similar findings have also been reported for the halotolerant cyanobacterium *Synechocystis* DUN52, where glycine betaine at 1.0 M and 1.2 M increased glutamine synthetase activity by 24% and 28% respectively at 2.0 M NaCl (Warr *et al.*, 1984).

The preceding discussion applies to salt-adapted cells, growing in saline conditions. However, a further aspect of protection against salt inactivation may also be considered. Recent research has shown that short-term changes in intracellular Na^+ levels may occur upon initial transfer to saline media. Thus Ehrenfeld and Cousin (1984) have shown for the euryhaline alga *Dunaliella tertiolecta* that intracellular Na^+ increased by approximately 0.13 M when cells were upshocked to a medium containing additional NaCl at 0.4 M. Rapid Na^+ entry (complete within 2 minutes) was followed by net Na^+ extrusion from the cell, and a new (low) steady state level was achieved within 30 minutes. Similar transient increases in cell Na^+ followed by net Na^+ extrusion have been reported for *Dunaliella parva* (Ginzburg, 1981) and the freshwater cyanobacterium *Synechococcus* PCC6311 (Blumwald *et al.*, 1983), although the time course for Na^+ extrusion was extended beyond 30 minutes. Reed *et al.* (1986a) have shown that transient increases in cell Na^+ may occur upon upshock in several cyanobacteria belonging to the genera *Synechococcus* and *Synechocystis*, and this response has been characterized for the euryhaline unicell *Synechocystis* PCC6714 (Reed *et al.*, 1985). Significant increases in cell Na^+ occurred when cells of *Synechocystis* PCC6714 were subjected to a change in external NaCl concentration of 0.2 M or greater, with maximum uptake of Na^+ being recorded upon transfer to the most saline medium (containing 1 M NaCl). Extrusion of Na^+

was found to be energy-dependent and sensitive to the external K^+ concentration (Reed *et al.*, 1985). These transient ion changes provide further evidence that intracellular organic osmotica may be involved in protection against NaCl inactivation, at least in the short term, following upshock. Thus organisms containing high levels of compatible solutes such as glycine betaine may well show less Na^+ inhibition upon transfer to a high-salt medium than organisms containing lower levels of organic osmotica. This phenomenon also provides some insight into possible mechanisms of acclimation to elevated salinities and the "training" of organisms to grow in hypersaline media (see, for example, Borowitzka *et al.*, 1977). Further implications of transient changes in intracellular solute status are discussed by Reed (1984b) and Reed *et al.* (1985, 1986).

In conclusion, it is clear that while the need for protection against Na^+ toxicity will be dependent upon the extent of insulation against NaCl entry at the outer membrane, compatible (organic) osmotica may also be involved in (i) the maintenance of cytoplasmic "water structure" (Schobert, 1977) and (ii) protection against transient increases in intracellular salt levels.

Modification

This category encompasses those features of cellular metabolism that have been modified to function in an environment of elevated salt concentration, showing no inhibition (optimal functioning) at high salt levels, in contrast with similar processes in salt-sensitive organisms. Because of the uniquely specialized nature of *Halobacterium* and *Halococcus*, with an absolute requirement for salt at more than 2 M (Table 1), the adaptations have been characterized to greatest extent in these microorganisms.

The cell envelope and outer membrane of *Halobacterium* are exposed to high salt levels at all times and show extreme modification. They are unstable in solutions of low ionic strength (less than 1–2 M NaCl); this disintegration is rapid and is not due solely to osmotic effects, since high concentrations of non-ionic solutes do not prevent disruption (Brown, 1983). This requirement for high levels of monovalent salts (or somewhat lower levels of divalent salts) appears to be due mainly to the acidic nature of the envelope proteins. The outer surface of each cell of *Halobacterium* consists of a single lipo-protein membrane with glycoprotein subunits on the outer surface and the proteins within this envelope complex contain a substantial proportion of aspartic acid and glutamic acid residues (Brown, 1965). Furthermore, the surface charge density is increased by acidic groups on the membrane lipids. It would seem that these excess negative charges are shielded by the presence of high concentrations of cations, thus preventing mutual repulsion. This charge-shielding effect has been clearly demonstrated by Brown (1964b) who showed that succinylation of the outer membrane and cell envelope of a salt-sensitive *Pseudomonas* sp. resulted in the substitution of —COO⁻ groups for −NH₃⁺. This led to the disaggregation of the succinylated membrane and envelope at low salt concentrations, as in *Halobacterium*. There is also some evidence that hydrophobic bonding within the membrane is weaker in *Halobacterium* than in non-halophilic bacteria, requiring higher salt concentrations to bring it to operational strength (see Brown, 1976, 1983).

No comparable data exist for halotolerant cyanobacteria although the growth of these organisms in a wide range of hyposaline and hypersaline media suggests that no analogous modifications of cell wall and/or plasma membrane exist, otherwise cellular integrity would not be maintained at low external salt concentrations. Clearly, any modification in the membrane composition of *Dunaliella* (see Evans and Kates, 1984) and/or halotolerant cyanobacteria must be related to insulation against NaCl entry (as discussed above) rather than to stability in high salt concentrations (and disaggregation at lower salt concentrations).

It is also applicable to consider the enzymology of *Halobacterium* within this category. Kushner (1978) has recognized three principal types of salt response for halobacterial enzymes, namely (1) enzymes requiring more than 1 M salt, with maximum activity at 2–4 M, (2) enzymes with maximum activity at 0.5–1.5 M, inhibited by more than 1.5 M NaCl, and (3) enzymes that are strongly inhibited by elevated NaCl concentrations, showing highest activity in the absence of salt. The last category includes the enzymes of fatty acid synthesis (Pugh *et al.*, 1971). It is noteworthy that the synthesis of esterified fatty acids is minimal in cells of *Halobacterium* and Brown (1983) has speculated that these type 3 enzyme systems must be vestigial (and may indicate a past ancestry including non-halophilic forms?). Of the remaining enzyme systems, those associated with the cell membrane fall into type 1 while soluble cytoplasmic enzymes generally conform to type 2. In general, both NaCl and KCl give similar stimulation–inhibition curves with type 1 and 2 enzymes (see Brown, 1976), although there are exceptions where KCl is less inhibitory than NaCl (Lanyi, 1974; Cazzulo, 1978). It seems that type 1 and type 2 enzyme systems have been modified to function in conditions of elevated ionic strength and Brown (1983) has suggested that these modifications serve to manipulate the conformation of salt-requiring enzymes at high salt concentrations to a state which broadly corresponds to that of non-halophile enzymes under low-salt conditions. Thus high concentrations of monovalent cations are required for enzyme stability as well as enzyme activity. This seems to apply to many enzymes from *Halobacterium* where stability, rather than activity, is critically dependent upon high salt concentrations (see Aitken *et al.*, 1970; Norberg *et al.*, 1973).

The modifications of enzymes in *Halobacterium* appear to involve an excess of acidic amino acid residues within the enzyme proteins, with charge shielding at high salt concentrations leading to long-term stability and, in some cases, conformational changes that serve to increase enzyme activity (Lanyi, 1974). In this respect, these enzymes share several features with the cell envelope, as described above.

The ribosomes of *Halobacterium* also show extreme modification when compared with those of non-halophilic microbes; they are unique in requiring high salt levels for stability and in their possession of large amounts of acidic proteins. High concentrations of KCl (3–4 M) and Mg^{2+} (0.1 M) are required to maintain stability (Bayley, 1976). These modifications can also be regarded as characteristic of *Halobacterium* (and, presumably, of *Halococcus*), resulting from the high concentrations of monovalent salts (KCl) in the intracellular fluid.

In *Dunaliella*, the salt relations of glycerol dehydrogenase and glucose-6-phosphate dehydrogenase are functionally identical in the marine *D. tertiolecta* and the halophilic *D. viridis* (Borowitzka and Brown, 1974) and there are no apparent specializations or modifications of these intracellular enzyme systems to function at high salt concentrations, presumably as a result of the presence of glycerol, rather than either KCl or NaCl, as the principal intracellular solute. To date, the only relevant study for cyanobacterial enzymes (Warr *et al.*, 1984) has shown that high concentrations of NaCl are also inhibitory to glutamine synthetase from the halotolerant isolate *Synechocystis* DUN52 (see Fig. 1b), although low NaCl concentrations permitted greater activity than in the marine isolate *N. harveyana* (Fig. 1a). Thus, modifications that enable intracellular enzymes to remain stable and functional at very high salt levels appear to be restricted to the halophilic bacteria *Halobacterium* and *Halococcus*, arising from their unique use of a monovalent salt, rather than an organic solute, as an intracellular osmoticum.

Another aspect of the osmotic physiology of *Halobacterium* and *Halococcus* that can be considered within this section concerns the use of the trans-plasma-membrane Na^+ gradient (generated by insulating the cell interior against Na^+ entry) for the cotransport (symport) or countertransport (antiport) of solutes and metabolites. Thus the extracellular–intracellular Na^+ gradient has been shown to drive the active uptake systems of all nutritionally important amino acids in *Halobacterium* (McDonald *et al.*, 1977; Lanyi, 1978). Somewhat similar findings have also been reported for amino acid uptake in *Halococcus* (Rodriguez-Valera *et al.*, 1982). While this feature is also observed in marine bacteria and microalgae (Hellebust, 1978; Rees *et al.*, 1980), it contrasts with the transport systems of freshwater microbes, which are mainly dependent upon H^+ cotransport (see Harold, 1977; Raven, 1980). An increased dependence upon the trans-plasmamembrane Na^+ gradient will also impose a requirement for elevated levels of Na^+ in the bathing medium and may influence the lower salinity limit for sustained growth.

It is clear from the above discussion that many of the modificiations exhibited by *Halobacterium* and *Halococcus* also lead to a requirement for high levels of salt, both for structural stability and for optimal physiological functioning. Modifications of this type can thus explain the *halophilic* rather than *halotolerant* nature of these organisms. The salt dependence of microbial cells may thus be related directly to their methods of dealing with salt toxicity. Insulation and/or protection may provide a means of limiting salt toxicity and thereby increasing the salinity range of an organism, while modificiation may lead to salt dependence and to a restriction of growth and/ or survival in response to low extracellular Na(Cl) concentration.

Acknowledgements

Research support for the author's studies of osmotic adjustment and osmoacclimation has been provided by the Royal Society and the Science and Engineering Research Council, UK. S. R. C. Warr provided helpful comments on the manuscript.

References

Aaronson, S. and Behrens, U. (1974). Ultrastructure of an unusual contractile vacuole in several Chrysomonad flagellates. *Journal of Cell Science* **14**, 1–9.

Aitken, D. M. and Brown, A. D. (1972). Properties of halophil nicotinamide-adenine dinucleotide phosphate-specific isocitrate dehydrogenase. *Biochemical Journal* **130**, 645–662.

Aitken, D. M., Wicken, A. J. and Brown, A. D. (1970). Properties of a halophil nicotinamide-adenine dinucleotide phosphate-specific isocitrate dehydrogenase. *Biochemical Journal* **116**, 125–134.

Batterton, J. C. and Van Baalen, C. (1971). Growth responses of blue–green algae to sodium chloride concentration. *Archives of Microbiology* **76**, 151–165.

Bayley, S. T. (1976). Information transfer. *In* "Extreme Environments: Mechanisms of Microbial Adaptation" (Ed. M. R. Heinrich), pp. 119–136. Academic Press, New York and London.

Bayley, S. T. and Morton, R. A. (1979). Biochemical evolution of halobacteria. *In* "Strategies of Microbial Life in Extreme Environments" (Ed. M. Shilo), pp. 109–124. Dahlem Konferenzen, Berlin.

Ben-Amotz, A. and Avron, M. (1973). The role of glycerol in osmotic regulation of the halophilic alga *Dunaliella parva*. *Plant Physiology* **51**, 875–878.

Ben-Amotz, A. and Avron, M. (1981). Glycerol and β-carotene metabolism in the halotolerant alga *Dunaliella*: a model system for biosolar energy conversion. *Trends in Biochemical Sciences* **6**, 297–299.

Ben-Amotz, A. and Avron, M. (1983). Accumulation of metabolites by halotolerant algae and its industrial potential. *Annual Review of Microbiology* **37**, 95–119.

Ben-Amotz, A. and Grunwald, T. (1981). Osmoregulation in the halotolerant alga *Asteromonas gracilis*. *Plant Physiology* **67**, 613–616.

Bisson, M. A. and Bartholemew, D. (1984). Osmoregulation or turgor regulation in *Chara*? *Plant Physiology* **74**, 252–255.

Bisson, M. A. and Gutknecht, J. (1980). Osmotic regulation in algae. *In* "Membrane Transport Phenomena: Current Conceptual Issues" (Eds W. J. Lucas, R. M. Spanswick and J. Dainty), pp. 131–142. North-Holland, Amsterdam.

Blumwald, E., Mehlhorn, R. J. and Packer, L. (1983). Ionic osmoregulation during salt adaptation of the cyanobacterium *Synechococcus* 6311. *Plant Physiology* **73**, 377–380.

Borowitzka, L. J. (1981). Solute accumulation and regulation of cell water activity. *In* "The Physiology and Biochemistry of Drought Resistance in Plants" (Eds L. G. Paleg and D. Aspinall), pp. 97–130. Academic Press, New York and Sydney.

Borowitzka, L. J. and Brown, A. D. (1974). The salt relations of marine and halophilic species of the unicellular green alga, *Dunaliella*. The role of glycerol as a compatible solute. *Archives of Microbiology* **96**, 37–52.

Borowitzka, L. J., Kessly, D. S. and Brown, A. D. (1977). The salt relations of *Dunaliella*. Further observations on glycerol production and its regulation. *Archives of Microbiology* **113**, 131–138.

Borowitzka, L. J., Demmerle, S., Mackay, M. A. and Norton, R. S. (1980). Carbon-13 nuclear magnetic resonance study of osmoregulation in a blue–green alga. *Science* **210**, 650–651.

Brock, T. D. (1976). Halophilic blue–green algae. *Archives of Microbiology* **107**, 109–111.

Brock, T. D. (1979). Ecology of saline lakes. *In* "Strategies of Microbial Life in Extreme Environments" (Ed. M. Shilo), pp. 29–47. Dahlem Konferenzen, Berlin.

Brown, A. D. (1964a). Aspects of bacterial response to the ionic environment. *Bacteriological Reviews* **28**, 296–329.

Brown, A. D. (1964b). The development of halophilic properties in bacterial membranes by acylation. *Biochimica et Biophysica Acta* **93**, 136–142.

Brown, A. D. (1965). Hydrogen ion titration of intact and dissolved lipoprotein membranes. *Journal of Molecular Biology* **12**, 491–508.

Brown, A. D. (1976). Microbial water stress. *Bacteriological Reviews* **40**, 803–846.

Brown, A. D. (1978). Compatible solutes and extreme water stress in eukaryotic microorganisms. *Advances in Microbial Physiology* **17**, 181–242.

Brown, A. D. (1983). Halophilic prokaryotes. *In* "Encyclopedia of Plant Physiology", Vol. 12C, "Physiological Plant Ecology III: Responses to the Chemical and Biological Environment" (Eds O. L. Lange, P. S. Nobel, C. B. Osmond and H. Ziegler), pp. 137–162. Springer, Berlin.

Brown, A. D. and Simpson, J. R. (1972). Water relations of sugar-tolerant yeasts: the role of intracellular polyols. *Journal of General Microbiology* **72**, 589–591.

Brown, A. D. and Sturtevant, J. M. (1980). The state of water in extremely halophilic bacteria: freezing transitions of *Halobacterium halobium* observed by differential scanning calorimentry. *Journal of Membrane Biology* **54**, 21–30.

Brown, L. M. and Hellebust, J. A. (1980). The contribution of organic solutes to osmotic balance in some green and eustigmatophyte algae. *Journal of Phycology* **16**, 265–270.

Caplan, S. R. and Ginzburg, M. (Eds) (1978). "Energetics and Structure of Halophilic Microorganisms". North-Holland, Amsterdam.

Cazzulo, J. J. (1978). Regulatory properties of enzymes from marine and extremely halophilic bacteria. *In* "Energetics and Structure of Halophilic Microorganisms" (Eds S. R. Caplan and M. Ginzburg), pp. 371–376. North-Holland, Amsterdam.

Christian, J. H. B. and Waltho, J. A. (1961). The sodium and potassium content of non-halophilic bacteria in relation to salt tolerance. *Journal of General Microbiology* **25**, 97–102.

Christian, J. H. B. and Waltho, J. A. (1962). Solute concentrations within cells of halophilic and non-halophilic bacteria. *Biochimica et Biophysica Acta* **65**, 506–508.

Cohen, Y., Padan, E. and Shilo, M. (1975). Facultative anoxygenic photosynthesis in the cyanobacterium *Oscillatoria limnetica*. *Journal of Bacteriology* **123**, 855–861.

Cohen, Y., Krumbein, W. E. and Shilo, M. (1977). Solar Lake (Sinai) 2. Distribution of photosynthetic organisms and primary production. *Limnology and Oceanography* **22**, 609–620.

Craigie, J. S. (1969). Some salinity-induced changes in growth, pigments and cyclohexanetetrol content of *Monochrysis lutheri*. *Journal of the Fisheries Research Board, Canada* **26**, 2959–2967.

Craigie, J. S. and McLachlan, J. (1964). Glycerol as a photosynthetic product in *Dunaliella tertiolecta* Butcher. *Canadian Journal of Botany* **42**, 777–778.

Cram, W. J. (1976). Negative feedback regulation of transport in cells. The maintenance of turgor, volume and nutrient supply. *In* "Encyclopedia of Plant Physiology", Vol. 2A, "Transport in Plants" (Eds U. Lüttge and M. G. Pitman), pp. 284–316. Springer, Berlin.

Dainty, J. (1976). Water relations of plant cells. *In* "Encyclopedia of Plant Physiology", Vol. 2A, "Transport in Plants" (Eds U. Lüttge and M. G. Pitman), pp. 12–35. Springer, Berlin.

Dewar, M. A. and Barber, J. (1973). Cation regulation in *Anacystis nidulans*. *Planta* **113**, 143–155.

Ehrenfeld, J. and Cousin, J. L. (1982). Ionic regulation of the unicellular green alga *Dunaliella tertiolecta*. *Journal of Membrane Biology* **70**, 47–57.

Ehrenfeld, J. and Cousin, J. L. (1984). Ionic regulation of the unicellular green alga *Dunaliella tertiolecta*: response to hypertonic shock. *Journal of Membrane Biology* **77**, 44–55.

Evans, R. W. and Kates, M. (1984). Lipid composition of halophilic species of *Dunaliella* from the Dead Sea. *Archives of Microbiology* **140**, 50–56.

Galinski, E. A. and Trüper, H. G. (1982). Betaine, a compatible solute in the extremely halophilic phototrophic bacterium *Ectothiorhodospira halochloris*. *FEMS Microbiology Letters* **13**, 357–360.

Ginzburg, B. Z. (1978). Regulation of cell volume and osmotic pressure in *Dunaliella*. *In* "Energetics and Structure of Halophilic Microorganisms" (Eds S. R. Caplan and M. Ginzburg), pp. 543–558. North-Holland, Amsterdam.

Ginzburg, M. (1969). The unusual membrane permeability of two halophilic unicellular organisms. *Biochimica et Biophysica Acta* **173**, 370–376.

Ginzburg, M. (1981). Measurements of ion concentrations in *Dunaliella parva* subjected to hypertonic shock. *Journal of Experimental Botany* **32**, 333–340.

Ginzburg, M., Sachs, L. and Ginzburg, B. Z. (1971). Ion metabolism in a *Halobacterium*. II, Ion concentrations in cells at different levels of metabolism. *Journal of Membrane Biology* **5**, 78–101.

Golubic, S. (1980). Halophily and halotolerance in cyanophytes. *Origins of Life* **10**, 169–183.

Harold, F. M. (1977). Ion currents and physiological functions in microorganisms. *Annual Review of Microbiology* **31**, 181–203.

Hastings, D. F. and Gutknecht, J. (1974). Turgor pressure regulation: modulation of active potassium transport by hydrostatic pressure gradients. *In* "Membrane Transport in Plants" (Eds U. Zimmermann and J. Dainty), pp. 79–83. Springer, Berlin.

Hastings, D. F. and Gutknecht, J. (1976). Ionic relations and the regulation of turgor pressure in the marine alga *Valonia macrophysa*. *Journal of Membrane Biology* **28**, 263–275.

Hellebust, J. A. (1976). Osmoregulation. *Annual Review of Plant Physiology* **27**, 485–505.

Hellebust, J. A. (1978). Uptake of organic substrates by *Cyclotella cryptica* (Bacillariophyceae): effect of ions, ionophores and metabolic and transport inhibitors. *Journal of Phycology* **14**, 79–83.

Horowitz, N. H. (1979). Biological water requirements. *In* "Strategies of Microbial Life in Extreme Environments" (Ed. M. Shilo), pp. 15–27. Dahlem Konferenzen, Berlin.

Imhoff, J. F. and Rodriguez-Valera, F. (1984). Betaine is the main compatible solute of halophilic eubacteria. *Journal of Bacteriology* **160**, 478–479.

Imhoff, J. F. and Trüper, H. G. (1977). *Ectothiorhodospira halochloris* sp. nov., a new extremely halophilic phototrophic bacterium containing bacteriochlorophyll b. *Archives of Microbiology* **114**, 115–121.

Jennings, D. H. (1983). Some aspects of the physiology and biochemistry of marine fungi. *Biological Reviews* **58**, 423–459.

Kao, O. H. W., Berns, D. S. and Town, W. R. (1973). The characterization of C-phycocyanin from an extremely halotolerant blue–green alga, *Coccochloris elabens*. *Biochemical Journal* **131**, 39–50.

Kauss, H. (1978). Osmotic regulation in algae. *Progress in Phytochemistry* **5**, 1–27.

Kirst, G. O. (1977). Co-ordination of ionic relations and mannitol concentrations in the euryhaline unicellular alga *Platymonas subcordiformis* (Hazen) after osmotic shocks. *Planta* **135**, 69–75.

Kremer, B. P. (1980). Taxonomic implications of algal photoassimilate patterns. *British Phycological Journal* **15**, 399–409.

Kushner, D. J. (1978). Life in high salt and solute concentrations: halophilic bacteria. *In* "Microbial Life in Extreme Environments" (Ed. D. J. Kushner), pp. 317–361. Academic Press, London and Orlando.

Lanyi, J. K. (1974). Salt-dependent properties of proteins from extremely halophilic bacteria. *Bacteriological Reviews* **38**, 272–290.

Lanyi, J. K. (1978). Light energy conversion in *Halobacterium halobium*. *Microbiological Reviews* **42**, 682–706.

Lanyi, J. K. (1979). The role of Na$^+$ in transport processes of bacterial membranes. *Biochimica et Biophysica Acta* **559**, 377–397.

Liu, M. S. and Hellebust, J. A. (1976). Effects of salinity and osmolarity of the medium on amino acid metabolism in *Cyclotella cryptica*. *Canadian Journal of Botany* **54**, 938–948.

Mackay, M. A., Norton, R. S. and Borowitzka, L. J. (1984). Organic osmoregulatory solutes in cyanobacteria. *Journal of General Microbiology* **130**, 2177–2191.

McDonald, R. E., Greene, R. V. and Lanyi, J. K. (1977). Light-activated amino acid transport systems in *Halobacterium halobium* envelope vesicles: role of chemical and electrical gradients. *Biochemistry* **16**, 3227–3235.

Measures, J. C. (1975). Role of amino acids in osmoregulation of nonhalophilic bacteria. *Nature, London* **257**, 398–400.

Miller, D. M., Jones, J. H., Yopp, J. H., Tindall, D. R. and Schmid, W. D. (1976). Ion metabolism in a halophilic blue–green alga, *Aphanothece halophytica*. *Archives of Microbiology* **111**, 145–149.

Mohammad, F. A. A., Reed, R. H. and Stewart, W. D. P. (1983). The halophilic cyanobacterium *Synechocystis* DUN52 and its osmotic responses. *FEMS Microbiology Letters* **16**, 287–290.

Nobel, P. S. (1974). "Introduction to Biophysical Plant Physiology". Freeman, San Francisco, California.

Nobel, P. S. (1983). "Biophysical Plant Physiology and Ecology". Freeman, San Francisco, California.

Norberg, P., Kaplan, J. G. and Kushner, D. J. (1973). Kinetics and regulation of the salt-dependent asparate transcarbamylase of *Halobacterium cutirubrum*. *Journal of Bacteriology* **113**, 680–686.

Paschinger, H. (1977). DCCD-induced sodium uptake in *Anacystis nidulans*. *Archives of Microbiology* **113**, 285–291.

Pollard, A. and Wyn Jones, R. G. (1979). Enzyme activities in concentrated solutions of glycine betaine and other solutes. *Planta* **144**, 291–298.

Pugh, E. L., Wassef, M. K. and Kates, M. (1971). Inhibition of fatty acid synthetase in *Halobacterium cutirubrum* and *Escherichia coli* by high salt concentrations. *Canadian Journal of Biochemistry* **49**, 953–958.

Raven, J. A. (1976). Transport in algal cells. *In* "Encyclopedia of Plant Physiology", Vol. 2A, "Transport in Plants" (Eds U. Lüttge and M. G. Pitman), pp. 129–188. Springer, Berlin.

Raven, J. A. (1980). Nutrient transport in microalgae. *Advances in Microbial Physiology* **21**, 47–226.

Reed, R. H. (1983). Taxonomic implications of osmoacclimation in *Cyanidium caldarium* (Tilden) Geitler. *Phycologia* **22**, 351–354.

Reed, R. H. (1984a). Use and abuse of osmo-terminology. *Plant, Cell and Environment* **7**, 165–170.

Reed, R. H. (1984b). Transient breakdown in the selective permeability of the plasma membrane of *Chlorella emersonii* in response to hyperosmotic shock: implications for cell water relations and osmotic adjustment. *Journal of Membrane Biology* **82**, 83–88.

Reed, R. H. (1985). Osmoacclimation in *Bangia atropurpurea* (Rhodophyta, Bangiales): the osmotic role of floridoside. *British Phycological Journal* **20**, 211–218.

Reed, R. H., Collins, J. C. and Russell, G. (1980). The effects of salinity upon galactosyl–glycerol content and concentration of the marine red alga *Porphyra purpurea* (Roth) C. Ag. *Journal of Experimental Botany* **31**, 1539–1554.

Reed, R. H., Chudek, J. A., Foster, R. and Stewart, W. D. P. (1984a). Osmotic adjustment in cyanobacteria from hypersaline environments. *Archives of Microbiology* **138**, 333–337.

Reed, R. H., Richardson, D. L., Warr, S. R. C. and Stewart, W. D. P. (1984b). Carbohydrate accumulation and osmotic stress in cyanobacteria. *Journal of General Microbiology* **130**, 1–4.

Reed, R. H., Richardson, D. L. and Stewart, W. D. P. (1985). Na$^+$ uptake and extrusion in the cyanobacterium *Synechocystis* PCC6714 in response to hypersaline treatment: evidence for transient changes in plasmalemma Na$^+$ permeability. *Biochimica et Biophysica Acta* **814**, 347–355.

Reed, R. H., Richardson, D. L. and Stewart, W. D. P. (1986). Osmotic responses of unicellular blue–green algae (cyanobacteria): changes in cell volume and intracellular solute levels in response to hyperosmotic treatment. *Plant, Cell and Environment* **9**, 25–31.

Rees, T. A. V., Cresswell, R. C. and Syrett, P. J. (1980). Sodium-dependent uptake of nitrate and urea by a marine diatom. *Biochimica et Biophysica Acta* **596**, 141–144.

Richardson, D. L., Reed, R. H. and Stewart, W. D. P. (1983). Glucosylglycerol accumulation in *Synechocystis* PCC6803 in response to osmotic stress. *British Phycological Journal* **18**, 209.

Rippka, R., Deruelles, J., Waterbury, J. B., Herdman, M. and Stanier, R. Y. (1979). Generic assignments, strain histories and properties of pure cultures of cyanobacteria. *Journal of General Microbiology* **111**, 1–61.

Rodriguez–Valera, F., Ventosa, A., Quesada, E. and Ruiz-Berraguero, F. (1982). Some physiological features of a *Halococcus* sp. at low salt concentrations. *FEMS Microbiology Letters* **15**, 249–252.

Schobert, B. (1977). Is there an osmotic regulatory mechanism in algae and higher plants? *Journal of Theoretical Biology* **68**, 17–26.

Schobert, B. (1980). Proline catabolism, relaxation of osmotic strain and membrane permeability in the diatom *Phaeodactylum tricornutum*. *Physiologia Plantarum* **50**, 37–42.

Schobert, B. and Marsh, D. (1982). Spin label studies on osmotically-induced changes in the aqueous cytoplasm of *Phaeodactylum tricornutum*. *Biochimica et Biophysica Acta* **720**, 87–95.

Setter, T. L. and Greenway, H. (1979). Growth and osmoregulation of *Chlorella emersonii* in NaCl and neutral osmotica. *Australian Journal of Plant Physiology* **6**, 47–60 (Corrigendum: *Australian Journal of Plant Physiology* **6**, 569–572).

Stewart, C. R. and Hanson, A. D. (1980). Proline accumulation as a metabolic response to water stress. *In* "Adaptation of Plants to Water and High Temperature Stress" (Eds N. C. Turner and P. J. Kramer), pp. 173–189. Wiley, New York.

Stewart, G. R. and Lee, J. A. (1974). The role of proline accumulation in halophytes. *Planta* **120**, 279–289.

Stoeckenius, W., Lozier, R. H. and Bogolmoni, R. A. (1979). Bacteriorhodopsin and the purple membrane of halobacteria. *Biochimica et Biophysica Acta* **505**, 215–278.

Storey, R. and Wyn Jones, R. G. (1977). Quaternary ammonium compounds in plants in relation to salt resistance. *Phytochemistry* **16**, 447–453.

Taiz, L. (1984). Plant cell expansion: regulation of cell wall mechanical properties. *Annual Review of Plant Physiology* **35**, 585–657.

Volcani, B. E. (1944). The microorganisms of the Dead Sea. *In* "Papers Collected to Commemorate the 70th Anniversary of Dr. Chaim Weizmann", pp. 71–85. Sieff Research Institute, Rehovot, Israel.

Walsby, A. E., van Rijn, J. and Cohen, Y. (1983). The biology of a new gas-vacuolate cyanobacterium, *Dactylococcopsis salina* sp. nov., in Solar Lake. *Proceedings of the Royal Society, London, Series B* **217**, 417–447.

Warr, S. R. C., Reed, R. H. and Stewart, W. D. P. (1984). Osmotic adjustment of cyanobacteria: the effects of NaCl, KCl, sucrose and glycine betaine on glutamine synthetase activity in a marine and a halotolerant strain. *Journal of General Microbiology* **130**, 2169–2175.

Wegmann, K. (1971). Osmotic regulation of photosynthetic glycerol production in *Dunaliella*. *Biochimica et Biophysica Acta* **234**, 317–323.

Wyn Jones, R. G. and Gorham, J. (1983). Osmoregulation. *In* "Encyclopedia of Plant Physiology", Vol. 12C, "Physiological Plant Ecology III: Responses to the Chemical and Biological Environment" (Eds O. L. Lange, P. S. Nobel, C. B. Osmond and H. Ziegler), pp. 35–58. Springer, Berlin.

Yamamoto, M. and Okamoto, H. (1967). Osmotic regulation in a halophilic *Chlamydomonas* cell. I. General feature of the response to the change in osmotic pressure. *Zeitschrift für Allgemeine Mikrobiologie* **7**, 143–150.

Yancey, P. H., Clark, M. E., Hand, S. C., Bowlus, R. D. and Somero, G. N. (1982). Living with water stress: evolution of osmolyte systems. *Science* **217**, 1214–1222.

Yopp, J. H., Tindall, D. R., Miller, D. R. and Schmid, W. E. (1977). Isolation, purification and evidence for a halophilic nature of the blue–green alga *Aphanothece halophytica* Fremy (Chroococcales). *Phycologia* **17**, 172–178.

Yopp, J. H., Miller, D. M. and Tindall, D. R. (1978). Regulation of intracellular water potential in the halophilic blue–green alga *Apanothece halophytica* (Chroococcales). *In* "Energetics and Structure of Halophilic Microorganisms" (Eds S. R. Caplan and M. Ginzburg), pp. 619–624. North-Holland, Amsterdam.

Zimmermann, U. (1978). Physics of turgor and osmoregulation. *Annual Review of Plant Physiology* **29**, 121–148.

Zimmermann, U., Steudle, E. and Lelkes, P. I. (1976). Turgor pressure regulation in *Valonia utricularis*: effect of cell wall elasticity and auxin. *Plant Physiology* **58**, 608–613.

Zimmermann, U., Beckers, F. and Steudle, E. (1977). Turgor sensing in plant cells by the electromechanical properties of the membrane. *In* "Transmembrane Ionic Exchanges in Plants" (Eds M. Thellier, A. Monnier, M. Demarty and J. Dainty), pp. 155–165. University of Rouen, Paris.

Zmiri, A., Wax, Y. and Ginzburg, B. Z. (1984). Electrical sizing of cells of the halophilic alga *Dunaliella parva*. *Plant, Cell and Environment* **7**, 229–237.

4

Fungal Responses Towards Heavy Metals

GEOFFREY MICHAEL GADD

Department of Biological Sciences, University of Dundee, Dundee DD1 4HN, UK

Introduction

The term "heavy metal" is imprecise: under biotic conditions the principal chemical species are cations (Hughes *et al.*, 1980), and arbitrary definitions based on density or position in the periodic table encompass a wide array of elements with diverse chemical and biological properties (Passow *et al.*, 1961; Bowen, 1966; Hughes *et al.*, 1980; Martin and Coughtrey, 1982). It has been proposed that the term heavy metals be abandoned in favour of a system which separates metal ions on the basis of their ability to form complexes with a variety of organic and inorganic ligands: class A (oxygen seeking), class B (nitrogen and/or sulphur seeking) and borderline or intermediate elements (Nieboer and Richardson, 1980). While this has some advantages, it is unwieldy and all the metals generally considered to be "heavy" occur in class B and borderline categories; that there is a borderline class shows that this system is also imprecise and again accentuates difficulties in attempting definitions. Most workers agree that, despite the problems, use of the term heavy metal will continue (Hughes *et al.*, 1980; Martin and Coughtrey, 1982).

In this review, the term "heavy metal" will be used in its broadest sense (Passow *et al.*, 1961; Hughes *et al.*, 1980; Martin and Coughtrey, 1982) without further attempts at definition and will include many of those metals encountered at elevated levels in the environment owing to natural and industrial activities, especially lead, cadmium, copper, mercury, manganese and zinc. Some, like copper and zinc, are essential for growth and metabolism at low concentrations (hence "trace elements") but it is at higher concentrations that these, and those metals with no essential biological functions like mercury, lead and cadmium, can be toxic to living organisms (Bowen, 1966). Cell death can result, or growth and metabolism be impaired, which can result in extensive damage to aquatic and terrestrial ecosystems.

Heavy metals exert toxicity by a number of methods. Their ability to bind to a variety of organic ligands can cause denaturation of proteins, including enzymes, disruption of cell membranes and decomposition of essential metabolites, and many can act as antimetabolites towards essential nutrients (Bowen, 1966). Partial correlation of toxicities with certain chemical properties like electronegativity, sulphide solubility or chelate stability has been demonstrated (Passow et al., 1961; Somers, 1961; Bowen, 1966) but there are considerable deviations, which is not surprising in view of the multiplicity of interactions that can occur between metals, molecules and cells. In this review, detailed attention to mechanisms of toxicity will not be given and readers are recommended to refer to other works (Passow et al., 1961; Bowen, 1966; Christie and Costa, 1984).

Heavy metal toxicity is the basis of many fungicidal preparations for the control of pathogens and preservation of natural and man-made materials (Foye, 1977). However, in spite of toxicity the ability of many fungi to survive and grow in polluted conditions is well known (Ashida, 1965; Ross, 1975). Fungi fulfil fundamental roles in natural environments, especially with regard to decomposition and nutrient cycling, and a knowledge of their responses not only is of scientific interest but also may be relevant to reclamation or detoxification of polluted habitats.

The ability of a fungus to survive and grow in the presence of high metal concentrations may depend on genetical adaptation, which is usually stable, and/or physiological adaptation, which is usually unstable (Ashida, 1965; Dekker, 1976). Experiments where fungi, particularly yeasts, are "trained" by successive transfers on media containing increasing concentrations of heavy metals may involve physiological adaptation (Macara, 1978) and/or the selection of genetically stable variants from the original population (Ashida, 1965). Adaptation can result in the expression of a resistance mechanism which can ensure avoidance or detoxification of the metal. However, adaptation may not be necessary if survival depends on intrinsic properties of the organism or metal detoxification by environmental factors (Gadd and Griffiths, 1978). Thus specific mechanisms of resistance involving adaptation should be clearly distinguished from those which do not. It should also be emphasized that the terms "tolerance" or "resistance" are arbitrary and whether a fungus is deemed "tolerant", "resistant" or "sensitive" often depends on subjective criteria such as the ability to grow on a certain metal concentration in media. Naturally, comparisons are easier between strains or mutants. Both "tolerance" and "resistance" are used interchangeably in the literature without clear distinction (Dekker, 1976) although it would perhaps be more appropriate to use "resistance" when an active mechanism produced as a direct response to a metal is initiated.

Another point worth mentioning here is that in the literature a variety of ways are used to express heavy metal concentrations. Some, like the outmoded "parts per million", can give rise to quite meaningless conclusions—similar parts per million values for different metal ions refer to quite different concentrations. In this chapter, concentrations are expressed in molarities whenever possible.

Environmental Influence on Heavy Metal Toxicity

Physical, organic and inorganic environmental components can markedly influence heavy metal availability and toxicity in aquatic and terrestrial environments (Gadd and Griffiths, 1978; Babich and Stotzky, 1978, 1980; Tyler, 1981).

A decreasing pH increases heavy metal availability whereas towards and above neutrality insoluble hydroxides and carbonates may form. However, while there are examples that fungal toxicity is unaffected by pH changes (Starkey, 1973) or is reduced at pH values above neutrality (Babich and Stotzky, 1979, 1982a), most evidence indicates that toxicity is reduced at acidic pH values (Babich and Stotzky, 1980) and the most extreme examples of fungal tolerance are confined to acidic conditions (Starkey, 1973; Stokes and Lindsay, 1979). This may be related to decreased metal uptake at low pH (see later).

In other instances, reduction of availability can result in reductions in toxicity. Clay minerals, particularly montmorillonite and vermiculite, can adsorb many heavy metals and reduce their toxicity (Babich and Stotzky, 1977a,b) and organic components, e.g. humic substances, proteins, amino acids, polysaccharides and chelating agents, can act in a similar fashion in aquatic and terrestrial habitats (Babich and Stotzky, 1980; Tyler, 1981). Other anions and cations can reduce toxicity: divalent cations like Ca^{2+} (Abel and Bärlocher, 1984), perhaps by competing with uptake systems (see later), and anions by the formation of complexes, as with chloride (Babich and Stotzky, 1982b, 1983a), or by the formation of insoluble compounds, as with sulphide, carbonate and phosphate (Gadd and Griffiths, 1978; Babich and Stotzky, 1980).

Similar interactions may obviously occur in laboratory media with anions and cations and where undefined substances like agar, peptones and yeast extract or complexing agents like ethylenediaminetetraacetic acid and citric acid can bind significant amounts of metals. In most cases, toxicity is reduced by such binding, and tolerance levels in such media are higher than in simple defined media.

Not always, however, is metal toxicity reduced by the kinds of interactions mentioned above. Some metal complexes are more toxic than the free ions (Gadd and Griffiths, 1978); marked inhibitory effects can still result in complex media even when few or no free ions can be detected (Ramamoorthy and Kushner, 1975) and synergistic effects of mixtures of metal ions can occur (Babich and Stotzky, 1983b).

Fungi in Heavy Metal-Polluted Habitats

Fungal numbers, as determined by dilution plating, were reduced and there were alterations in species composition in highly polluted soil near a zinc smelter (Jordan and Lechevalier, 1975). However, there was little difference in the zinc tolerance of fungi isolated from either site and most were capable of 50% growth at 700 μM Zn^{2+}. At control sites, zinc-tolerant genera included *Bdellospora, Verticillium* and *Paecilomyces* with *Penicillium, Torula* and *Aureobasidium* at polluted sites. In another study, fungal populations in soil contaminated with nickel, copper, iron and cobalt were not significantly different from those at control sites (Freedman and Hutchinson, 1980). Copper- and nickel-tolerant fungi (defined as being capable of growth at approximately 1.6 mM Cu^{2+} and/or Ni^{2+}) were isolated from both control and contaminated sites, the predominant genera being *Penicillium* (60%) followed by *Trichoderma, Rhodotorula, Oidiodendron, Mortierella* and *Mucor*. Along a steep copper and zinc gradient towards a brass mill, fungal biomass decreased by about 75% (Nordgren *et al.*, 1983). Below 100 μg Cu g^{-1}, no obvious effects were observed apart from some decrease in total mycelial length and, although there was no decrease in numbers of colony-forming units along the gradient, *Penicillium* and *Oidiodendron* declined at polluted sites whereas *Geomyces, Paecilomyces* and some sterile forms increased. Changes in species composition were also observed in an organo-mercurial-treated golf green as compared with the untreated fairway. *Chaetomium, Fusarium, Penicillium* and *Paecilomyces* species, *Gliomastix murorum* var *felinum* and *Myrothecium striatisporum* were greatly reduced whereas *Trichocladium asperum, Trichoderma hamatum, Zygorrhynchus moelleri* and *Chrysosporium pannorum* were isolated more frequently from the treated green (Pugh and Williams, 1971; Williams and Pugh, 1975). *Penicillium lilacinum* constituted 23% of all the fungi isolated from soil polluted by mine drainage and was cadmium tolerant up to 0.09 M. Other cadmium-tolerant fungi included *Paecilomyces* sp., *Synnematium, Penicillium waksmanni* and *Trichoderma* sp. (Tatsuyama *et al.*, 1975).

Filamentous fungi are frequently isolated from industrial and laboratory solutions of low pH containing high metal concentrations (Starkey, 1973).

Some, like *Penicillium ochro-chloron*, can grow in copper concentrations up to saturation values (Stokes and Lindsay, 1979).

Heavy metal pollution of plant surfaces is widespread and while there may be significant decreases in total microbial numbers (including bacteria) on phylloplanes, generally numbers of filamentous fungi and non-pigmented yeasts are little affected (Gingwell *et al.*, 1976; Bewley, 1979, 1980; Bewley and Campbell, 1980; Mowll and Gadd, 1985) and if metal-supplemented isolation media are used only fungi may be isolated (Bewley, 1980). On polluted oak leaves, *Aureobasidium pullulans* and *Cladosporium* species were the most numerous organisms and a greater proportion were metal-tolerant when compared with control isolates (Bewley and Campbell, 1978; Bewley, 1980). In fact, numbers of *A. pullulans* show good positive correlation with lead concentrations whether smelter or vehicle derived and on polluted leaves it can become the dominant organism (Bewley and Campbell, 1980; Mowll and Gadd, 1985). However, adaptation was unnecessary for growth of *A. pullulans* on polluted leaf surfaces and the ability to tolerate high lead levels occurred in isolates from the unpolluted site (Mowll and Gadd, 1985). Non-pigmented yeasts on phylloplanes, particularly *Cryptococcus* sp., also exhibit metal tolerance *in vivo* and *in vitro* (Gingell *et al.*, 1976; Bewley, 1979) in contrast with pigmented yeasts like *Sporobolomyces roseus* which is extremely sensitive to heavy metals (Bewley, 1980; Bewley and Campbell, 1980; Gadd, 1983; Mowll and Gadd, 1985). Smith (1977) also studied heavy metal effects on phylloplane fungi and found that *A. pullulans*, *Epicoccum* sp. and *Phialophora verrucosa* were tolerant, *Gnomia platani*, *Cladosporium* sp. and *Pleurophomella* were of intermediate tolerance while *Pestalotiopsis* and *Chaetomium* were sensitive.

Mercury-tolerant fungi have been isolated from the surfaces of seeds treated with mercury compounds. These include *Pyrenophora avenae*, *Penicillium crustosum*, *Cladosporium cladosporoides*, *Syncephalastrum racemosum* and *Ulocladium atrum* (Noble *et al.*, 1966; Greenaway, 1972, 1973; Ross and Old, 1973a,b; Greenaway *et al.*, 1974). However, mercury tolerance was not stable and populations were not uniformly composed of tolerant or susceptible strains (Noble *et al.*, 1966; Old, 1968).

Thus, fungi can exhibit considerable tolerance towards heavy metals and may become dominant organisms in some polluted habitats. While there is evidence that species diversity may be reduced in metal-polluted habitats, tolerance can be exhibited by fungi from polluted and unpolluted habitats. This indicates that, in many cases, survival must be dependent on intrinsic properties, "gratuitous mechanisms of resistance" (Gadd and Griffiths, 1978), rather than adaptive changes and also that other environmental changes associated with the metal pollution, e.g. altered pH and nutrient status, may be important in altering species composition.

Mycorrhizal Interactions with Heavy Metals

A knowledge of the responses of mycorrhizal fungi to heavy metals is of importance in view of the interest in the reclamation of polluted sites; their involvement in the revegetation of spoil heaps has been described (Daft *et al.*, 1975; Allen and Allen, 1980).

Hyphae of vesicular–arbuscular mycorrhizas can translocate zinc and copper to the host plant (Benson and Covey, 1976; Cooper and Tinker, 1978; Gildon and Tinker, 1983b). This has implications for growth in polluted environments since toxicity may result towards the host plants or, if the fungus is inhibited, there could be a reduction in the phosphorus supply to the plant (McIlveen and Cole, 1979).

Isolates of the endophyte *Glomus mosseae* from polluted sites have been shown to exhibit marked tolerance to zinc, copper, nickel and cadmium in contrast with isolates from unpolluted soils (Gildon and Tinker, 1981, 1983a). In the presence of zinc or cadmium, infection with a sensitive strain was much lower than with a tolerant strain and, in zinc-amended soil, growth with the tolerant endophyte was greater although a clear benefit only resulted at metal levels sufficient almost to eliminate the susceptible strain (Gildon and Tinker, 1981). Further studies have confirmed the protective influence of mycorrhizal infection against heavy metals (Gildon and Tinker, 1983a).

The protective influence of mycorrhizas towards heavy metals in ericaceous plants has also been demonstrated; little or no growth occurs in mycorrhiza-free plants in the presence of copper and zinc (Bradley *et al.*, 1981, 1982). The endophytes all grew well in pure culture at copper concentrations up to 1.2 mM and zinc concentrations up to 2.3 mM. Mycorrhizal plants had lower concentrations in their shoots than non-mycorrhizal plants but higher metal concentrations in their roots. It was suggested that the hyphal complexes of the endophyte absorbed the metals which resulted in exclusion from the shoot and avoidance of toxicity. This function of ericoid mycorrhizas is of ecological significance and tolerance levels of mycorrhizal ericaceous plants were considerably greater than those of metal-tolerant grasses like *Agrostis*. This explains why *Calluna* plants, for example, show greater colonization powers on contaminated soil than *Agrostis* and on some sites are the only significant plant cover (Bradley *et al.*, 1981, 1982).

Heavy Metal Uptake by Fungi

Heavy metal uptake by fungi is of fundamental importance to organisms growing in polluted habitats since tolerance may be determined by the ability to prevent cellular entry of a potentially toxic metal or the ability to compartmentalize or detoxify it within the cell. Heavy metal uptake should not be considered in isolation from the considerable amount of information on K^+, Ca^{2+} and Mg^{2+} (Borst-Pauwels, 1981) since such ions may compete with heavy metal uptake or, like K^+, be released from cells during influx. In addition, divalent cations may enter cells by a common mechanism.

Characteristics and Mechanisms

Metal uptake by fungi generally comprises two phases. The first is metabolism-independent surface binding while the second is energy-dependent intracellular influx and, in cases of ambiguity, these have been distinguished in this section. However, a variety of terminology is found in the literature and "uptake" or "accumulation" may refer to either or both mechanisms. In this chapter, uptake is defined as the sum of both mechanisms in living cells.

Wall binding is rapid, often taking less than a few minutes (Rothstein and Hayes, 1956; Khovrychev, 1973), and amounts bound can be large. *Saccharomyces cerevisiae* walls can bind their own weight of Hg^{2+} (Murray and Kidby, 1975), 30–40% of total Co^{2+} uptake by *Neurospora crassa* was due to surface binding (Venkateswerlu and Sastry, 1980) and 90% of Ag^{2+} taken up by *Candida utilis* was associated with surfaces (Golubovich *et al.*, 1976). Dilute acids and complexing agents may remove a large proportion of bound metals (Rothstein and Hayes, 1956; Venkateswerlu and Sastry, 1970; Golubovich *et al.*, 1976). Differences in binding capacity occur between different species which can reflect different wall compositions. Spore walls of *Penicillium italicum* accumulated more copper than spore walls of *N. crassa* (Somers, 1963) and greater wall binding of zinc by *S. roseus* than *S. cerevisiae* could be accounted for by differences in wall composition (Mowll and Gadd, 1983). Chlamydospores of *A. pullulans* bound much more metal ions than hyaline yeast-like cells or mycelium (Mowll and Gadd, 1984) and all the metal was localized in the thick, melanized cell wall (Gadd, 1984; Mowll and Gadd, 1984). The surface binding of metals can be expressed using Freundlich, Scatchard or Langmuir plots (Paton and Budd, 1972; Brown *et al.*, 1974; Murray and Kidby, 1975; Duddridge and Wainwright, 1980; Tsezos and Volesky, 1981; Mowll and Gadd, 1983, 1984). These can illustrate differences between organisms and/or different cell types of the same organism (Duddridge and Wainwright, 1980; Mowll and Gadd, 1983, 1984). However, there are a number of assumptions including that fitment of accumulation data to a Freundlich or Langmuir isotherm indicates a uniform unimolecular layer and that each ion is adsorbed onto an adsorption site (Brown *et al.*, 1974). Binding may be complex and of more than one type and manipulation of data in this way may not be very meaningful (Passow and Rothstein, 1960). Walls of dead fungal cells can bind metals and their capacity can be greater than (Somers, 1963; Ponta and Broda, 1970), equal to (Golubovich *et al.*, 1976) or less than (Duddridge and Wainwright, 1980) that of living cells depending on alterations to wall structure caused by the killing process used. Competition between metal ions can occur for binding sites on cell walls. K^+, Mg^{2+} and Ca^{2+} competed for Cu^{2+}-binding sites on fungal spores (Somers, 1963); Pb^{2+} and Mg^{2+} influenced Cu^{2+} absorption to *C. utilis* (Khovrychev, 1973); Hg^{2+} inhibited Co^{2+} binding to *N. crassa* (Venkateswerlu and Sastry, 1970); Hg^{2+}, Co^{2+}, Mg^{2+}, Mn^{2+} and Cu^{2+} inhibited Cd^{2+} binding by yeast-like cells of *A. pullulans* although for chlamydospores only Cu^{2+} had a significant effect (Mowll and Gadd, 1984). It should be mentioned that metal-binding sites on cell surfaces may not be involved in subsequent influx; the surface binding of metal ions can be considerably reduced, or even absent (Mohan *et al.*, 1984), without affecting influx rates and, conversely, influx rates can be reduced when surface binding is unaffected (see Borst-Pauwels (1981) and later sections).

Energy-dependent influx proceeds at a slower rate than surface binding and can be inhibited by low temperatures, glucose analogues and metabolic inhibitors and uncouplers (Fuhrmann and Rothstein, 1968; Norris and Kelly, 1977; Mowll and Gadd, 1983, 1984) and phosphate needs to be present in cells prior to or during incubation with a given metal (Fuhrmann and Rothstein, 1968; Okorokov *et al.*, 1975; Roomans *et al.*, 1979).

In yeasts energy-dependent influx of Cu^{2+}, Cd^{2+}, Zn^{2+}, Co^{2+}, Ni^{2+}, Mg^{2+}, Ca^{2+} and Sr^{2+} has been demonstrated with an apparent affinity series of Mg^{2+}, Co^{2+}, Zn^{2+}, $> Mn^{2+} > Ni^{2+} > Ca^{2+} > Sr^{2+}$ (Fuhrmann and Rothstein, 1968). That there is a similar sequence for divalent cation activation of the plasma membrane ATPase implied ATPase involvement in divalent cation influx (Fuhrmann, 1974). However, a direct role has been discounted since no activation of the ATPase by Ca^{2+} or Sr^{2+} was observed (Peters and Borst-Pauwels, 1979) and initial rates of Sr^{2+} influx were not significantly different from those for Ca^{2+} and Mn^{2+}, the latter ion being capable of strong stimulation of ATPase activity (Roomans *et al.*, 1979; Nieuwenhuis *et al.*, 1981). Even though influx rates of Mn^{2+} and Sr^{2+} were similar, about 10 times more Mn^{2+} than Sr^{2+} was subsequently accumulated, which suggested efflux systems of different rate and the involvement of Mn^{2+} compartmentalization within the vacuole (Nieuwenhuis *et al.*, 1981).

While most work has been carried out on yeasts, energy-dependent intracellular influx of Ni^{2+}, Co^{2+}, Zn^{2+}, Cd^{2+} and Cu^{2+} has been demonstrated in several fungi including *N. crassa* (Mohan and Sastry, 1983a), *Neocosmospora vasinfecta* (Paton and Budd, 1972) and *A. pullulans* (Gadd,

1981; Mowll and Gadd, 1984). Reasons why there is less work on filamentous fungi include problems arising from the mycelial growth habit. High wall binding can mask low rates of intracellular influx and there may be difficulties in removing free ions trapped in interhyphal spaces. In fungi other than yeasts, influx systems seem to be of low affinity and, in *A. pullulans*, sensitive radioisotopic methods proved best for analysis (Gadd, 1981; Mowll and Gadd, 1984). By applying accumulation data to adsorption isotherms it was concluded that Pb^{2+}, Cd^{2+} and Zn^{2+} uptake by *Pythium* sp., *Scytalidium lignicola* and *Dictyuchus sterile* was largely an absorptive process (Duddridge and Wainwright, 1980). However, metal concentrations were low (less than 0.15 mM) and there may have been some of the problems described above. The use of protoplasts may eliminate some of the difficulties associated with mycelium and their application has been demonstrated in yeast (Gadd *et al.*, 1984a). It should be stressed that there must be some kind of influx mechanism in filamentous fungi for essential ions like Cu^{2+} and Zn^{2+}.

Divalent cation influx may be driven by the membrane potential and be reduced or halted when the cell membrane is depolarized by K^+ or substrate–proton cotransport (Roomans *et al.*, 1979) or where the membrane potential is abolished by uncouplers (Lichko *et al.*, 1980). Enhanced K^+ efflux from cells, which may result if the membrane is hyperpolarized by ethidium bromide or ionophores, may increase rates of cation influx because of the resulting negative diffusion potential (Boutry *et al.*, 1977; Peña, 1978). Although divalent cation efflux may not always be detected when the membrane is depolarized, this does not necessarily exclude membrane potential involvement in transport because of internal compartmentalization and/or complexation of metal ions (see later). It should also be mentioned that the monovalent ion transport system can also transport Ca^{2+} and Mg^{2+} in low amounts (Borst-Pauwels, 1981).

In fungi, the influx of several metal ions can be described by the Michaelis–Menten equation which may indicate that there are saturable sites on the cell membrane. However, there are wide variations in reported K_m and V_{max} values even for the same metal and the same organism (Borst-Pauwels, 1981). While there are several reasons for this including strain differences, complex formation and adsorption phenomena (which may reduce the external free ion concentration), it may also indicate the existence of two transport systems, one of high affinity (low K_m) and one of low affinity (high K_m). Two affinity constants were found for Co^{2+} influx in *S. cerevisiae* (Norris and Kelly, 1977; Heldwein *et al.*, 1977) and deviations from Michaelis-Menten kinetics were also observed for Ca^{2+} and Sr^{2+} (Roomans *et al.*, 1979). At low divalent cation concentrations, where sites are not saturated, deviations from Michaelis–Menten kinetics may be due to surface potential effects (Roomans *et al.*, 1979; Borst-Pauwels, 1981). However, studies with heavy metals may be concerned with high concentrations where surface potential effects may be minimized.

It should be borne in mind that different kinetic parameters may be obtained using cells of different physiological states. The V_{max} for Zn^{2+} influx by *C. utilis* was 17 times greater in late exponential phase cells than in cells at other stages of growth (Failla and Weinberg, 1977), while for Cu^{2+} influx rates were maximal in early exponential phase cells (Khovrychev, 1973). Mn^{2+} influx into *S. cerevisiae* was maximal in mid-exponential phase cells (Okorokov *et al.*, 1979).

Release of Monovalent Cations

After metal influx, electroneutrality may be maintained by K^+ efflux and a stoichiometric relationship of $1M^{2+}_{in}:2K^+_{out}$ has often been observed (Fuhrmann and Rothstein, 1968; Norris and Kelly, 1977; Okorokov *et al.*, 1979; Lichko *et al.*, 1980; Mowll and Gadd, 1984). In fact, good rates of metal influx can depend on prior loading of cells with K^+ and increasing levels of external K^+ can inhibit influx (Rothstein *et al.*, 1958; Okorokov *et al.*, 1979; Lichko *et al.*, 1980). However, K^+ efflux may not always be a feature of metal influx; Na^+-loaded yeasts release Na^+ (Fuhrmann and Rothstein, 1968) and there was no K^+ efflux during Co^{2+} influx into K^+-deficient yeast (Norris and Kelly, 1977), Cu^{2+} influx into *Rhodotorula mucilaginosa* (Norris and Kelly, 1979) or Zn^{2+} influx into *S. roseus* (Mowll and Gadd, 1983). It is possible that K^+ efflux may not be detected if it is at low concentration and completely adsorbed onto cell walls or retaken up into cells (Borst-Pauwels, 1981).

An important consideration is whether K^+ efflux is an integral part of a metal influx system or represents membrane damage by the heavy metal and leakage of intracellular contents. This can be a combination of an "all-or-none" effect for individual cells, so that with increasing metal concentration a greater proportion of the population is affected, together with gradual K^+ loss (Passow and Rothstein, 1960; Kuypers and Roomans, 1979). Thus, a simple stoichiometric relationship between metal influx and K^+ efflux may not occur for cadmium, mercury, copper and silver (Norris and Kelly, 1977) nor is such a relationship likely for toxic concentrations of Co^{2+} or Mn^{2+}. For *S. cerevisiae*, energy-dependent Cd^{2+} influx/K^+ efflux was apparent at low concentrations (Norris and Kelly, 1977; Gadd and Mowll, 1983) but, at 0.5 mM, over 90% of cell K^+ (and Mg^{2+}) was released after 5 minutes, whether glucose was present or absent, and this mirrored exactly the pattern of cell death (Gadd and Mowll, 1983). A similar relationship between K^+

release and cell death occurred at Zn^{2+} concentrations above 0.1 mM (Mowll and Gadd, 1983). However, there was no detectable Zn^{2+}-induced K^+ efflux from *S. roseus* despite great Zn^{2+} sensitivity which suggested a mechanism of toxicity other than membrane disruption (Mowll and Gadd, 1983).

Intracellular Compartmentation of Metal Ions

Internal compartmentation not only ensures a pool of ions for the maintenance of essential functions but may also be a means of detoxification and localization of potentially toxic metal ions (Gadd and Griffiths, 1978). A major proportion of Mn^{2+}, Mg^{2+} and K^+ taken in by *S. cerevisiae* and *Saccharomyces carlsbergensis* is located in the vacuoles (Okorokov *et al.*, 1977, 1980; Lichko *et al.*, 1980, 1982) and bound to low molecular weight polyphosphates, the biosynthesis of which accompanies Mg^{2+} or Mn^{2+} accumulation (Okorokov *et al.*, 1980; Lichko *et al.*, 1982). A decrease in vacuolar K^+ accompanies Mn^{2+} accumulation in the vacuole (Lichko *et al.*, 1982) and the vacuolar membrane is assumed to have a transport system for cation transfer from cytoplasm to vacuole. Thus, intracellular metal ion concentrations may be regulated by both vacuolar and plasma membrane transport systems and also polyphosphate levels (Lichko *et al.*, 1982). Ca^{2+} transport by Ca^{2+}/H^+ antiport has been demonstrated in intact yeast vacuoles and right-side-out vacuolar membrane vesicles (Ohsumi and Anraku, 1983). Other data exist for Fe^{3+} and Ca^{2+} accumulation in vacuoles of *Penicillium chrysogenum* (cited in Okorokov *et al.*, 1980). Other work has demonstrated localization of Mg^{2+}, Sr^{2+} and Ca^{2+} in cytoplasmic granules, possibly polyphosphate, in yeast cells lacking vacuoles (Roomans, 1980).

Certain yeasts can precipitate thallium as thallium oxide within mitochondria, which may be subsequently excreted from cells, by a process termed oxidative detoxification (Lindegren and Lindegren, 1973). Localized silver deposition around cell walls and within vacuoles has been observed in *Cryptococcus albidus* (Brown and Smith, 1976), mercury precipitation in electron-dense bodies occurs in mercury-exposed hyphae of *C. pannorum* (Williams and Pugh, 1975) and similar bodies, presumed to contain zinc, occurred in *N. vasinfecta* after Zn^{2+} influx (Paton and Budd, 1972).

Influence of pH on Metal Uptake

In general, low external pH reduces both surface binding and intracellular influx (Fuhrmann and Rothstein, 1968; Venkateswerlu and Sastry, 1970; Paton and Budd, 1972; Okorokov *et al.*, 1979; Roomans *et al.*, 1979; Tsezos and Volesky, 1981; Mowll and Gadd, 1984). There may be an optimal pH value or range for maximal rates of influx below or above which a decrease occurs: pH 4–7 for Cu^{2+} influx into *C. utilis* (Khovrychev, 1973), pH 4 for Ni^{2+} and *N. crassa* (Mohan *et al.*, 1984), pH 6.5 for Zn^{2+} and *N. vasinfecta* (Paton and Budd, 1972), pH 5.0 for Mn^{2+} and *S. cerevisiae* (Okorokov *et al.*, 1979) and pH 4.8 for Zn^{2+} and *C. utilis* (Failla *et al.*, 1976). At higher pH values than these, metal availability may be reduced because of precipitation as hydroxides for example (Gadd and Griffiths, 1978).

Intracellular pH may also affect influx rates: Sr^{2+} influx rates into *S. cerevisiae* decreased with increasing cell pH (Roomans *et al.*, 1979). There is little information for other metals and organisms.

Relationship between Metal Uptake and Toxicity

A relationship between metal uptake and sensitivity has been observed. Decreased influx or impermeability may result in tolerance and those environmental factors that decrease uptake like low external pH may also confer tolerance (Gadd and Griffiths, 1978).

Increased Cu^{2+} influx by *S. cerevisiae* was associated with increased toxicity; both influx and toxicity were reduced at low temperatures (Ross, 1977). A similar close connection between metal influx and toxicity in yeast has been shown for Zn^{2+} and Cd^{2+} (Gadd and Mowll, 1983; Mowll and Gadd, 1983). A decreased influx of Cd^{2+}, Cu^{2+} and Li^{+} by metal-tolerant yeast mutants or strains, when compared with sensitive parent strains, has been described (Asensio *et al.*, 1976; Ross and Walsh, 1981; Joho *et al.*, 1983; Gadd *et al.*, 1984a): a copper-tolerant strain of *A. pullulans* took up less copper than a sensitive strain (Gadd and Griffiths, 1980a), and there was no energy-dependent Co^{2+} influx into a resistant strain of *N. crassa* in contrast with the parent strain (Venkateswerlu and Sastry, 1979). The use of protoplasts demonstrated that decreased Cu^{2+} influx by a tolerant strain of *S. cerevisiae* was due to a change in membrane transport properties and not, for example, altered wall permeability (Gadd *et al.*, 1984a). Chlamydospores of *A. pullulans* are impermeable to heavy metals and were more tolerant than hyaline yeast-like cells or mycelium (Gadd and Griffiths, 1980b; Gadd, 1981; Mowll and Gadd, 1984).

Naturally, the relationship between influx and toxicity is best ascertained using parent and mutant strains of the same species. Different species will differ in sensitivity and amounts of metal taken up. Cd^{2+} uptake was similar in both *S. cerevisiae* and *Rhodotorula rubra* but the former was more sensitive (Berthe-Corti *et al.*, 1984); *S. roseus* took up less zinc than *S. cerevisiae* but was much more sensitive (Mowll and Gadd, 1983). Another deviation from the apparently clear relationship between uptake and toxicity in yeasts was where Mn^{2+}-resistant strains accumulated more Mn^{2+} than the sensitive wild type (Bianchi et al., 1981). However, other properties which may have enabled enhanced uptake by the resistant strain like increased sulphide production, synthesis of metal-binding proteins or efficient internal compartmentation may also have been involved.

As described previously, low external pH generally decreases both surface binding and rates of intracellular uptake and thus tolerance to a given metal may increase at low pH. Scytalidium can grow in saturated $CuSO_4$ at pH 0.3–2.0 but is sensitive to concentrations as low as 4×10^{-5} M near neutrality (Starkey, 1973). For P. ochro-chloron, a fungus also capable of growth in high Cu^{2+} concentrations, a constant copper uptake occurs above external concentrations of approximately 16 mM (Okamoto et al., 1977). However, at pH 6.0 and above, copper uptake and toxicity markedly increased, even at low external concentrations (Gadd and White, 1985). For Scytalidium, the pH range of 4.2–5.0 appeared critical for copper toxicity (Starkey, 1973) and, for a strain of A. pullulans, copper uptake and toxicity were markedly reduced below pH 4.0 (Gadd and Griffiths, 1980a). Further experiments with a copper-tolerant strain of A. pullulans showed that a pH decrease in the medium could explain the kinetics of copper uptake through growth and the observed alleviation of copper toxicity (Gadd and Griffiths, 1980a).

The connection between reduced uptake at low pH and low toxicity is only clear in a limited number of cases. A toxicity reduction with increased acidity has not been observed for all fungi (Starkey, 1973; Babich and Stotzky, 1981) although some pH effects can be accounted for by the chemical behaviour of the metal ions involved, e.g. the formation of insoluble hydroxides at pH values around neutrality.

Influence of Other Cations on Metal Influx and Toxicity

In N. crassa, the alleviation of Cu^{2+} influx and toxicity by Fe^{2+} and Mn^{2+} has been shown (Rao et al., 1984), as has suppression of Ni^{2+} and Co^{2+} influx in certain strains by Mg^{2+} and Mn^{2+} (Venkateswerlu and Sastry, 1970; Mohan and Sastry, 1983a,b; Mohan et al., 1984). Similarly, Mn^{2+} and Mg^{2+} inhibited Zn^{2+} influx in N. vasinfecta (Paton and Budd, 1972) although Zn^{2+} influx in C. utilis was only inhibited by Cd^{2+} (Failla et al., 1976). Mg^{2+} was a competitive inhibitor of Co^{2+} influx in S. cerevisiae (Norris and Kelly, 1977) and Ca^{2+} depressed rates of Cd^{2+} influx in S. cerevisiae (Norris and Kelly, 1977) and A. pullulans (Mowll and Gadd, 1984). The inhibition of metal cation influx by others may be related to ionic radii (Norris and Kelly, 1977) although the situation may be complicated when toxic symptoms like membrane disruption occur, influx rates being reduced because of this or sometimes being enhanced because of increased cell permeability (Borst-Pauwels, 1981). Competition for influx between various metal ions may indicate that there is a general mechanism of uptake for divalent cations of low specificity (Furhmann and Rothstein, 1968; Borst-Pauwels, 1981).

Metallothioneins

Metallothioneins are low molecular weight, cysteine-rich proteins that can bind metals like cadmium, zinc and/or copper. They occur widely in biological systems, their most obvious functions being detoxification and storage and regulation of intracellular metal ions (Webb, 1979; Christie and Costa, 1984).

Fungal metallothioneins exclusively contain copper (Prinz and Weser, 1975; Weser *et al.*, 1977; Lerch, 1980; Beltramini and Lerch, 1983) in contrast with those of higher eukaryotic organisms which contain copper and/or zinc under physiological conditions (Webb, 1979).

The occurrence of a copper–thionein (cuprodoxin) in *S. cerevisiae* (molecular weight 9500 ± 500) was first demonstrated by Prinz and Weser (1975) and there are further studies on the structure of this compound (Weser *et al.*, 1977; Rupp *et al.*, 1979). Low molecular weight (approximately 8000) copper–proteins were also isolated from yeast by others and showed a similarity with rat liver copper–chelatin (Premakumar *et al.*, 1975), a metalloprotein with properties different from those of metallothioneins (Webb, 1979). Weser *et al.* (1977) have considered copper–chelatin to be a similar type of protein to their copper–thionein (cuprodoxin) although a direct relationship was not initially obvious. Indeed, in metallothionein research there is controversy arising from methodology and data interpretation (Webb, 1979), much undoubtedly being due to the reactive crosslinking properties of metallothioneins which can result in mistaken identification of unique metal-binding proteins (Christie and Costa, 1984). Three types of copper-binding proteins, all of molecular weight approximately 10 000 and resembling copper–thionein, were extracted from a copper-resistant strain of *S. cerevisiae* after growth in copper-containing medium (Naiki and Yamagata, 1976). A detoxification mechanism was ascribed and, in fact, such proteins were also detected in the sensitive parent strain when it was grown at sublethal copper concentrations (Naiki and Yamagata, 1976). In *C. utilis*, using an extraction procedure relatively specific for metallothioneins, evidence was presented for the synthesis of a metallothionein-like protein which may have been involved in storage and regulation of intracellular zinc (Failla *et al.*, 1976) although no metallothionein-like protein was found in Cd^{2+}-adapted *S. cerevisiae* (Macara, 1978).

N. crassa, when grown in the presence of copper, synthesizes a low molecular weight (2200) copper-binding metallothionein with some structural features common to cadmium or zinc metallothioneins from man, horse and mouse (Lerch, 1980; Beltramini and Lerch, 1983).

At low physiological copper concentrations, there was complete copper uptake by *N. crassa* and over 40% was associated with metallothionein which indicated a storage function. That there was tolerance to copper over a wide concentration range and a constant copper metallothionein content emphasized a detoxification function (Lerch, 1980).

While at low exposure levels detoxification and/or intracellular regulation of metal ions by metallothioneins seems apparent, this may not be their normal physiological function. It has been suggested in other eukaryotic cells that the free protein may be biologically active with functions in the maintenance of redox potential, ion transport and as a metabolic pool of cysteine residues (Webb, 1979). Prinz and Weser (1975) proposed a role for the reversibly oxidizable cuprodoxin from *S. cerevisiae* in bioenergetic systems while Lerch (1980) suggested that *Neurospora* metallothionein may act as a metal donor to active sites of metal-containing enzymes.

Although it has been stated that resistance to heavy metal toxicity in yeast is mediated by the synthesis of metallothionein (Fogel and Welch, 1982), many other aspects of structure and physiology can be involved (see other sections). For organisms suddenly exposed to potentially lethal concentrations, survival in the first instance probably depends on other properties. Metallothioneins are synthesized over a time period and, by definition, such synthesis must occur at sublethal concentrations.

Industrial Applications of Fungal Heavy Metal Uptake

The uptake of metal ions by fungi is of industrial relevance for the removal of metals from waste waters for environmental protection and/or subsequent recovery of the metal (Brierley and Brierley, 1983). Fungi may be better suited for this purpose than other microbial groups because of their often great tolerance towards metals and other adverse external conditions such as low pH, their high wall binding capacity and often high intracellular uptake values for living cells. Yeasts were more effective for Ni^{2+} and Cu^{2+} uptake than several bacteria (Norris and Kelly, 1979). Filamentous fungi have received more attention than yeasts and most work has been carried out with dead cells. As described previously, affinities and rates of intracellular influx may be low for filamentous fungi and wall binding high. Use of dead cells obviates possible toxicity problems and any irreversible internal compartmentation of metal ions. So far, most published work has been concerned with uranium and thorium which can enter the environment by means of contaminated effluents from the nuclear energy industry. As for metals, rates and extent of adsorption are governed by a variety of external factors like pH, temperature, other anions and cations and ionic radii (Strandberg et al., 1981; Tsezos and Volesky, 1981, 1982; Tsezos, 1983; Tobin et al., 1984).

S. cerevisiae can accumulate up to 10–15% of its dry weight as uranium on the cell surface (Strandberg et al., 1981). Rhizopus arrhizus walls can absorb up to 18% uranium and thorium (Tsezos and Volesky, 1981, 1982) while Penicillium sp. takes up at least 0.5% (Galun et al., 1983). Binding is to a variety of ligands including phosphate, carboxylate, amine, hydroxyl and sulphydryl groups (Rothstein and Meier, 1951; Strandberg et al., 1981; Tobin et al., 1984). Non-living biomass of R. arrhizus absorbs a variety of other metal cations like Cu^{2+}, Zn^{2+}, Hg^{2+} and Cd^{2+} (but not those of alkali metals, i.e. Na^+, K^+, Rb^+ and Cs^+) and uptake was directly related to ionic radii, larger ions being absorbed more strongly than smaller ones (Tobin et al., 1984). The chitin component of R. arrhizus cell walls was most important in metal complexation (Tseozs and Volesky, 1982; Tsezos, 1983) and the chelating ability of a variety of other kinds of fungal biomass was also related to chitin and chitosan content (Muzzarelli and Tanfani, 1982).

Industrial interest in metal accumulation by fungi is sure to increase not only because of high uptake capacity and low cost, fungal biomass being often available as industrial waste (Tsezos and Volesky, 1981; Muzzarelli and Tanfani, 1982), but also because after removal of complexed metals, by sodium bicarbonate for example (Tsezos, 1984), the biomass can be reused as biosorbent (Strandberg et al., 1981; Galun et al., 1983).

Morphological Changes in Response to Heavy Metals

Cell morphology, the production of fruiting structures and yeast–mycelial (Y–M) transitions can all be affected by heavy metals. This section deals with such effects, largely omitting the considerable amount of data pertaining to essential metals like calcium and magnesium (see Pitt and Ugalde, 1984; Walker et al., 1984). Obviously, elements like copper and zinc are essential micronutrients and may have specific roles in morphogenesis at low non-toxic concentrations but this section is mainly concerned with morphological changes that occur at concentrations where some toxicity may be manifest. This is a difficult distinction to make because many studies have not considered toxicity which can make it unclear whether observed morphological changes do reflect a specific physiological role for the metal or are non-specific and the same as those changes which occur in response to other adverse external conditions.

The Y–M transition of *Candida albicans* can be affected by certain metal ions. In general, zinc suppresses the mycelial phase, which results in maximal yeast cell production (Widra, 1964). Yamaguchi (1975) found that maximal yeast cell development occurred at 9 μM Zn^{2+}; concentrations above 10 μM were inhibitory. This zinc effect was influenced by temperature, itself an important determinant of the Y–M transition. In short-term experiments, micromolar zinc concentrations inhibited mycelial development at 25 °C but not at 37 °C even at millimolar concentrations (Bedell and Soll, 1979). However, in longer-term experiments, zinc at 400 and 600 μM (concentrations which caused up to a 136% increase in generation time) did suppress mycelial development but had no effect on germination (Ross, 1982). Copper was also found to inhibit mycelial development at 37 °C and, over the range 15–250 μM Cu^{2+}, populations consisted of yeast cells (Vaughn and Weinberg, 1978). However, Ross (1982) found that although copper almost completely inhibited germination at 7.5 and 10 μM there were inhibitory effects and significant increases in generation time. It was concluded that this copper effect was due to general copper toxicity and not to specific inhibition of mycelial development (Ross, 1982). Cadmium, at 1.5 and 2.5 μM, also inhibited germination of *C. albicans* but was similarily toxic. Thus, the effects of copper and cadmium appeared to be different from those of zinc which did not affect germination but prevented the further production of mycelium (Ross, 1982). Zinc has also been shown to favour yeast-like development and to suppress mycelial formation in *Sporothrix schenckii* (Alsina and Rodriguez-del Valle, 1984), and in some strains of *Histoplasma capsulatum* (Pine and Peacock, 1958). Some recent studies have shown that cobalt, at 10^{-5} M, enhanced the activity of a morphogenic autoregulatory substance (MARS) produced by *C. albicans* which suppressed the Y–M transition, but no other metal ions, including Zn^{2+}, influenced MARS activity (Hazen and Cutler, 1983).

Manganese is reported to stimulate yeast-phase development in *C. albicans* (Widra, 1964) and the number of conidia-bearing structures (coremia) and conidia produced by *Penicillium claviforme* and *Penicillium clavigerum* (Tinnell *et al.*, 1977). Manganese concentrations of less than 10^{-7} M resulted in the abnormal development of squat, bulbous hyphae and swollen spores in *Aspergillus niger* whereas above 10^{-6} M only filamentous hyphae resulted (Kisser *et al.*, 1980). Although these manganese concentrations were unlikely to be toxic, further additions of copper (0.1–1.0 mM depending on the Mn^{2+} concentration) reproduced the same morphological effects as manganese deficiency (Kisser *et al.*, 1980).

In *A. pullulans*, a variety of heavy metals were found to induce or accelerate the formation of melanin-pigmented chlamydospores and hyphae on solid and in liquid media (Gadd and Griffiths, 1980b; Gadd, 1981). Since several of these metals have no apparent biological function, like aluminium, lead, cadmium and mercury, it seemed that this morphological change was a general response to unfavourable conditions and did in fact resemble those changes that occurred in normal metal-free populations owing to nutrient limitation and alterations in growth rate (Gadd, 1980).

Other Mechanisms of Heavy Metal Detoxification

Organic and inorganic substances, produced inside or exterior to fungal cells, may detoxify metals by complexation or precipitation although whether many of these are primary resistance mechanisms remains uncertain.

Citric acid is an efficient chelating agent while oxalic acid interacts with iron and copper to form insoluble oxalate crystals (Levi, 1969; Englander and Corden, 1971). The latter mechanism can be effective against copper–chrome–arsenate wood preservatives and has been observed in *Poria* sp. (Levi, 1969), as well as *A. niger, Penicillium spinulosum* and *Verticillium psalliotae* (Murphy and Levy, 1983). Compounds implicated in the mercury resistance of *P. avenae* include a red anthraquinone pigment (Greenaway, 1971) and non-protein thiols (Ross and Old, 1973b). However, the significance of the pigment is dubious (Ross and Old, 1973a) and there were no significant differences in amounts of non-protein thiols between resistant and sensitive strains (Ross, 1974). Similarly, extracellular thiols were detected after growth of *Fusarium solani* in media supplemented with sodium thiosulphate but both in the absence and presence of Hg^{2+}; exogenously added thiols could confer Hg^{2+} protection (Wainwright and Grayston, 1983). Other postulated mechanisms are via intracellular pools of sulphydryl compounds (Ashworth and Amin, 1964; Kikuchi, 1964) and links between cysteine and methionine metabolism have been described (Singh and Sherman, 1974; Wainwright and Grayston, 1983). One copper-resistant yeast strain was found to contain a large amount of superoxide dismutase when repeatedly cultured in the presence of copper which aided survival (Naiki, 1980). In *P. ochro-chloron*, high levels of intracellular glycerol, induced by high external Cu^{2+}, were involved in the exclusion of Cu^{2+} (Gadd et al., 1984b).

The production of H_2S can result in the precipitation of heavy metals as insoluble sulphides (Ashida, 1965). In yeasts, copper sulphide is deposited in and around the cell wall which can make colonies appear dark brown (Ashida et al., 1963). However, as described in other sections, this is not the sole explanation for observed tolerance: H_2S production by lag and early logarithmic phase cells of a copper-resistant strain did not differ from that of a sensitive strain, and copper tolerance can be exhibited by yeasts with no H_2S-producing ability (Ashida et al., 1963).

Metal transformation by fungi has been demonstrated but there is little work in this area. Hg^{2+} can be reduced by yeast to the elemental state (Hg^0) which can then be lost from the medium (Brunker and Bott, 1974). Methylation of mercury can also be catalysed by yeasts and fungi (Landner, 1971; Vonk and Sijpesteijn, 1973). Although methylation products may be more toxic than the free cations, they are often volatilized from the medium (Jernelöv and Martin, 1975). Certain fungi have been implicated in the oxidation of Mn^{2+} to form MnO_2 which precipitates in and around oxidizing colonies (Konetzka, 1977).

Concluding Remarks

Fungi are ubiquitous, important and sometimes dominant in metal-polluted habitats and they exhibit a variety of responses towards heavy metals which can determine survival. Some may be specific mechanisms of resistance which are induced by the presence of metal ions while others are non-specific and depend on intrinsic properties, but all may be influenced to varying extents by environmental factors. Of special importance is heavy metal uptake and again specific and non-specific mechanisms occur which can decrease or prevent entry or detoxify a metal within the cell. Apart from the occasional use of mutants, detailed genetical studies have received little attention except for the work relating to metallothionein and copper resistance in yeast (Fogel and Welch, 1982; Butt et al., 1984). Of course it is difficult to carry out genetic analysis with many fungi so perhaps future work on those organisms with well-understood genetics should be encouraged.

Clearly, interactions of heavy metals with fungi are embraced within wider biological contexts from the cellular level such as transport and morphogenesis to ecology and industrial applications. I hope that this perspective is not lost in this chapter and that the great potential of fungi for fundamental studies is indicated.

References

Abel, T. H. and Bärlocher, F. (1984). Effects of cadmium on aquatic hyphomycetes. *Applied and Environmental Microbiology* **48**, 245–251.

Allen, E. B. and Allen, M. F. (1980). Natural re-establishment of vesicular–arbuscular mycorrhizae following strip-mine reclamation in Wyoming. *Journal of Applied Ecology* **17**, 139–147.

Alsina, A. and Rodriguez-del Valle, N. (1984). Effects of divalent cations and functionally related substances on the yeast to mycelium transition in *Sporothrix schenckii*. *Sabouraudia: Journal of Medical and Veterinary Mycology* **22**, 1–5.

Asensio, J., Ruiz-Argüeso, T. and Rodriguez-Navarro, A. (1976). Sensitivity of yeasts to lithium. *Antonie van Leeuwenhoek; Journal of Microbiology and Serology* **42**, 1–8.

Ashida, J. (1965). Adaptation of fungi to metal toxicants. *Annual Review of Phytopathology* **3**, 153–174.

Ashida, J., Higashi, N. and Kikuchi, T. (1963). An electronmicrocopic study on copper precipitation by copper-resistant yeast cells. *Protoplasma* **57**, 27–32.

Ashworth, L. J. and Amin, J. V. (1964). A mechanism for mercury tolerance in fungi. *Phytopathology* **54**, 1459–1463.

Babich, H. and Stotzky, G. (1977a). Reductions in the toxicity of cadmium to microorganisms by clay minerals. *Applied and Environmental Microbiology* **33**, 696–705.

Babich, H. and Stotzky, G. (1977b). Effect of cadmium on fungi and on interactions between fungi and bacteria in soil: influence of clay minerals and pH. *Applied and Environmental Microbiology* **33**, 1059–1066.

Babich, H. and Stotzky, G. (1978). Effects of cadmium on the biota: influence of environmental factors. *Advances in Applied Microbiology* **23**, 55–117.

Babich, H. and Stotzky, G. (1979). Abiotic factors affecting the toxicity of lead to fungi. *Applied and Environmental Microbiology* **38**, 506–513.

Babich, H. and Stotzky, G. (1980). Environmental factors that influence the toxicity of heavy metal and gaseous pollutants to microorganisms. *Critical Reviews in Microbiology* **8**, 99–145.

Babich, H. and Stotzky, G. (1981). Manganese toxicity to fungi: influence of pH. *Bulletin of Environmental Contamination and Toxicology* **27**, 474–480.

Babich, H. and Stotzky, G. (1982a). Nickel toxicity to fungi: influence of environmental factors. *Ecotoxicology and Environmental Safety* **6**, 577–589.

Babich, H. and Stotzky, G. (1982b). Influence of chloride ions on the toxicity of cadmium to fungi. *Zentralblatt für Bakteriologie und Hygiene* **C3**, 421–426.

Babich, H. and Stotzky, G. (1983a). Nickel toxicity to estuarine/marine fungi and its amelioration by magnesium in sea water. *Water, Air and Soil Pollution* **19**, 193–202.

Babich, H. and Stotzky, G. (1983b). Synergism between nickel and copper in their toxicity to microbes: mediation by pH. *Ecotoxicology and Environmental Safety* **7**, 576–587.

Bedell, G. W. and Soll, D. R. (1979). Effects of low concentrations of zinc on the growth and dimorphism of *Candida albicans*: evidence for zinc-resistant and -sensitive pathways for mycelium formation. *Infection and Immunity* **26**, 348–354.

Beltramini, M. and Lerch, K. (1983). Spectroscopic studies on *Neurospora* copper metallothionein. *Biochemistry* **22**, 2043–2048.

Benson, N. R. and Covey, R. P. (1976). Response of apple seedlings to zinc fertilisation and mycorrhizal inoculation. *Hortscience* **11**, 252–253.

Berthe-Corti, L., Pietsch, I., Mangir, M., Ehrlich, W. and Lochmann, E.-R. (1984). Die Wirkung von Cadmium auf Wachstum und Stoffwechsel von *Saccharomyces*- und *Rhodotorula*-zellen. *Chemosphere* **13**, 107–119.

Bewley, R. J. F. (1979). The effects of zinc, lead and cadmium pollution on the leaf surface microflora of *Lolium perenne* L. *Journal of General Microbiology* **110**, 247–254.

Bewley, R. J. F. (1980). Effects of heavy metal pollution on oak leaf microorganisms. *Applied and Environmental Microbiology* **40**, 1053–1059.

Bewley, R. J. F. and Campbell, R. (1978). Scanning electron microscopy of oak leaves contaminated with heavy metals. *Transactions of the British Mycological Society* **71**, 508–511.

Bewley, R. J. F. and Campbell, R. (1980). Influence of zinc, lead and cadmium pollutants on the microflora of hawthorn leaves. *Microbial Ecology* **6**, 227–240.

Bianchi, M. E., Carbone, M. L. and Lucchini, G. (1981). Mn^{2+} and Mg^{2+} uptake in Mn-sensitive and Mn-resistant yeast strains. *Plant Science Letters* **22**, 345–352.

Borst-Pauwels, G. W. F. H. (1981). Ion transport in yeast. *Biochimica et Biophysica Acta* **650**, 88–127.

Boutry, M., Foury, F. and Goffeau, A. (1977). Energy-dependent uptake of calcium by the yeast *Schizosaccharomyces pombe*. *Biochimica et Biophysica Acta* **464**, 602–612.

Bowen, H. J. M. (1966). "Trace Elements in Biochemistry". Academic Press, London.

Bradley, R., Burt, A. J. and Read, D. J. (1981). Mycorrhizal infection and resistance to heavy metal toxicity in *Calluna vulgaris*. *Nature, London* **292**, 335–337.

Bradley, R., Burt, A. J. and Read, D. J. (1982). The biology of mycorrhiza in the Ericaceae. VIII. The role of mycorrhizal infection in heavy metal resistance. *New Phytologist* **91**, 197–209.

Brierley, J. A. and Brierley, C. L. (1983). Biological accumulation of some heavy metals—biotechnological applications. *In* "Biomineralization and Biological Metal Accumulation" (Eds P. Westbroek and E. W. de Jong), pp. 499–509. Reidel, Dordrecht.

Brown, R. B., Vairo, M. L. R. and Borzani, W. (1974). Quantitative study of labelled Hg^{2+} adsorption by live yeast cells. Evaluation of the number of glucose penetration sites. *Journal of Fermentation Technology* **52**, 536–541.

Brown, T. A. and Smith, D. G. (1976). The effects of silver nitrate on the growth and ultrastructure of the yeast *Cryptococcus albidus*. *Microbios Letters* **3**, 155–162.

Brunker, R L. and Bott, T. L. (1974). Reduction of mercury to the elemental state by a yeast. *Applied Microbiology* **27**, 870–873.

Butt, T. R., Sternberg, E. J., Gorman, J. A., Clark, P., Hamer, D., Rosenberg, M. and Crooke, S. T. (1984). Copper metallothionein of yeast, structure of the gene, and regulation of expression. *Proceedings of the National Academy of Sciences of the United States of America* **81**, 3332–3337.

Christie, N. T. and Costa, M. (1984). *In vitro* assessment of the toxicity of metal compounds. IV. Disposition of metals in cells: interactions with membranes, glutathione, metallothionein and DNA. *Biological Trace Element Research* **6**, 139–158.

Cooper, K. M. and Tinker, P. B. (1978). Translocation and transfer of nutrients in vesicular–arbuscular mycorrhizas. II. Uptake and translocation of phosphorus, zinc and sulphur. *New Phytologist* **81**, 43–52.

Daft, M. J., Hackscaylo, E. and Nicolson, T. H. (1975). Arbuscular mycorrhizas in plants colonising coal spoils in Scotland and Pennsylvania. *In* "Endomycorrhizas" (Eds. F. E. Sanders, B. Mosse and P. B. Tinker), pp. 561–580. Academic Press, London.

Dekker, J. (1976). Acquired resistance to fungicides. *Annual Review of Phytopathology* **14**, 405–428.

Duddridge, J. E. and Wainwright, M. (1980). Heavy metal accumulation by aquatic fungi and reduction in viability of *Gammarus pulex* fed Cd^{2+} contaminated mycelium. *Water Research* **14**, 1605–1611.

Englander, C. M. and Corden, M. E. (1971). Stimulation of mycelial growth of *Endothia parasitica* by heavy metals. *Applied Microbiology* **22**, 1012–1016.

Failla, M. L. and Weinberg, E. D. (1977). Cyclic accumulation of zinc by *Candida utilis* during growth in batch culture. *Journal of General Microbiology* **99**, 85–97.

Failla, M. L., Benedict, C. D. and Weinberg, E. D. (1976). Accumulation and storage of Zn^{2+} by *Candida utilis*. *Journal of General Microbiology* **94**, 23–36.

Fogel, S. and Welch, J. W. (1982). Tandem gene amplification mediates copper resistance in yeast. *Proceedings of the National Academy of Sciences of the United States of America* **79**, 5342–5346.

Foye, W. O. (1977). Antimicrobial activities of mineral elements. *In* "Microorganisms and Minerals" (Ed. E. D. Weinberg), pp. 387–419. Dekker, New York.

Freedman, B. and Hutchinson, T. C. (1980). Effects of smelter pollutants on forest leaf litter decomposition near a nickel–copper smelter at Sudbury, Ontario. *Canadian Journal of Botany* **58**, 1722–1736.

Fuhrmann, G. F. (1974). Relation between divalent cation transport and divalent cation activated ATPase in yeast plasma membranes. *Experientia* **30**, 686.

Fuhrmann, G. F. and Rothstein, A. (1968). The transport of Zn^{2+}, Co^{2+} and Ni^{2+} into yeast cells. *Biochimica et Biophysica Acta* **163**, 325–330.

Gadd, G. M. (1980). Melanin production and differentiation in batch cultures of the polymorphic fungus *Aureobasidium pullulans*. *FEMS Microbiology Letters* **9**, 237–240.

Gadd, G. M. (1981). Mechanisms implicated in the ecological success of polymorphic fungi in metal-polluted habitats. *Environmental Technology Letters* **2**, 531–536.

Gadd, G. M. (1983). The use of solid medium to study effects of cadmium, copper and zinc on yeasts and yeast-like fungi: applicability and limitations. *Journal of Applied Bacteriology* **54**, 57–62.

Gadd, G. M. (1984). Effect of copper on *Aureobasidium pullulans* in solid medium: adaptation not necessary for tolerant behaviour. *Transactions of the British Mycological Society* **82**, 546–549.

Gadd, G. M. and Griffiths, A. J. (1978). Microorganisms and heavy metal toxicity. *Microbial Ecology* **4**, 303–317.

Gadd, G. M. and Griffiths, A. J. (1980a). Influence of pH on toxicity and uptake of copper in *Aureobasidium pullulans*. *Transactions of the British Mycological Society* **75**, 91–96.

Gadd, G. M. and Griffiths, A. J. (1980b). Effect of copper on morphology of *Aureobasidium pullulans*. *Transactions of the British Mycological Society* **74**, 387–392.

Gadd, G. M. and Mowll, J. L. (1983). The relationship between cadmium uptake, potassium release and viability in *Saccharomyces cerevisiae*. *FEMS Microbiology Letters* **16**, 45–48.

Gadd, G. M. and White, C. (1985). Copper uptake by *Penicillium ochro-chloron*: influence of pH on toxicity and demonstration of energy-dependent copper influx using protoplasts. *Journal of General Microbiology* **131**, 1875–1879.

Gadd, G. M., Stewart, A., White, C. and Mowll, J. L. (1984a). Copper uptake by whole cells and protoplasts of a wild-type and copper-resistant strain of *Saccharomyces cerevisiae*. *FEMS Microbiology Letters* **24**, 231–234.

Gadd, G. M., Chudek, J. A., Foster, R. and Reed, R. H. (1984b). The osmotic responses of *Penicillium ochro-chloron*: changes in internal solute levels in response to copper and salt stress. *Journal of General Microbiology* **130**, 1969–1975.

Galun, M., Keller, P., Feldstein, H., Galun, E., Siegel, S. and Siegel, B. (1983). Recovery of uranium(VI) from solution using fungi. II. Release from uranium-loaded *Penicillium* biomass. *Water, Air and Soil Pollution* **20**, 277–283.

Gildon, A. and Tinker, P. B. (1981). A heavy metal-tolerant strain of a mycorrhizal fungus. *Transactions of the British Mycological Society* **77**, 648–649.

Gildon, A. and Tinker, P. B. (1983a). Interactions of vesicular–arbuscular mycorrhizal infection and heavy metals in plants. I. The effects of heavy metals on the development of vesicular–arbuscular mycorrhizas. *New Phytologist* **95**, 247–261.

Gildon, A. and Tinker, P. B. (1983b). Interactions of vesicular–arbuscular mycorrhizal infections and heavy metals in plants. II. The effects of infection on uptake of copper. *New Phytologist* **95**, 263–268.

Gingell, S. M., Campbell, R. and Martin, M. H. (1976). The effect of zinc, lead and cadmium pollution on the leaf surface microflora. *Environmental Pollution* **11**, 25–37.

Golubovich, V. N., Khovrychev, M. P. and Rabotnova, I. L. (1976). Binding of silver ions by cells of *Candida utilis*. *Microbiology* **45**, 105–107.

Greenaway, W. (1971). Relationship between mercury resistance and pigment production in *Pyrenophora avenae*. *Transactions of the British Mycological Society* **56**, 37–44.

Greenaway, W. (1972). Epiflora of oat seed treated with organo-mercury. *Transactions of the British Mycological Society* **58**, 321–327.

Greenaway, W. (1973). Resistance of *Ulocladium atrum* to phenyl-mercury. *Transactions of the British Mycological Society* **60**, 359–360.

Greenaway, W., Cripps, A. and Ward, S. (1974). Resistance to organo-mercury by *Penicillia* isolated from cereal seed. *Transactions of the British Mycological Society* **63**, 137–141.

Hazen, K. C. and Cutler, J. E. (1983). Effect of cobalt and morphogenic autoregulatory substance (MARS) on morphogenesis of *Candida albicans*. *Experimental Mycology* **7**, 182–187.

Heldwein, R., Tromballa, H. W. and Broda, E. (1977). Aufnahme von Cobalt, Blei und Cadmium durch Bäckerhefe. *Zeitschrift für Allgemeine Mikrobiologie* **17**, 299–308.

Hughes, M. K., Lepp, N. W. and Phipps, D. A. (1980). Aerial heavy metal pollution and terrestrial ecosystems. *In* "Advances in Ecological Research" (Ed. A. Macfadyen), Vol. 11, pp. 217–237. Academic Press, London.

Jernelöv, A. and Martin, A. L. (1975). Ecological implications of metal metabolism by microorganisms. *Annual Review of Microbiology* **29**, 61–77.

Joho, M., Sukenobu, Y., Egashira, E. and Murayama, T. (1983). The correlation between Cd^{2+} sensitivity and Cd^{2+} uptake in the strains of *Saccharomyces cerevisiae*. *Plant and Cell Physiology* **24**, 389–394.

Jordan, M. J. and Lechevalier, M. P. (1975). Effects of zinc-smelter emissions on forest soil microflora. *Canadian Journal of Microbiology* **21**, 1855–1865.

Khovrychev, M. P. (1973). Absorption of copper ions by cells of *Candida utilis*. *Microbiology* **42**, 745–749.

Kikuchi, T. (1964). Comparison of original and secondarily developed copper resistance of yeast strains. *Botanical Magazine, Tokyo* **77**, 395–402.

Kisser, M., Kubicek, C. P. and Rohr, M. (1980). Influence of manganese on morphology and cell wall composition of *Aspergillus niger* during citric acid fermentation. *Archives of Microbiology* **128**, 26–33.

Konetzka, W. A. (1977). Microbiology of metal transformations. *In* "Microorganisms and minerals" (Ed. E. D. Weinberg), pp. 317–342. Dekker, New York.

Kuypers, G. A. J. and Roomans, G. M. (1979). Mercury-induced loss of K$^+$ from yeast cells investigated by electron probe X-ray microanalysis. *Journal of General Microbiology* **115**, 13–18.

Landner, L. (1971). Biochemical model for the biological methylation of mercury suggested from methylation studies *in vivo* with *Neurospora crassa*. *Nature, London* **230**, 452–453.

Lerch, K. (1980). Copper metallothionein, a copper-binding protein from *Neurospora crassa*. *Nature, London* **284**, 368–370.

Levi, M. P. (1969). The mechanisms of action of copper–chrome–arsenate preservatives against wood-destroying fungi. *Record of the Annual Convention of the British Wood Preserving Association*, 113–127.

Lichko, L. P., Okorokov, L. A. and Kulaev, I. S. (1980). Role of vacuolar ion pool in *Saccharomyces carlsbergensis*: potassium efflux from vacuoles is coupled with manganese or magnesium influx. *Journal of Bacteriology* **144**, 666–671.

Lichko, L. P., Okorokov, L. A. and Kulaev, I. S. (1982). Participation of vacuoles in regulation of levels of K$^+$, Mg^{2+} and orthophosphate ions in cytoplasm of the yeast *Saccharomyces carlsbergensis*. *Archives of Microbiology* **132**, 289–293.

Lindegren, C. C. and Lindegren, G. (1973). Oxidative detoxification of thallium in the yeast mitochondria. *Antonie van Leeuwenhoek; Journal of Microbiology and Serology* **39**, 351–353.

Macara, I. G. (1978). Accommodation of yeast to toxic levels of cadmium ions. *Journal of General Microbiology* **104**, 321–324.

Martin, M. H. and Coughtrey, P. J. (1982). "Biological Monitoring of Heavy Metal Pollution". Applied Science Publishers, London.

McIlveen, W. D. and Cole, H. (1979). Influence of zinc on development of the endomycorrhizal fungus *Glomus mosseae* and its mediation of phosphorus uptake by *Glycine max* "Amsoy 71". *Agricultural Environment* **4**, 245–256.

Mohan, P. M. and Sastry, K. S. (1983a). Interrelationships in trace-element metabolism in metal toxicities in nickel-resistant strains of *Neurospora crassa*. *Biochemical Journal* **212**, 205–210.

Mohan, P. M. and Sastry, K. S. (1983b). Studies on copper toxicity in nickel-resistant strains of *Neurospora crassa*. *Current Microbiology* **9**, 127–132.

Mohan, P. M., Rudra, M. P. P. and Sastry, K. S. (1984). Nickel transport in nickel-resistant strains of *Neurospora crassa*. *Current Microbiology* **10**, 125–128.

Mowll, J. L. and Gadd, G. M. (1983). Zinc uptake and toxicity in the yeasts *Sporobolomyces roseus* and *Saccharomyces cerevisiae*. *Journal of General Microbiology* **129**, 3421–3425.

Mowll, J. L. and Gadd, G. M. (1984). Cadmium uptake by *Aureobasidium pullulans*. *Journal of General Microbiology* **130**, 279–284.

Mowll, J. L. and Gadd, G. M. (1985). Effect of vehicular lead pollution on phylloplane mycoflora. *Transactions of the British Mycological Society* **84**, 685–689.

Murphy, R. J. and Levy, J. F. (1983). Production of copper oxalate by some copper tolerant fungi. *Transactions of the British Mycological Society* **81**, 165–168.

Murray, A. D. and Kidby, D. K. (1975). Sub-cellular location of mercury in yeast grown in the presence of mercuric chloride. *Journal of General Microbiology* **86**, 66–74.

Muzzarelli, R. A. A. and Tanfani, F. (1982). The chelating ability of chitinous materials from *Aspergillus niger, Streptomyces, Mucor rouxii, Phycomyces blakeseanus* and *Choanephora curcurbitarum. In* "Chitin and Chitosan" (Eds S. Mirano and S. Tokura), pp. 183–186. The Japanese Society of Chitin and Chitosan, Tottori.

Naiki, N. (1980). Role of superoxide dismutase in a copper-resistant strain of yeast. *Plant and Cell Physiology* **21**, 775–783.

Naiki, N. and Yamagata, S. (1976). Isolation and some properties of copper-binding proteins found in a copper-resistant strain of yeast. *Plant and Cell Physiology* **17**, 1281–1295.

Nieboer, E. and Richardson, D. H. S. (1980). The replacement of the nondescript term "heavy metals" by a biologically and chemically significant classification of metal ions. *Environmental Pollution* **1**, 3–26.

Nieuwenhuis, B. J. W. M., Weijers, A. G. M. and Borst-Pauwels, G. W. F. H. (1981). Uptake and accumulation of Mn^{2+} and Sr^{2+} in *Saccharomyces cerevisiae*. *Biochimica et Biophysica Acta* **649**, 83–88.

Noble, M., MacGarvie, Q. D., Hams, A. F. and Leafe, E. L. (1966). Resistance to mercury of *Pyrenophora avenae* in Scottish seed oats. *Plant Pathology* **15**, 23–28.

Nordgren, A., Bååth, E. and Söderström, B. (1983). Microfungi and microbial activity along a heavy metal gradient. *Applied and Environmental Microbiology* **45**, 1829–1837.

Norris, P. R. and Kelly, D. P. (1977). Accumulation of cadmium and cobalt by *Saccharomyces cerevisiae*. *Journal of General Microbiology* **99**, 317–324.

Norris, P. R. and Kelly, D. P. (1979). Accumulation of metals by bacteria and yeasts. *Developments in Industrial Microbiology* **20**, 299–308.

Ohsumi, Y. and Anraku, Y. (1983). Calcium transport driven by a proton motive force in vacuolar membrane vesicles of *Saccharomyces cerevisiae*. *Journal of Biological Chemistry* **258**, 5614–5617.

Okamoto, K., Suzuki, M., Fukami, M., Toda, S. and Fuwa, K. (1977). Uptake of heavy metals by a copper-tolerant fungus, *Penicillium ochro-chloron. Agricultural Biology and Chemistry* **41**, 17–22.

Okorokov, L. A., Lichko, L. P., Kholodenko, V. P., Kadomtseva, V. M., Petrikevich, S. B., Zaichkin, E. and Karimova, A. M. (1975). Free and bound magnesium in fungi and yeasts. *Folia Microbiologica* **20**, 460–466.

Okorokov, L. A., Lichko, L. P., Kadomtseva, V. M., Kholodenko, V. P., Titovsky, V. T. and Kulaev, I. S. (1977). Energy-dependent transport of manganese into yeast cells and distribution of accumulated ions. *European Journal of Biochemistry* **75**, 373–377.

Okorokov, L. A., Kadomtseva, V. M. and Titovskii, B. I. (1979). Transport of manganese into *Saccharomyces cerevisiae. Folia Microbiologica* **24**, 240–246.

Okorokov, L. A., Lichko, L. P. and Kulaev, I. S. (1980). Vacuoles: main compartments of potassium, magnesium and phosphate ions in *Saccharomyces carlsbergensis* cells. *Journal of Bacteriology* **144**, 661–665.

Old, K. M. (1968). Mercury-tolerant *Pyrenophora avenae* in seed oats. *Transactions of the British Mycological Society* **51**, 525–534.

Passow, H. and Rothstein, A. (1960). The binding of mercury by the yeast cell in relation to changes in permeability. *Journal of General Physiology* **43**, 621–633.

Passow, H., Rothstein, A. and Clarkson, T. W. (1961). The general pharmacology of the heavy metals. *Pharmacological Reviews* **13**, 185–224.

Paton, W. H. N. and Budd, K. (1972). Zinc uptake in *Neocosmospora vasinfecta. Journal of General Microbiology* **72**, 173–184.

Peña, A. (1978). Effect of ethidium bromide on Ca^{2+} uptake by yeast. *Journal of Membrane Biology* **42**, 199–213.

Peters, P. H. J. and Borst-Pauwels, G. W. F. H. (1979). Properties of plasmamembrane ATPase and mitochondrial ATPase of *Saccharomyces cerevisiae*. *Physiologia Plantarum* **46**, 330–337.

Pine, L. and Peacock, C. L. (1958). Studies on the growth of *Histoplasma capsulatum* IV. Factors influencing conversion of the mycelial phase to the yeast phase. *Journal of Bacteriology* **75**, 167–174.

Pitt, D. and Ugalde, U. O. (1984). Calcium in fungi. *Plant, Cell and Environment* **7**, 467–475.

Ponta, H. and Broda, E. (1970). Mechanismen der Aufnahme von Zink durch Bäckerhefe. *Planta* **95**, 18–26.

Premakumar, R., Winge, D. R., Wiley, R. D. and Rajagopalan, K. V. (1975). Copper–chelatin: isolation from various eucaryotic sources. *Archives of Biochemistry and Biophysics* **170**, 278–288.

Prinz, R. and Weser, U. (1975). Cuprodoxin. *FEBS Letters* **54**, 224–228.

Pugh, G. J. F. and Williams, J. I. (1971). Effect of an organo-mercury fungicide on saprophytic fungi and on litter decomposition. *Transactions of the British Mycological Society* **57**, 164–166.

Ramamoorthy, S. and Kushner, D. J. (1975). Binding of mercuric and other heavy metal ions by microbial growth media. *Microbial Ecology* **2**, 162–176.

Rao, S., Subramanyam, C. and Venkateswerlu, G. (1984). Nitrogen metabolism in the blue mycelia of *Neurospora crassa* isolated from copper toxic cultures. *Current Microbiology* **10**, 79–84.

Roomans, G. M. (1980). Localization of divalent cations in phosphate-rich cytoplasmic granules in yeast. *Physiologia Plantarum* **48**, 47–50.

Roomans, G. M., Theuvent, A. P. R., Van den Berg, T. P. R. and Borst-Pauwels, G. W. F. H. (1979). Kinetics of Ca^{2+} and Sr^{2+} uptake by yeast. Effect of pH, cations and phosphate. *Biochimica et Biophysica Acta* **551**, 187–196.

Ross, I. S. (1974). Non-protein thiols and mercury resistance of *Pyrenophora avenae*. *Transactions of the British Mycological Society* **63**, 77–83.

Ross, I. S. (1975). Some effects of heavy metals on fungal cells. *Transactions of the British Mycological Society* **64**, 175–193.

Ross, I. S. (1977). Effect of glucose on copper uptake and toxicity in *Saccharomyces cerevisiae*. *Transactions of the British Mycological Society* **69**, 77–81.

Ross, I. S. (1982). Effect of copper, cadmium and zinc on germination and mycelial growth in *Candida albicans*. *Transactions of the British Mycological Society* **78**, 543–545.

Ross, I. S. and Old, K. M. (1973a). Mercuric chloride resistance of *Pyrenophora avenae*. *Transactions of the British Mycological Society* **60**, 293–300.

Ross, I. S. and Old, K. M. (1973b). Thiol compounds and resistance of *Pyrenophora avenae* to mercury. *Transactions of the British Mycological Society* **60**, 301–310.

Ross, I. S. and Walsh, A. L. (1981). Resistance to copper in *Saccharomyces cerevisiae*. *Transactions of the British Mycological Society* **77**, 27–32.

Rothstein, A. and Hayes, A. D. (1956). The relationship of the cell surface to metabolism. XIII. The cation-binding properties of the yeast cell surface. *Archives of Biochemistry and Biophysics* **63**, 87–99.

Rothstein, A. and Meier, R. (1951). The relationship of the cell surface to metabolism. VI. The chemical nature of uranium-complexing groups of the cell surface. *Journal of Cellular and Comparative Physiology* **38**, 245–270.

Rothstein, A., Hayes, A. D., Jennings, D. and Hooper, D. (1958). The active transport of Mg^{2+} and Mn^{2+} into the yeast cell. *Journal of General Physiology* **41**, 585–594.

Rupp, H., Cammack, R., Hartmann, H.-J. and Weser, U. (1979). Oxidation–reduction reactions of copper-thiolate centres in Cu-thionein. *Biochimica et Biophysica Acta* **578**, 462–475.

Singh, A. and Sherman, F. (1974). Association of methionine requirement with methyl mercury resistant mutants of yeast. *Nature, London* **247**, 227–229.

Smith, W. H. (1977). Influence of heavy metal leaf contaminants on the *in vitro* growth of urban-tree phylloplane fungi. *Microbial Ecology* **3**, 231–239.

Somers, E. (1961). The fungitoxicity of metal ions. *Annals of Applied Biology* **49**, 246–253.

Somers, E. (1963). The uptake of copper by fungal cells. *Annals of Applied Biology* **51**, 425–437.

Starkey, R. L. (1973). Effect of pH on toxicity of copper to *Scytalidium* sp., a copper-tolerant fungus, and some other fungi. *Journal of General Microbiology* **78**, 217–225.

Stokes, P. M. and Lindsay, J. E. (1979). Copper tolerance and accumulation in *Penicillium ochro-chloron* isolated from copper-plating solution. *Mycologia* **71**, 796–806.

Strandberg, G. W., Shumate, S. E. and Parrott, J. R. (1981). Microbial cell as biosorbents for heavy metals: accumulation of uranium by *Saccharomyces cerevisiae* and *Pseudomonas aeruginosa*. *Applied and Environmental Microbiology* **41**, 237–245.

Tatsuyama, K., Egawa, H., Senmaru, H., Yamamoto, H., Ishioka, S., Tamatsukuri, T. and Saito, K. (1975). *Penicillium lilacinum*: its tolerance to cadmium. *Experientia* **31**, 1037–1038.

Tinnell, W. H., Jefferson, B. L. and Benoit, R. E. (1977). Manganese-mediated morphogenesis in *Penicillium claviforme* and *Penicillium clavigerum*. *Canadian Journal of Microbiology* **23**, 209–212.

Tobin, J. M., Cooper, D. G. and Neufeld, R. J. (1984). Uptake of metal ions by *Rhizopus arrhizus* biomass. *Applied and Environmental Microbiology* **47**, 821–824.

Tsezos, M. (1983). The role of chitin in uranium adsorption by *Rhizopus aarhizus*. *Biotechnology and Bioengineering* **25**, 2025–2040.

Tsezos, M. (1984). Recovery of uranium from biological adsorbents—desorption equilibrium. *Biotechnology and Bioengineering* **26**, 973–981.

Tsezos, M. and Volesky, B. (1981). Biosorption of uranium and thorium. *Biotechnology and Bioengineering* **23**, 583–604.

Tsezos, M. and Volesky, B. (1982). The mechanism of uranium biosorption by *Rhizopus aarhizus*. *Biotechnology and Bioengineering* **24**, 385–401.

Tyler, G. (1981). Heavy metals in soil biology and biochemistry. *In* "Soil Biochemistry" (Eds E. A. Paul and J. N. Ladd), Vol. 5, pp. 371–414. Dekker, New York.

Vaughn, V. J. and Weinberg, E. D. (1978). *Candida albicans* dimorphism and virulence: role of copper. *Mycopathologia* **64**, 39–42.

Venkateswerlu, G. and Sastry, K. S. (1970). The mechanism of uptake of cobalt ions by *Neurospora crassa*. *Biochemical Journal* **118**, 497–503.

Venkateswerlu, G. and Sastry, K. S. (1979). Cobalt transport in a cobalt-resistant strain of *Neurospora crassa*. *Journal of Biosciences* **1**, 433–439.

Vonk, J. W. and Sijpesteijn, A. K. (1973). Studies on the methylation of mercuric chloride by pure cultures of bacteria and fungi. *Antonie van Leeuwenhoek; Journal of Microbiology and Serology* **39,** 505–513.

Wainwright, M. and Grayston, S. J. (1983). Reduction in heavy metal toxicity towards fungi by addition to media of sodium thiosulphate and sodium tetrathionate. *Transactions of the British Mycological Society* **81,** 541–546.

Walker, G. M., Sullivan, P. A. and Shepherd, M. G. (1984). Magnesium and the regulation of germ-tube formation in *Candida albicans. Journal of General Microbiology* **130,** 1941–1945.

Webb, M. (1979). The metallothioneins. *In* "The Chemistry, Biochemistry and Biology of Cadmium" (Ed. M. Webb), pp. 195–266. Elsevier–North-Holland, Amsterdam.

Weser, U., Hartmann, H.-J., Fretzdorff, A. and Strobel, G. J. (1977). Homologous copper(I)-(thiolate)₂-chromophores in yeast copper thionein. *Biochimica et Biophysica Acta* **493,** 465–477.

Widra, A. (1964). Phosphate directed Y–M variation in *Candida albicans. Mycopathologia et Mycologia Applicata* **23,** 197–202.

Williams, J. I. and Pugh, G. J. F. (1975). Resistance of *Chrysosporium pannorum* to an organomercury fungicide. *Transactions of the British Mycological Society* **64,** 255–263.

Yamaguchi, H. (1975). Control of dimorphism in *Candida albicans* by zinc: effect on cell morphology and composition. *Journal of General Microbiology* **86,** 370–372.

5

Strategies of Phototrophic Bacteria in Sulphide-containing Environments

H. VAN GEMERDEN and R. DE WIT

*Department of Microbiology, University of Groningen, Kerklaan 30,
NL-9751 NN Haren, The Netherlands*

The presence of sulphide is a common phenomenon in aqueous ecosystems. In particular in eutrophic and stratified lakes of some depth, a hypolimnion is formed during the summer in which oxygen is utilized as the result of respiration by a wide variety of organisms, eventually resulting in anaerobic conditions. In the deeper parts of such lakes, and even more markedly in the sediments, the conditions then become favourable for the obligate anaerobic sulphate-reducing bacteria, resulting in the production of sulphide. The sulphide also reaches higher water strata, either by diffusion or by internal water movements, where it may be taken up by various organisms, including phototrophic bacteria.

In most natural fresh waters the sulphate ion is the second or third most abundant anion; nevertheless, its concentration rarely exceeds 0.3 mM. Consequently, in most fresh waters the production of sulphide by sulphate-reducing bacteria is often limited by the availability of sulphate. Because of the decomposition of organic matter, somewhat increased sulphide concentrations may be encountered locally. However, in general the concentration of sulphide does not exceed 0.3 mM, which is not considered inhibitory for anoxygenic phototrophic bacteria, although appreciable differences exist in this respect between different representatives. For other organisms the presence of sulphide, even at such relatively low concentrations, may well be lethal. In natural fresh waters the impact of the concentration of sulphide on the selection of anoxygenic phototrophic bacteria thus emphasizes the affinity for sulphide, rather than the inhibition by sulphide. Although this is of relevance as well in marine and brackish waters and sediments, it is obvious from the sulphate content of such ecosystems that here the concentration of sulphide—provided that sufficient degradable substrates are available—may well exceed a concentration of 10 mM. For comparison, the sulphate concentration in sea water is about 28 mM. Such high concentrations of sulphide create serious problems even for those phototrophic bacteria that utilize sulphide as an electron donor in photosynthesis. Likewise, the presence of sulphide in high concentrations as a result of geological activities (volcanos, deep-sea wells, etc.) has a similar impact. In the latter cases, however, phototrophic bacteria are less commonly involved.

Different groups of phototrophic bacteria are involved in the utilization of various reduced forms of sulphur. The purple sulphur bacteria (Chromatiaceae and Ectothiorhodospiraceae) and the green sulphur bacteria (Chlorobiaceae) are well known for their ability to utilize sulphide as an electron donor in anoxygenic photosynthesis. The number of cyanobacteria studied in this respect is gradually increasing (Garlick et al., 1977). The data show that many of them are also able to fix carbon dioxide at the expense of sulphide oxidation, in addition to having the ability to perform an oxygenic photosynthesis resulting in the production of oxygen from water.

Judged from the types of ecosystems inhabited by purple and green sulphur bacteria and cyanobacteria, it becomes evident that high sulphide concentrations at least incidentally, but presumably frequently, are encountered by these microbes. The fact that such organisms thrive in the aforementioned habitats indicates that they are able to cope with elevated concentrations of sulphide and have developed or maintained certain "strategies" to do so. When discussing strategies of phototrophic bacteria, ways of life are meant which by selection have shown survival value during evolution.

Sulphides of many trace elements are extremely insoluble: the solubility product of FeS is 3.7×10^{-19} mol 1^{-1} whereas the corresponding value for cobalt sulphide is 3×10^{-29} mol 1^{-1}. One of the adaptation mechanisms reported for fungi which may explain their ability to grow in the presence of high concentrations of heavy metals is the production of sulphide as an extracellular detoxifying compound (see Chapter 4 by Gadd, this volume). Conceivably, the toxicity of sulphide could thus be explained on the basis of an inaccessibility of trace elements.

It has indeed been shown in the laboratory that continuous cultures of sulphate-reducing bacteria—intended to be sulphate limited—were in fact limited by the concentration of soluble iron. In order to maintain anaerobic conditions and a sufficiently low redox potential, in these experiments sulphide (final concentration 1–2 mM) was added to the reservoir solution. As a consequence, a precipitate of FeS was formed in the reservoir bottle and very little iron was actually added to the organisms inside the culture vessel. There the concentration of sulphide was not reduced but was even somewhat higher than in the reservoir solution owing to the organisms' (limited) activities. The organisms were found to be neither sulphate limited nor limited by availability of organic carbon. Instead, iron was the growth-rate-limiting factor. Had the reservoir solution been stirred, FeS would have been added to the culture. However, owing to the presence of excess sulphide, still very little iron would have been available for incorporation into cellular material.

In a comparable experimental design, used for the cultivation of anoxygenic phototrophic bacteria, no such problems were encountered. Even when FeS which was formed in the reservoir bottle was added to the culture, the organisms grew well and it was established that sulphide was the limiting factor. The explanation for these differences is that *Chromatium* oxidizes sulphide to either elemental sulphur or sulphate, thereby establishing an extremely low concentration of free sulphide in the culture solution which, in turn, results in a dissociation of the added FeS. As a result, both the sulphide and the iron are fully available to the organisms. However, in natural waters the phototrophic bacteria may not always be able to reduce the concentration of free sulphide sufficiently. Then problems similar to those described for the sulphate-reducing bacteria (or any other type of sulphide-producing bacterium) will be encountered by sulphide-utilizing phototrophs. Although experimental support in this respect is very limited, it is generally believed that anoxygenic phototrophic bacteria in natural waters are limited either by sulphide (or more precisely by the electron donors) or by light. In a few cases this has been demonstrated to be correct (Sorokin, 1970; Parkin and Brock, 1980a,b; Guerrero *et al.*, 1985). However, the unavailability of sufficient concentrations of trace metals in solution inevitably would result in even more elevated concentrations of sulphide, and it is of interest to envisage possible useful strategies to cope with such conditions.

It appears that of the three forms in which free sulphide exists (i.e. H_2S, HS^- and S^{2-}) the undissociated hydrogen sulphide is the most toxic. The equilibrium between the sulphide ion, the bisulphide ion and the undissociated hydrogen sulphide is very much controlled by the environmental pH. At neutral pH the ratio $H_2S:HS^-:S^{2-}$ is $1.1:1:10^{-5}$ and shifts towards the dissociated forms with increasing pH values ($pK_1 = 7.04$, $pK_2 = 11.96$). Already in 1931 Van Niel showed that various species of purple sulphur bacteria were able to grow at elevated concentrations of free sulphide, provided that the alkalinity was increased. The higher the total free sulphide concentration, the more narrow the pH range was at which growth was observed (Fig. 1).

The reason why sulphide is such an inhibitory compound is not exactly known, and it may very well be different for different organisms or groups of organisms. Yet it appears that the explanation for the more serious toxicity of the undissociated free sulphide compared with the other forms of free sulphide is that organisms are unable to prevent the undissociated form from

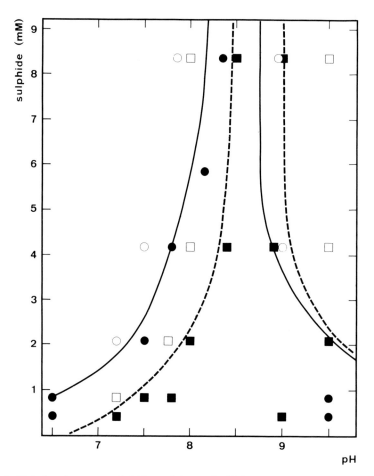

Fig. 1. Relation between sulphide toxicity and pH in two purple sulphur bacteria
(■ □, ● ○). Combinations of pH and initial sulphide concentration which
supported growth are indicated by closed symbols (■, ●); no growth after 16 days is
indicated by open symbols (□, ○). The rate of growth at different sulphide
concentrations cannot be read from the figure. (Redrawn after van Niel (1931).)

entering the cell. Bisulphide (HS^-) and sulphide (S^{2-}) ions, being charged,
are probably taken up by active transport which can be regulated. However,
there is no way to prevent the diffusion of the undissociated H_2S through the
cell membrane. In this respect, the inhibitory effect of sulphide is comparable
with the bacteriostatic effects of undissociated organic acids in, for example,
yoghurt, sauerkraut and silage. Of course, this is unlikely to be the full
explanation. Conceivably, the presence of sulphide inside the cell could result
in an insufficient availability of certain trace elements.

It is obvious that the environmental pH is of influence on the sulphide equilibrium, but the pH affects all kinds of reactions in the cell as well. In order to differentiate between such gross effects of pH and the effect of pH on the sulphide equilibrium, a comparison is made between the impact of pH on the specific growth rates of a *Chromatium* species in a sulphide–carbon dioxide medium and in an acetate medium. Acetate can be considered a non-inhibitory substrate, since no reduction of the specific growth rate is observed at concentrations far exceeding the saturating concentration. The relation between the specific growth rate of *Chromatium* sp. and the concentration of sulphide and acetate at different pH values is schematically presented in Fig. 2. Major differences are observed on two points. The first is that the maximal specific growth rate at optimal pH is higher in acetate media than in sulphide media. The second difference is that the optimal pH for growth in sulphide-containing media is higher than that in acetate-containing media. As shown, there is not much difference between the growth rates in sulphide and acetate media once the pH is higher than 9. At such high pH values, very little of the total free sulphide is present in the undissociated form. The reduction of the maximal specific growth rate at elevated pH values, being very similar in sulphide and acetate media, can thus be taken as reflecting effects not related to the equilibrium of sulphide; further specifications appear to be irrelevant.

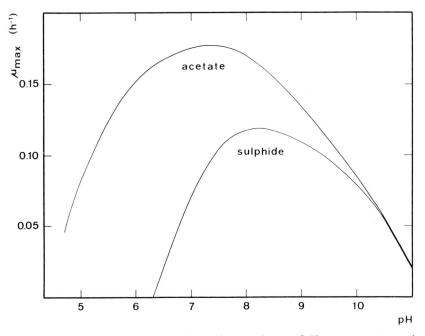

Fig. 2. Relation between the maximal specific growth rate of *Chromatium vinosum* in media containing 1 mM sulphide or 1 mM acetate and the pH.

The arguments mentioned above point to the same conclusion as that drawn by Van Niel (1931), i.e. the toxic effects of free sulphide can be explained by the presence of the undissociated form. Howsley and Pierson (1979) reached the same conclusion for oxygenic photosynthesis in the cyanobacteria *Anabaena* sp., *Synechocystis* sp. and *Oscillatoria* sp. At equal total free sulphide concentrations in the medium there was an increase in the degree of repression of oxygenic photosynthesis with decreasing pH. The degree of repression correlated with the concentration of undissociated sulphide.

To cope with the presence of inhibitory concentrations of sulphide, "strategies" have developed which can be grouped into two categories.

(a) In the presence of sulphide, processes are stimulated which result in the oxidation of sulphide. Such reactions can be either biotic or abiotic: the common result is that the concentration of sulphide is reduced.

(b) The contact with the undissociated sulphide is avoided altogether, or at least drastically reduced. Examples of such strategies are illustrated below with known examples of growth or activities of anoxygenic and oxygenic phototrophic bacteria.

The compromise between the fact that sulphide has an inhibitory effect but at the same time is required for anoxygenic photosynthesis is to utilize it actively. In addition, photosynthesis preferably should not be reduced at elevated concentrations of sulphide, regardless of the impact of the presence of sulphide on the specific growth rate. Once sulphide has entered the cell, undissociated sulphide may be formed as a consequence of the internal pH. Even in the complete absence of active transport, the build-up of an intracellular pool will continue owing to the diffusion of undissociated hydrogen sulphide. The only solution to prevent this is to oxidize the sulphide to less obnoxious compounds such as elemental sulphur or sulphate or others. One way to reduce the concentration is to grow, i.e. to fix carbon dioxide at the expense of sulphide oxidation. The faster the growth rate, the more effectively is sulphide removed. Also, the better the sulphide affinity, the more effectively is the concentration of sulphide reduced. However, it appears that the sulphide affinity of all anoxygenic phototrophic bacteria is high and at best only slightly related to the toxicity of sulphide (Van Gemerden, 1984). Data on the sulphide affinity of cyanobacteria are lacking but, judged from carbon dioxide uptake rates (Garlick *et al.*, 1977), it appears that the sulphide affinity in this group of organisms is not nearly as good as that found in the purple sulphur bacteria (Fig. 3).

High growth rates are thus of importance. However, the relation between the environmental concentration of sulphide and the specific growth rate of phototrophic bacteria is not a saturating type of curve as predicted by

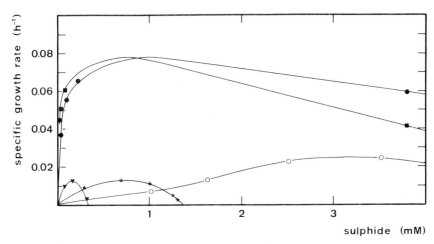

Fig. 3. Relation between the specific growth rate of phototrophic bacteria and the concentration of sulphide. For the purple sulphur bacteria *Thiocapsa roseopersicina* (●) and *Ectothiorhodospira shaposhnikovii* (■) growth rates actually observed are shown. For the cyanobacteria *Lyngbya* sp. (▼), *Aphanothece halophytica* ((★) and *Oscillatoria limnetica* (○) it has been assumed that the short-term carbon dioxide fixation in the presence of sulphide (data from Garlick *et al.*, 1977) results in full growth. If not, the specific growth rates shown for the cyanobacteria are overestimates.

Monod (1950) kinetics, but is better described by other formulae, e.g. that of Haldane (see Andrews, 1968). A comparison between the different kinetics is shown in Fig. 4. In the Monod perception of growth there is one concentration of the substrate at which the specific growth rate is half the maximal specific growth rate. That concentration is referred to as the saturation constant K_s. However, in the utilization of inhibitory substrates, there are two concentrations of the substrate at which the specific growth rate is half the maximal specific growth rate. The lower concentration is referred to as K_s, whereas the upper concentration is described as the inhibition constant K_i. The magnitude of K_i can be used to compare the toxicity of the substate for different organisms, unlike affinity where the ratio μ_{max}/K_s can be used to compare substrate affinities (see Healey, 1980; Zevenboom, 1980; Van Gemerden, 1983).

For the purple and green sulphur bacteria an average sulphide concentration of about 2.9 mM (0.7–8 mM) is required to reduce the specific growth rate to half its maximal value. The inhibition appeared to be somewhat more pronounced in those organisms that deposit elemental sulphur inside the cells and it has been postulated that in those organisms sulphide is taken up as

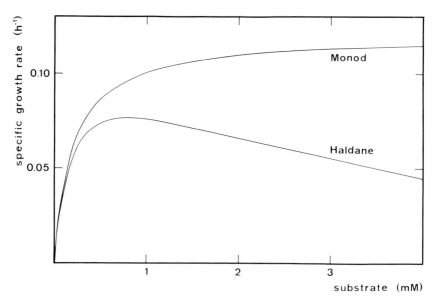

Fig. 4. Effect of the utilization of an inhibitory substrate on the specific rate of growth of a microorganism. The curves shown have been calculated using the following kinetic parameters: Monod, $\mu_{max} = 0.12$ hour^{-1}, $K_s = 0.2$ mM; Haldane, μ_{max} (theoretical) $= 0.12$ hour^{-1}, $K_s = 0.2$ mM, $K_i = 3$ mM.

such and oxidized therafter, in contrast with organisms that deposit the elemental sulphur outside the cell (Van Gemerden, 1984).

The reduction in the specific growth rate at higher sulphide concentrations would result in slower sulphide oxidation rates if growth were the only way of utilizing sulphide. However, the diffusion of undissociated sulphide from outside to inside the cells would be faster. As a consequence, the internal concentration of sulphide would increase, unless the organisms have means to augment the rate of sulphide oxidation. For most phototrophic bacteria this has not been studied yet. In *Chromatium* sp. several strategies are effective in this respect. The first is that at higher sulphide concentrations a shift can be observed from a sulphide-to-sulphate oxidation to a sulphide-to-sulphur oxidation. In the latter step two electrons are released, whereas in the oxidation of sulphide to sulphate eight electrons are transferred. In other words, if the rate of growth were to be reduced to one-quarter of its original value the full rate of sulphide oxidation could still be maintained by oxidizing sulphide to sulphur only. Some organisms have even completely lost (or maybe never gained) the capacity to oxidize the sulphur formed during the oxidation of sulphide. This could be a precautionary measure to prevent the build-up of too high a sulphide concentration. The phenomenon is observed not only in all cyanobacteria that are able to utilize sulphide as electron donor but also in some representatives of the purple non-sulphur bacteria (Rhodospirillaceae), e.g. *Rhodobacter capsulatus* (*Rhodoseudomonas capsulata*) (Hansen and Van Gemerden, 1972). Of itself, the oxidation of sulphide to sulphur could be an effective measure; however, mixed culture studies (to be discussed later) have shown the opposite.

The second strategy observed in *Chromatium* is that the oxidation of sulphide and the concurrent fixation of carbon dioxide do not always result in balanced growth, i.e. in the synthesis of proteins, carbohydrates, nucleic acids and other major cell constituents in proportion to each other. In particular, carbohydrates may be accumulated inside the cell at elevated concentrations of sulphide. Thus, at higher sulphide concentrations the specific growth rate may be reduced indeed, but not—or at least to a far lesser extent—the fixation of carbon dioxide through the Calvin cycle. Actually, the extent to which *Chromatium* spp. are able to reduce the external sulphide concentration cannot be judged at all from the plot of specific growth rate *versus* the sulphide concentration. Likewise, the limits of growth of an organism cannot be deduced from the limits of carbon dioxide fixation. Growth of *Chromatium vinosum* ceases completely long before the environmental concentration of sulphide has become 30 mM. However, at such concentrations the organisms are still able to fix carbon dioxide at the expense of sulphide oxidation, resulting in the intracellular deposition of glycogen. The ratio between growth and glycogen formation at increasing concentrations of sulphide is schematically presented in Fig. 5. It should be realized, however, that these are to be considred as short-term strategies. The organisms cannot continue to increase the intracellular concentration of elemental sulphur and glycogen indefinitely, because of the lack of internal space.

A somewhat similar strategy can be observed on decreasing the environmental value of the pH. At constant total free sulphide concentration, a shift in the pH towards lower values results in an increased concentration of the undissociated form. The time courses of total free sulphide, elemental sulphur and glycogen under identical conditions except for pH showed virtually identical rates of glycogen synthesis. However, at pH 5 far more elemental sulphur was formed than at pH 8. As a result of this, the concentration of sulphide decreased more rapidly in the pH 5 culture (Fig. 6).

The concentration of sulphide encountered in natural waters and sediments inhabited by phototrophic bacteria as a rule shows variations. These variations often extend to the complete absence of sulphide for a time period of some hours or more. In part, this can be explained by the organisms' own photosynthetic activities. If phototrophic bacteria under these conditions

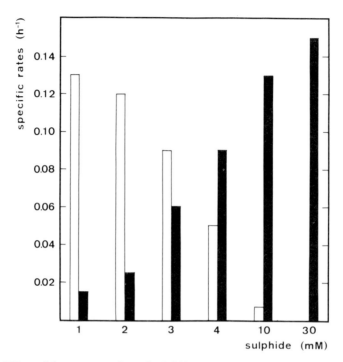

Fig. 5. Effect of the concentration of sulphide on the specific rate of growth (□) and the specific rate of glycogen synthesis (■) of *C. vinosum*.

could react by accurately adjusting the enzyme levels of the Calvin cycle to the low need, an immediate reaction to a sudden reappearance of sulphide would be impossible. An effective strategy thus appears to be to maintain sufficient levels of certain enzymes in the absence of the substrates. It has indeed been observed that, although the level of Calvin enzymes is somewhat reduced under heterotrophic conditions (Hurlbert and Lascelles, 1963), acetate-grown *Chromatium* cells are able to oxidize sulphide and to fix carbon dioxide as well as sulphide-grown cells can (Beeftink and Van Gemerden, 1979). Conceivably, this enables the organisms to react rapidly under fluctuating conditions, resulting in the elimination of competitors that do not show this apparently superfluous synthesis of enzymes. It is unknown at present whether similar strategies occur among the cyanobacteria.

Cyanobacteria, having both photosystems I and II, exhibit an oxygenic photosynthesis. Many of them are also able to perform an anoxygenic photosynthesis in which sulphide acts as the electron donor. This was first reported by Cohen *et al.* (1975) for *O. limnetica* but later research showed it

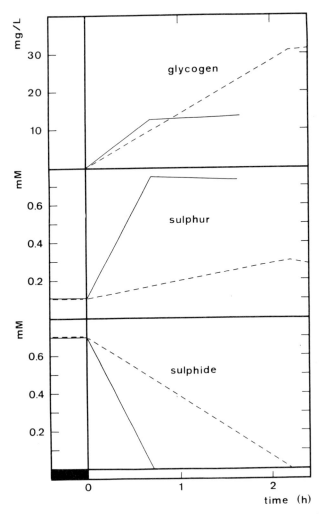

Fig. 6. Time course of sulphide, sulphur and glycogen in batch cultures of *C. vinosum* at pH 5 (———) and pH 8 (– – – –). Cultures devoid of intracellular elemental sulphur were supplied with sulphide in the dark. From zero time onwards, saturating light intensities were provided. In the time period shown, growth (increase in protein) was not observed.

to be true for many other strains as well (Garlick *et al.*, 1977). In some strains the involvement of photosystem II ceases completely at sulphide concentrations as low as 0.1 mM (Oren *et al.*, 1979). However, only after an induction period of several hours is the sulphide-dependent anoxygenic photosynthesis fully induced. Chloramphenicol inhibits the initiation of anoxygenic photosynthesis (Oren and Padan, 1978), indicating that a *de novo* synthesis of enzymes is required.

The examples discussed above all relate to the utilization of sulphide as an electron donor in anoxygenic photosynthesis. Another way to reduce the environmental concentration of sulphide is to produce sulphide-binding compounds. It appears that such strategies are predominantly encountered among the cyanobacteria (Cohen, 1984a,b).

Cyanobacteria are able to perform an oxygenic photosynthesis resulting in the production of oxygen. Subsequently, the oxygen reacts with sulphide; the reaction products can be elemental sulphur, thiosulphate or sulphate depending on the concentration of the reactants (Chen and Morris, 1972). The usefulness of this strategy is illustrated by the behaviour of an *Oscillatoria* species isolated from Wilbur Hot Springs. The sulphide concentration encountered in the organism's habitat is about 0.006 mM. When the isolate was exposed to a sulphide concentration of 0.8 mM, the rate of oxygenic photosynthesis increased by a factor of 4.5 (Cohen, 1984a). As a consequence of the stoichiometric relationship between the fixation of carbon dioxide and the production of oxygen the latter rate also increased drastically. The abiotic oxidation of sulphide by oxygen is not an extremely fast reaction: oxygen uptake rates of filtered Solar Lake water have been reported to be 45 μmol l^{-1} hour^{-1} (Jørgensen, *et al.*, 1979). Elemental sulphur is one of the reaction products of sulphide and oxygen (Chen and Morris, 1972). Owing to the subsequent abiotic reaction between the residual sulphide and the elemental sulphur formed—resulting in the production of polysulphides (Chen and Gupta, 1973)—the actual decrease in the sulphide concentration can therefore be expected to exceed the rate of oxygen uptake. It thus appears that the *Oscillatoria* species thriving in Lake Wilbur is able to prevent a build-up of sulphide in its vicinity.

A sulphide-stimulated enhancement of oxygenic photosynthesis has also been observed in *Microcoleus chtonoplastes* (Cohen, 1984a). This cyanobacterium is commonly found as the most dominant mat-forming organism in many marine and hypersaline habitats (Bauld, 1984; Stolz, 1984; Stal *et al.*, 1985). However, in this case the stimulation of oxygen production is observed only at low sulphide concentrations. At higher concentrations of sulphide the organism gradually shifts to anoxygenic photosynthesis with sulphide as the electron donor. Between 0.2 and 8 mM sulphide, both oxygenic and anoxygenic photosynthesis can be observed (Cohen, 1984b). The *Oscillatoria* strain isolated from Wilbur Hot Springs cannot do this. In this organism anoxygenic photosynthesis has not been observed, despite the inhibition of oxygenic carbon dioxide fixation (Cohen, 1984a).

In the presence of sulphide, the photosynthesis of cyanobacteria resembles that of the purple sulphur bacteria, the differences being that sulphur is deposited outside the cells and that sulphur cannot be oxidized to sulphate. In a similar manner as described above, the formation of polysulphides from sulphide and elemental sulphur could further reduce the environmental concentration of sulphide. However, it is obvious that organisms which have the possibility to perform both oxygenic and anoxygenic photosynthesis do not depend as severely on a reduction in the concentration of sulphide as organisms which cannot photosynthesize anoxygenically. No strategy whatsoever is observed in, for example, *Anacystis nidulans* where complete and irreversible cessation of photoassimilation of carbon dioxide is observed upon brief exposure to sulphide (Cohen, 1984a). In general, there is good agreement between the strategy exhibited by cyanobacteria and the sulphide concentrations encountered under natural conditions (see Cohen, 1984a).

The deposition of elemental sulphur outside the cell is a feature observed not only in cyanobacteria but also in *Chlorobium* species and *Ectothiorhodospira* species. Therefore, it is to be expected that in these cases also the environmental concentration of sulphide is lowered owing to the formation of polysulphides. It appears that the extent to which the concentration of sulphide is reduced in this way is about 0.5 mM. In a continuous culture of *Chlorobium limicola* growing on sulphide at high dilution rates, the polysulphide concentration (as S_3^{2-}) was found to be about 0.2 mM. On the assumption that it is formed from elemental sulphur and sulphide, the latter compound will have been reduced by about 0.6 mM. Compared with the effectiveness of other strategies (see below), the effect of polysulphide formation thus appears to be marginal.

In the examples discussed above, the concentration of sulphide is reduced to less inhibitory levels by oxidation, either biotically or abiotically. Another way to reduce the toxic effect of sulphide is to avoid the contact with undissociated sulphide. Among the phototrophic bacteria, conceivably two strategies are encountered in this respect. The first is to produce extracellular sulphur and the second is to develop a high pH optimum for growth.

There is some circumstantial evidence to explain why certain phototrophic bacteria deposit the elemental sulphur formed during the oxidation of sulphide outside the cell. It appears that the initial acceptor of the electrons released is situated on the outside of the cellular membrane (Then and Trüper, 1983: see Van Gemerden, 1984). The consequence is that sulphide is oxidized on the outside of the membrane. The effectiveness of this strategy is enhanced by the fact that the sulphide affinity of all phototrophic bacteria producing extracellular elemental sulphur is very high compared with that of species that store the elemental sulphur intracellularly (Van Gemerden, 1984). However, the cyanobacteria appear to behave differently in this respect (Garlick *et al.*, 1977) (see Fig. 3).

As mentioned earlier, the undissociated form of sulphide is the most toxic. With increasing pH the environmental concentration of undissociated sulphide decreases. Thus, a relatively high pH optimum for growth reduces the chances of sulphide poisoning. Data on the intracellular concentration of sulphide are not known, but it is obvious that maintaining a fairly neutral internal pH would result in the opposite effect, namely an internal shift from S^{2-} and HS^- to H_2S. In this respect it is of interest that the internal pH of alkalophilic bacteria is balanced at an average value of 9.5 (see Padan et al., 1981; Konings and Veldkamp, 1983). *Ectothiorhodospira* species combine two strategies: in general these organisms have high pH optima, and also they deposit the elemental sulphur formed in the oxidation of sulphide extracellularly.

It cannot be denied that there is some subjectivity in the description of a strategy as being effective or not. Moreover, it should be stated quite clearly what is considered to be a positive effect. With respect to the toxicity of sulphide, an effective strategy could be interpreted as resulting in a rapid decrease in the environmental concentration of sulphide. However, in general terms an effective strategy would be interpreted as resulting in higher specific growth rates compared with other organisms that utilize the same substrate, thus resulting in the competitive exclusion of the other species. These two interpretations do not necessarily have identical results. This will be illustrated with an example concerning the competition for sulphide between anoxygenic bacteria. At present no published examples are known that enable a comparison between strategies encountered in the oxygenic cyanobacteria and in the anoxygenic purple and green bacteria.

Despite trivially being described as a non-sulphur purple bacterium, *R. capsulatus* has the ability to utilize sulphide as an electron donor in anoxygenic photosynthesis (Hansen and Van Gemerden, 1972). It appears that the bacterium has an extremely high affinity for sulphide, far higher than that exhibited by the purple sulphur bacterium *C. vinosum* (Van Gemerden, 1984). The exclusive product of sulphide oxidation was found to be elemental sulphur (Hansen and Van Gemerden, 1972). The combination of these characteristics can be expected to result in a rapid lowering of the concentration of sulphide, and at first sight this seems to be an effective strategy. However, the sulphur produced by *R. capsulatus* is deposited extracellularly and it is well known that purple sulphur bacteria like *C. vinosum* have the ability to utilize elemental sulphur as an electron donor. In the oxidation of sulphide to elemental sulphur two electrons are released, compared with six in the oxidation of sulphur to sulphate. The yields per millimole of reducing power are very similar in the two organisms. Consequently, with sulphide serving as the primary electron donor, the biomass of *Chromatium* can be expected to be at least three times that of *R. capsulatus*. Actually, the latter organism was found to contribute no more than 5% of the total biomass. These data show that a high affinity for sulphide combined with the production of elemental sulphur and no sulphate indeed result in a rapid lowering of the environmental sulphide concentration but still cannot be considered advantageous from an ecological point of view. This example may serve to illustrate that the definition of *R. capsulatus* as being a "non-sulphur" purple bacterium is well chosen, despite its ability to utilize sulphide as an electron donor.

On the basis of the previous information one may wonder how the description of Chlorobiaceae as being "sulphur" bacteria is justified. Chlorobiaceae are known to deposit the elemental sulphur formed in the oxidation of sulphide outside the cells. Therefore, one could expect that the competition for sulphide between *Chlorobium* sp. and *Chromatium* sp. would proceed in a similar fashion as described for the competition between *R. capsulatus* and *Chromatium*. The experimental data have shown otherwise. At all dilution rates, the two competing organisms were found to coexist: low dilution rates were in favour of *Chlorobium* whereas high dilution rates were in favour of *Chromatium* (Van Gemerden and Beeftink, 1981). It appears that the elemental sulphur formed by *Chlorobium* remains somehow attached to the cells and thus is not freely available for *Chromatium* (unpublished data).

In summary, it can be concluded that in different groups of organisms different strategies to reduce the contact with undissociated sulphide are encountered. Some of the strategies are effective in the sense that they enable an organism to thrive under conditions that otherwise would be fatal. Other strategies, however, do result in a reduction in the concentration of sulphide but are completely ineffective in the sense of being advantageous in the competition with other organisms in the same habitat.

A topic which deserves future interest is how the strategies encountered in the purple and green sulphur bacteria compare with the strategies observed in the cyanobacteria. Numerous examples show that cyanobacteria and purple sulphur bacteria thrive in the same habitat, and an understanding of the overall dynamics of such habitats requires a sound knowledge of the effectiveness of the strategies of the different (groups of) organisms.

Acknowledgements

We should like to thank P. T. Visscher for unpublished data on polysulphide formation in cultures of phototrophic bacteria and P. A. G. Hofman for his help in drawing the figures.

References

Andrews, J. F. (1968). A mathematical model for the continuous cultivation of microorganisms utilizing inhibitory substrates. *Biotechnology and Bioengineering* **10**, 707–723.

Bauld, J. (1984). Microbial mats in marginal marine environments: Shark Bay, Western Australia, and Spencer Gulf, South Australia. *In* "Microbial Mats: Stromatolites" (Eds Y. Cohen, R. W. Castenholz and H. O. Halvarson), pp. 39–58. Liss, New York.

Beeftink, H. H. and Van Gemerden, H. (1979). Actual and potential rates of substrate oxidation and product formation in continuous cultures of *Chromatium vinosum*. *Archives of Microbiology* **121**, 161–167.

Chen, K. Y. and Gupta, S. K. (1973). Formation of polysulfides in aqueous solution. *Environmental Letters* **4**, 187–200.

Chen, K. Y. and Morris, J. C. (1972). Kinetics of oxidation of aqueous sulfide by O_2. *Environmental Science and Technology* **6**, 529–537.

Cohen, Y. (1984a). The Solar Lake cyanobacterial mats: strategies of photosynthetic life under sulfide. *In* "Microbial Mats: Stromatolites" (Eds Y. Cohen, R. W. Castenholz and H. O. Halvarson), pp. 133–148. Liss, New York.

Cohen, Y. (1984b). Oxygenic photosynthesis, anoxygenic photosynthesis, and sulfate reduction in cyanobacterial mats. *In* "Current Perspectives in Microbial Ecology" (Eds M. J. Klug and C. A. Reddy), pp. 435–441. American Society for Microbiology, Washington, District of Columbia.

Cohen, Y., Jørgensen, B. B., Padan, E. and Shilo, M. (1975). Sulfide dependent anoxygenic photosynthesis in the cyanobacterium *Oscillatoria limnetica*. *Nature, London*, **257**, 489–492.

Garlick, S., Oren, A. and Padan, E. (1977). Occurrence of facultative anoxygenic photosynthesis among filamentous and unicellular cyanobacteria. *Journal of Bacteriology* **129**, 623–629.

Van Gemerden, H. (1983). Physiological ecology of purple and green bacteria. *Annales de Microbiologie (Institut Pasteur)* **134B**, 73–92.

Van Gemerden, H. (1984). The sulfide affinity of phototrophic bacteria in relation to the location of elemental sulfur. *Archives of Microbiology* **139**, 289–294.

Van Gemerden, H. and Beeftink, H. H. (1981). Coexistence of *Chlorobium* and *Chromatium* in a sulfide-limited continuous culture. *Archives of Microbiology* **129**, 32–34.

Guerrero, R., Montesinos, E., Pedros-Alio, C., Esteve, I., Mas, J., van Gemerden, H., Hofman, P. A. G. and Bakker, J. F. (1985). Phototrophic sulfur bacteria in two Spanish lakes: vertical distribution and limiting factors. *Limnology and Oceanography* **30**, 932–943.

Hansen, T. A. and van Gemerden, H. (1972). Sulfide oxidation by purple non-sulfur bacteria. *Archives of Microbiology* **86**, 49–56.

Healey, F. P. (1980). Slope of the Monod equation as an indicator of advantage in nutrient competition. *Microbial Ecology* **5**, 281–286.

Howsley, R. and Pearson, H. W. (1979). pH dependent sulphide toxicity to oxygenic photosynthesis in cyanobacteria. *FEMS Microbiology Letters* **6**, 287–292.

Hurlbert, R. E. and Lascelles, J. (1963). Ribulose diphosphate carboxylase in Thiorhodaceae. *Journal of General Microbiology* **33**, 445–458.

Jørgensen, B. B., Kuenen, J. G. and Cohen, Y. (1979). Microbial transformations of sulfur compounds in a stratified lake (Solar Lake). *Limnology and Oceanography* **24**, 799–822.

Konings, W. N. and Veldkamp, H. (1983). Energy transductions and solute transport mechanisms in relation to environments occupied by microorganisms. *In* "Microves in their Natural Environments (Eds J. H. Slater, R. Whittenbury and J. W. T. Wimpenny), pp. 153–186. SGM Symposium 34.

Monod, J. (1950). La technique de culture continue, théorie et applications. *Annales de l'Instut Pasteur* **77**, 390–410.

Oren, A. and Padan, E. (1978). Induction of anaerobic photoautotrophic growth in the cyanobacterium *Oscillatoria limnetica*. *Journal of Bacteriology* **133**, 558–563.

Oren, A., Padan, E. and Malkin, S. (1979). Sulfide inhibition of photosystem II in cyanobacteria (blue-green algae) and tobacco chloroplasts. *Biochimica et Biophysica Acta* **546**, 270–279.

Padan, E., Zilberstein, D. and Schuldiner, S. (1981). pH homeostasis in bacteria. *Biochimica et Biophysica Acta* **650**, 151–166.

Parkin, T. B. and Brock, T. D. (1980a). Photosynthetic bacterial production in lakes: the effects of light intensity. *Limnology and Oceanoagraphy* **25**, 711–718.

Parkin, T. B. and Brock, T. D. (1980b). The effect of light quality on the growth of phototrophic bacteria in lakes. *Archives of Microbiology* **125**, 19–27.

Sorokin, Y. I. (1970). Interrelations between sulphur and carbon turnover in meromictic lakes. *Archives of Hydrobiology* **66**, 391–446.

Stal, L. J., van Gemerden, H. and Krumbein, W. E. (1985). Structure and development of a benthic marine microbial mat. *FEMS Microbiology and Ecology* **31**, 111–125.

Soltz, J. F. (1984). Fine structure of the stratified microbial community at Laguna Figueroa, Baja California, Mexico: II. Transmission electron microscopy as a diagnostic tool in studying microbial communities *in situ*. *In* "Microbial Mats: Stromatolites" (Eds Y. Cohen, R. W. Castenholz and H. O. Halvarson), pp. 23–38. Liss, New York.

Then, J. and Trüper, H.-G. (1983). Sulfide oxidation in *Ectothiorhodospira abdelmalekii*. Evidence for the catalytic role of cytochrome *c*-551. *Archives of Microbiology* **135**, 254–258.

van Niel, C. B. (1931). On the morphology and physiology of the purple and green sulphur bacteria. *Archiv für Mikrobiologie* **3**, 1–112.

Zevenboom, W. (1980). Growth and nutrient uptake kinetics of *Oscillatoria agardhii*. Ph.D. Thesis, University of Amsterdam.

6

Damaging Effects of Light on Microorganisms

G. C. WHITELAM

Department of Botany, University of Leicester, Leicester LE1 7RH, UK

and

G. A. CODD

Department of Biological Sciences, University of Dundee DD1 4HN, UK

Introduction

The damaging effects of light on microbial cells have been recognized in general terms for many years and stress imposed by high intensity light is a feature of many extreme environments. Damage can be initiated by ultraviolet (UV) and visible wavelengths and the organisms affected include bacteria, algae, fungi and protozoa, in addition to viruses. Examples of such harmful effects on microbial cells and their components or activities, together with accounts of damage to animal and plant cells, can be found in several earlier reviews (Spikes and Livingstone, 1969; Krinsky, 1976; Harris, 1978; Thorington, 1980; Codd, 1981).

Appreciation of the ability of microorganisms to survive, adapt and thrive in extreme environments is increased by our recognition of a photodynamic, photooxidative or photoinhibitory syndrome in many natural and man-made environments. A review of microbial life under conditions of high irradiation was provided by Nasim and James (1978) in a previous anthology on microorganisms and extreme environments (Kushner, 1978). Both physiological and ecological aspects were covered although the review mainly centred on ionizing and UV irradiation and their effects on chemoheterotrophic microbes (Nasim and James, 1978). We have reviewed recent aspects of UV-mediated damage in this chapter and have additionally considered advances in the understanding of the harmful effects of visible wavelengths. Damage to chemoheterotrophic microbes continues to attract much interest in terms of basic mechanisms, ecophysiology and applied aspects. However, the chemoheterotrophic mode of nutrition permits the microbe to obtain its energy requirements for growth from organic carbon compounds alone and such organisms can thus grow in the dark. The microbial phototrophs, by contrast, must be able to absorb light and to use the energy obtained, to permit growth on inorganic carbon (photoautotrophs) or on organic carbon compounds which cannot by themselves provide all the energy needed for growth (photoheterotrophs). The phototrophs, including the purple and green photosynthetic bacteria, the cyanobacteria and the microalgae, are thus required to be able to grow in a light climate and are presented with the conflicting problems of too little *versus* too much light. Interest in the harmful effects of light on phototrophic microbes has increased rapidly over recent years in parallel with research into photooxidation and photoinhibition in higher plants (see Codd, 1981; Powles, 1984).

In this review, we consider damaging effects of light on both phototrophic and chemoheterotrophic microbes in terms of the ecology of the organisms and the nature of the damage at the molecular and physiological levels. The strategies used by microbes to avoid, or tolerate, the photooxidative or photoinhibitory environment are examined and some areas where light damage may be important in applied or economic activities are also considered.

Ecological Aspects

Injury to microbial cells is mainly studied in the laboratory, rather than in the field, although sufficient evidence exists to indicate that the harmful effects of solar UV and visible wavelengths are significant ecological factors in aquatic and terrestrial environments.

The literature contains several terms used to describe the quality of light which can cause cellular damage. Here, we define the regions of the electromagnetic spectrum responsible for initiating light-induced damage to microorganisms as follows:

UV-C (far UV), 200–290 nm
UV-B (mid UV), 290–320 nm
UV-A (near UV), 320–400 nm
visible light, 400–750 nm

The UV-A and UV-B regions are often combined in the literature and referred to as near UV.

Aquatic Environments

UV and visible wavelengths have been implicated in the death of bacteria and algae and the inhibition of photosynthesis and respiration in fresh, estuarine and marine waters and in extreme environments, including hot springs and hypersaline lakes. The pioneering work of Shilo and coworkers established that a photooxidative syndrome of high irradiance, high oxygen levels and reduced carbon dioxide levels can cause an impairment of cyanobacterial photosynthesis followed by the bleaching of photosynthetic pigments and subsequent death of the cells (Abeliovich and Shilo, 1972; Abeliovich et al., 1974). At the peak of their development, cyanobacterial blooms frequently form heavy scums at the water surface, buoyancy being due to the presence of gas vacuoles, a feature of several planktonic cyanobacteria (Walsby, 1972; van Liere and Walsby, 1982). At the water surface, this dense accumulation of cyanobacteria may be exposed to light for many hours, the intensive photosynthetic activity leading to the production of supersaturation dissolved oxygen levels and a depletion of carbon dioxide. Such natural conditions lead to the photooxidative death of several cyanobacterial strains and may explain, in part, the sudden die-off of cyanobacterial blooms in nature (Eloff et al., 1976). Coulombe and Robinson (1981) have shown further that photooxidative death contributes to the collapse of *Aphanizomenon flos-aquae* blooms during summer periods of thermal stratification in Canadian waters.

The water surface microlayer provides a complex extreme environment where photoinhibition of the microbiota can occur. Phytoplankton populations in marine and estuarine surface microlayers can be considerably smaller, or larger, than bulk-water populations (e.g. Manzi et al., 1977; Hardy and Valett, 1981), although large numbers of the microlayer population may be dead or show reduced metabolic activity (Marumo et al., 1971; Albright, 1980). Diurnal periodicities in microlayer chlorophyll a concentrations in the Damariscotta estuary (Maine, USA), with nocturnal enrichment, indicate that photoinhibition is a dominant inhibitory factor (Carlson, 1982). Diatom species characteristic of the surface microlayer phytoplankton have previously been shown to be killed by exposure to UV-B levels typical of the water surface (Calkins and Thordardottir, 1980; Thomson et al., 1980). Photoinhibition of chemoheterotrophic bacterial metabolism also occurs in water surface microlayers (Horrigan et al., 1981) and the killing effect of light on these organisms in microlayers and aerosols should be taken into account in sampling procedures for the water treatment industry (see later).

Subsurface photoinhibition of planktonic photosynthesis has been recorded widely in *in situ* bottle incubations at various depths (see Harris, 1978; Codd, 1981). Photoinhibition is a function of surface irradiance I_0, and photon fluence rates I_0 of 100–400 μmol photons m^{-2} second^{-1} (400–700 nm) are sufficient to cause inhibition in bottle tests. These threshold values are considerably less than full sunlight (about 2000 μmol photons m^{-2} second^{-1})

and may be exceeded for most of the day. Diurnal and seasonal patterns of phytoplankton photoinhibition in surface waters occur in temperate latitudes (e.g. Harris, 1978; Codd, 1981; Hammer, 1983). Several objections have been raised against the evidence for phytoplankton photoinhibition observed in *in situ* bottle incubation experiments, including the prevention of cell circulation, in what may otherwise be a well-mixed water, cell sedimentation and nutrient limitation in the containers. These factors may account for the photoinhibition observed in bottles held at fixed depths and the higher rates of photosynthesis measured in replicate samples vertically circulated during incubation (see Talling, 1971). However, "bottle effects" have been discounted in other long-term studies (see Antia *et al.* (1963) and references in Codd (1981)). Clearly, the onset and progress of photoinhibition is influenced by many opposing factors in addition to the basic inhibition threshold, including nutrient availability and cellular repair and adaptive or avoidance strategies (see later).

Photoinhibition of photosynthesis and of growth of members of all the major classes of microalgae has been reported in field and laboratory experiments. Although comparisons are constrained by differences in the previous history of the cells and incubation conditions during photoinhibitory treatment, some generalizations may be drawn: reports of photoinhibition of dinoflagellates (Dinophyceae) are more numerous than for diatoms (Bacillariophyceae). The sensitivity of diatoms to damage by light may be broadly similar to that of cyanobacteria (blue–green algae, Cyanophyceae), whilst the green algae (Chlorophyceae), as a class, may be less susceptible to damage. These generalizations are in agreement with the typical photon fluence rate preferences of the algal classes for photosynthesis and growth in marine and fresh waters, namely, in ascending rank, dinoflagellates and other flagellates < diatoms ≈ cyanobacteria < green algae (for a review see Richardson *et al.* (1983)).

Photosynthetic bacteria were among the first microbes to be used in laboratory studies on photodynamic action and the protective role of carotenoids (Griffiths *et al.*, 1955; Sistrom *et al.*, 1956). These organisms remain as useful model systems in which light- plus oxygen-dependent inhibition of carbon dioxide fixation and photophosphorylation can be investigated (Slooten and Sybesma, 1976; Akazawa *et al.*, 1978; Asami and Akazawa, 1978). Ecological evidence for the damaging effects of light on these phototrophs in the water column is lacking. In stratified lakes, purple and green bacteria typically occur as bands below the thermocline in a microaerobic or an anaerobic, sulphide-containing, environment, where photon fluence rates may be only 0.015% to 3% of surface values and are subsaturating for photosynthesis (e.g. Parkin and Brock, 1980a,b). Whether photoinhibition of photosynthetic bacteria occurs during or after periods of turbulence which would extend the oxygenated epilimnetic waters below the previous thermocline and may raise the organisms into a higher light climate requires investigation.

The viability of halobacteria (*Halobacterium salinarium, H. halobium* and *H. cutirubrum*) is also markedly reduced by laboratory exposure to UV and visible wavelengths (Brock and Petersen, 1976; Simon, 1980). During starvation, loss of *H. cutirubrum* viability only occurred in the light if oxygen was present. An oxygen requirement was also apparent for the photokilling of *H. halobium* (more than 90% loss of viability over 7 days), although a 50% reduction in viability was also recorded during the anaerobic irradiation. The hypersaline lakes inhabited by these rhodopsin-containing bacteria, which can perform a light-driven ATP synthesis but are incapable of autotrophic growth, are typically low in dissolved oxygen (Brock and Petersen, 1976). However, the hypersaline soda lakes in Africa and elsewhere which support massive blooms of halobacteria (Grant and Tindall, Chapter 2, this volume) receive high irradiance and photoinhibition in the upper layers probably occurs.

Of the toxic forms of oxygen known to be involved in photooxidative damage (see later), several have recently been demonstrated in natural waters. Absorption of visible and UV-A wavelengths by the brown substances present in humic waters results in the generation of superoxide anions (O_2^-) and singlet oxygen ($^1O_2$) (Baxter and Carey, 1982, 1983). The disproportionation of superoxide to hydrogen peroxide, which is also involved in cellular damage, has been demonstrated in surface and ground waters exposed to sunlight (Cooper and Zika, 1983). Hydrogen peroxide levels in excess of 30 µM have been recorded in eutrophic waters after short exposure to sunlight (Draper and Crosby, 1983).

Evidence of photoinhibition in benthic microbial mats, largely consisting of cyanobacteria, has been obtained in different extreme environments where undisturbed mat development is permitted. Wickstrom (1984) has observed a depression of the nitrogenase activity of the filamentous thermophilic mat former *Mastigocladus laminosus* when exposed to high *versus* low irradiances (712 *versus* 252 µmol photons m^{-2} second^{-1}), in a 50 °C stream in Yellowstone National Park, USA. Maximal nitrogenase activity (measured as acetylene reduction) occurred at 44% of full mid-day irradiance in August. These findings, plus earlier reports of a depression of planktonic cyanobacterial nitrogenase around mid-day (e.g. Peterson *et al.*, 1977), can be said to constitute a photoinhibition of nitrogen fixation. Differences occur between mat-forming cyanobacteria in the response of nitrogenase to irradiance. In contrast with *M. laminosus*, which forms thick streamers, the thin mats of nitrogen-fixing *Calothrix* spp. found in unshaded locations show an increase in nitrogenase activity with increasing irradiance with no indication of photoinhibition (Wickstrom and Castenholz, 1978; Wickstrom, 1984). The properties of nitrogenases, including their sensitivity to oxygen and the dependence of the majority of nitrogen-fixing cyanobacteria upon photosynthesis to supply the energy and reductant needed to drive nitrogen fixation, must clearly be taken into account in considering the events involved in the observed photoinhibition of nitrogen fixation.

Evidence of photoinhibition of photosynthesis in benthic cyanobacterial mats, consisting mainly of *Microcoleus* and *Phormidium* associated with flexibacteria, has been obtained in the hypersaline Solar Lake in Sinai (Jørgensen *et al.*, 1983). The laminated gelatinous mats occur at 0.2–1.0 m water depth and receive 1000–1200 µmol photons m^{-2} second^{-1} at noon on sunny days. Probing the layers at 0.2 mm intervals with microelectrodes has revealed a submat surface maximum of specific photosynthetic oxygen evolution. The manipulative techniques being developed by these workers, using microelectrodes for the measurement of pO_2, pS^{2-} and pH with tip diameters of 1–6 µm, will undoubtedly enable rapid advances to be made in many areas of microbial ecology.

Terrestrial Environments

The damaging effects of light on the activities and viability of microbes in terrestrial habitats have not received the widespread attention given to microbes in aquatic environments. Field studies have usually involved depth profiles of light penetration and incubations *versus* depth or at the water surface. Laboratory studies typically involve the suspension of microbes in aqueous media; these approaches clearly facilitate experimental design, manipulation and sampling. However, terrestrial habitats exposed to sunlight, e.g. soil, rock and plant surfaces, offer many prospects for research into the damaging effects of UV and visible wavelengths in microbes.

Biogenic rock varnishes, which have attracted scientific attention for almost two centuries (see Krumbein and Jens, 1981), are thin (5 µm–1 mm) brown, black or grey coatings on rock surfaces. They contain iron and manganese oxides and organic material which typically overlie a layer of cyanobacteria or algae, with lichens, fungi and chemoheterotrophic bacteria. The biological roles of the lacquer-like rock varnish, which often occurs in deserts receiving high irradiance, may protect the microbial communities from desiccation and from damage by UV (Krumbein and Jens, 1981) and visible light.

Physiochemical Aspects

Light Wavelengths Responsible

All the wavelength regions as defined above, have been shown to be effective in eliciting lethal or sublethal light damage in microorganisms. Emphasis will be placed on the action of wavelengths greater than 290 nm since the UV-C region does not represent a significant component of the solar spectrum which reaches the surface of the Earth. The atmospheric ozone layer effectively absorbs light from this wavelength region as well as significantly attenuating light in the UV-B region (Johnson *et al.*, 1976). Thus, microorganisms are not subjected to these shorter wavelengths under natural conditions, although the damaging effects of UV-C laboratory conditions have been extensively studied. Indeed, all cell types are susceptible to UV-C-induced damage since many important biological molecules, including nucleic acids and proteins, absorb strongly at these wavelengths.

With the obvious exception of the photosynthetic pigments of phototrophic organisms, the major cellular consituents of most cells do not show appreciable absorbance of longer wavelength UV and visible light. Nevertheless, significant damage to nucleic acids and other cellular components is elicited by exposure to light in the UV-B, UV-A and visible wavebands. In contrast with the destructive action of UV-C, light effects at these longer wavelengths are mediated by sensitizers rather than by direct absorption of the energy by the light-sensitive cellular components.

Increasingly, attention is being focused on the deleterious effects of UV-B radiation because of the increased transmission of this wavelength region resulting from the anthropogenic reduction of stratospheric ozone (Caldwell, 1979). Microorganisms contain a range of chromophores capable of absorbing UV-B radiation and initiating cellular damage. Similarly, a variety of chromophores exist which absorb in the UV-A and visible regions and are able to sensitize cells to light. Most notably, the photosynthetic pigments of the phototropic microorganisms, although efficient energy-transducing photoreceptors, can undergo a range of side reactions ultimately leading to cellular damage.

The Role of Oxygen

Oxygen is not involved in the damaging effects of UV-C radiation, but for wavelengths longer than this, where damage is mediated by a photosensitizer, the reactions almost always involve the participation of molecular oxygen. These reactions are sensitized photooxidation processes and are often referred to as photodynamic actions.

Photosensitized reactions involve the electronically excited states of the sensitizing molecules, produced by excitation with light. The absorption of a photon of light raises the ground state (S_0) of the sensitizer to an extremely short-lived excited state, referred to as the singlet state (^1S). The energy of the singlet state can be dissipated through a number of mechanisms including interactions with the surrounding solvent or by the emission of a photon of light as fluorescence. For the most efficient photosensitizers, however, the excited singlet state may undergo intersystem crossing to yield a longer-lived (10^{-3}–10 seconds) metastable excited state, the triplet state (^3S). The excited triplet state has a sufficiently long lifetime to allow interactions with other molecules, thereby initiating photochemical reactions, before returning to the ground state. These reactions can be summarized thus:

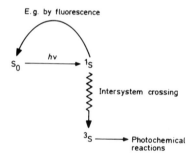

The excited triplet sensitizer can undergo reactions with molecules in its vicinity either by an energy transfer process or by electron or hydrogen atom transfer. Triplet sensitizers may also return to the ground state by radiation-less decay or by the action of physical quenchers. However, it is the reactions with neighbouring molecules which lead to photodynamic action, owing to the production of highly reactive oxygen species. Reactions of triplet sensitizers by electron or hydrogen atom transfer are termed type I or free radical reactions (Spikes, 1977). Electron transfer from the triplet sensitizer to a suitable substrate leads to the production of a semireduced sensitizer (S^-) and a semioxidized form of the substrate (A^+), i.e. free radicals. The semireduced sensitizer can then react with oxygen to regenerate the ground state sensitizer with the production of the oxygen superoxide radical (O_2^-).

The oxygen superoxide radical undergoes a rapid, spontaneous dismutation to produce ground state oxygen and hydrogen peroxide:

$$2O_2^- + 2H^+ \rightarrow O_2 + H_2O_2 \tag{1}$$

However, all aerobic organisms and many anaerobes contain the enzyme superoxide dismutase which catalyses this reaction. The existence of an enzyme to speed up a spontaneously rapid reaction is an indiciation of the potential cytotoxicity of the superoxide radical. Indeed, superoxide dismutase has been shown to play a vital role in the protection of cells from the toxic effects of oxygen (Fridovich, 1975; Halliwell, 1978). Furthermore, systems generating superoxide are known to kill cells, to inactivate enzymes and to degrade cellular constituents, including DNA and membranes (Fridovich, 1975; Halliwell, 1978). However, in aqueous solution, superoxide is only poorly reactive, acting chiefly as a reducing agent. This has led to suggestions that superoxide itself is not toxic (Fee, 1982). It now seems likely that the toxic effects of superoxide are due to the superoxide-dependent formation of highly reactive hydroxyl radicals (OH·), a reaction which requires hydrogen peroxide and traces of free iron (Halliwell, 1978; McCord and Day, 1978) but which can also be mediated by ascorbate and thiol compounds (Rowley and Halliwell, 1982).

$$O_2^- + H_2O_2 \xrightarrow{\text{Fe, ascorbate, thiols}} O_2 + OH· + OH^- \qquad (2)$$

The most common reactions of triplet state sensitizers by energy transfer (type II) are with ground state oxygen. This reaction proceeds efficiently since oxygen is paramagnetic and so exists as a triplet in its ground state. Hence the interaction of an excited state triplet sensitizer with ground state triplet oxygen is spin conserved. The products of the reaction are ground state sensitizer and the highly reactive singlet state of oxygen, $^1O_2$:

$$_3S + _3O_2 \longrightarrow S_0 + {}^1O_2 \qquad (3)$$

Singlet oxygen is much more reactive than ground state oxygen and can interact with a wide range of biological molecules (Spikes, 1977).

The type of pathway followed in a particular sensitized oxidation will depend upon the chemical nature of the sensitizer, the reaction conditions and the availability of oxygen. In many cases sensitized photooxidations can proceed by more than one pathway, the overall process being more complicated than described above.

In addition to these generalized pathways for sensitized photooxidation other pathways may operate in phototrophic organisms, involving electron transport mediated by the photosynthetic pigments. For example, it is well established that, in photosystem I of higher-plant-type photosynthetic electron transport, the univalent reduction of oxygen can occur at the expense of reduced ferredoxin (Asada and Kiso, 1973; Allen, 1977). The immediate product of this electron donation to molecular oxygen is the superoxide radical and thence hydrogen peroxide.

Sensitizers

Almost every biological molecule that absorbs light in the wavelength region between 290 and 750 nm has been proposed as a possible photosensitizer. In many instances of light-induced cellular damage in microorganisms it is possible to identify precisely the endogenous photoreceptor responsible for the photooxidative effect. Nevertheless, a few endogenous photosensitizers have been identified as mediating photooxidative reactions. For instance, it is well established that mutant strains of photosynthetic bacteria that lack coloured carotenoids are rapidly killed by visible light in the presence of oxygen; the action spectrum for this lethal effect of light is essentially identical with the absorption spectrum of bacteriochlorophyll, strongly suggesting that the chlorophyll is the photosensitizer (Giese, 1971; Asami and Akazawa, 1978). Chlorophyll also photosensitizes destructive reactions in cyanobacteria and in eukaryotic microalgae (Abeliovich *et al.*, 1974; Elstner and Oswald, 1980). Particularly under conditions of limiting electron transport through the photosystems, energy dissipation by normal charge separation may be decreased in favour of reactions which involve intersystem crossing and the formation of singlet oxygen:

$$_3Chl + {}^3O_2 \longrightarrow Chl_0 + {}^1O_2 \tag{4}$$

Although there are many similarities between the effects of UV-B and UV-A and the effects of visible light and a photosensitizer, e.g. the requirement for oxygen, the nature of the endogenous sensitizers mediating UV-A- and UV-B-induced damage are largely unknown. To some extent this may be due to the fact that a vast array of photoreceptors can function in this wavelength region including cytochromes, cytochrome oxidase, haem proteins, flavins, NAD, NADH, porphyrins, quinones and tryptophan. The possible roles of these compounds in the damaging effects of UV-B and UV-A have been discussed in reviews by Knowles (1975) and Krinsky (1976). Quinones, porphyrins and flavins are also known to photosensitize a variety of *in vitro* systems. For example, FMN in the presence of oxygen will sensitize a range of enzymes to blue light (Codd, 1972; Whitelam and Codd, 1983b). The inactivation of haem proteins by UV-A and visible light is not surprising since the haem moieties of these proteins absorb strongly in this wavelength region. Furthermore, it has been shown that catalase, an enzyme composed of four haem-containing subunits, is readily photoinactivated by visible and UV-A light (Mitchell and Anderson, 1965; Cheng *et al.*, 1981).

In a study of the induction of tryptophanase in *Escherichia coli*, Swenson and Setlow (1970) observed that this process was inhibited by broad-band UV radiation. The action spectrum for the inhibition showed a peak at 334 nm which corresponds to the action maximum for growth delay.

Pyridoxal phosphate was suggested to be the responsible chromophore since it absorbs strongly in this region of the spectrum as well as functioning as a cofactor for tryptophanase. Recently, however, a substantial body of evidence has been presented which suggests that the rare RNA nucleoside 4-thiouridine is involved in the UV-A-induced inhibition of tryptophanase and growth delay in *E. coli* (Jagger, 1981); Sharma *et al.*, 1981). This tRNA nucleoside has also been implicated in the 334 nm UV-induced single-strand breaks in DNA in the same organism. The rate of accumulation of DNA backbone breakage is more than twice as fast in wild-type strains than in strains of *E. coli* lacking 4-thiouridine (Peak *et al.*, 1983). Furthermore, in a recent *in vitro* study of the induction of single-strand breaks in DNA from *Bacillus subtilis*, Peak *et al.* (1984) have shown that free 4-thiouridine, as well as a range of other nucleic acid components, can function as sensitizers of 334 nm irradiation.

The tRNA for tyrosine in *B. subtilis* contains 4-thiouridine (Keith *et al.*, 1976); however, growth delay in this organism in response to UV-A appears to involve menaquinone as the target molecule (Taber *et al.*, 1978).

In addition to the range of endogenous photosensitizers a wide variety of naturally occurring exogenous sensitizers may also contribute to lethal and sublethal light damage. For example, it has become apparent that in some instances irradiation of the growth medium will lead to inhibition in several microorganisms. Yoakum and Eisenstark (1972) demonstrated that photoproducts especially toxic to mutants of *E. coli* and *B. subtilis* are produced by the aerobic irradiation of L-tryptophan with broad-band UV. This tryptophan photoproduct has been shown to be mutagenic and lethal to repair-deficient bacterial mutants and to inhibit repair-deficient single-strand DNA breaks. McCormick *et al.* (1976) identified hydrogen peroxide as the toxic product and the biological effects have been attributed to its presence. It is possible that tryptophan present in cellular proteins or in metabolic pools during UV exposure could also result in the generation of hydrogen peroxide.

It is well established that the lethality of UV-B and UV-A radiation in *E. coli* and in coliphage T7 is enhanced by the presence of hydrogen peroxide (Ananthaswamy and Eisenstark, 1976; Hartman and Eisenstark, 1978). Hydrogen peroxide is frequently generated during the production of bacteriological growth media (Carlsson *et al.*, 1978). Marler and Van Baalen (1965) have shown that hydrogen peroxide can arise in mineral media from a reaction between citrate and Mn^{2+} promoted by autoclaving and also by irradiation, UV-A wavelengths around 365 nm being especially effective. The action spectrum for the synergism of action of UV and hydrogen peroxide for inactivation of phage T7 shows a peak at 340 nm. Hydrogen peroxide absorbs only poorly in this region of the UV spectrum and it has been assumed that it is not acting as a photosensitizer (Jagger, 1981). However, Ahmad (1981) has presented evidence that broad-band near-UV photolysis of hydrogen peroxide leads to generation of the superoxide radical. Furthermore, since T7 inactivation by near UV and hydrogen peroxide is significantly reduced in the presence of superoxide dismutase, Ahmad (1981) has suggested that O_2^- or one of its reactive products is involved in phage inactivation.

Many natural surface waters have been shown to contain photosensitizers capable of producing strong oxidants upon absorption of sunlight. Zepp *et al.* (1977) have shown that singlet oxygen may be produced as a result of sensitization by dissolved organic materials in a number of samples of naturally coloured waters from the southern United States. These observations have been confirmed recently for humic water samples from a variety of sources (Baxter and Carey, 1982; Haag *et al.*, 1984). The formation of so toxic and reactive a species as singlet oxygen in natural waters is of environmental interest since it seems likely that it will have an effect on the organisms of coloured waters, although this has yet to be demonstrated.

Biochemical and Physiological Aspects

Processes and Components Affected in Heterotrophic Organisms

Nucleic acids and viability. The action spectra for inactivation or killing of a wide variety of microorganisms show close agreement with the absorbance spectrum of DNA, suggesting that direct absorption of light by DNA is responsible for the observed effects (Giese, 1968). If these action spectra are extended to wavelengths longer than 290 nm, then high fluences of UV-B, UV-A and visible radiation will also lead to killing of bacteria, in the absence of added sensitizers (Webb, 1977). In one of the earliest quantitative studies of photoinactivation, Hollaender (1943) demonstrated that *E. coli* could be readily inactivated with high fluences of radiation in the wavelength region between 350 and 490 nm. The efficiency of near-UV- and visible-light-induced killing of bacteria is much reduced compared with the lethal effects of UV-C radiation. Furthermore, as the wavelength of actinic irradiation increases, then increasing photon fluence rates are required to achieve the same degree of killing (Webb, 1977).

The damaging or lethal effects of UV and visible light appear to be largely due to an inhibition of the reproductive capacity of the cells through the direct or indirect action of DNA (Webb, 1977; Jagger, 1981; Peak and Peak, 1982). Since the nucleic acids have little or no absorption at wavelengths greater than 320 nm, it is reasonable to suppose that the lethal effects of these longer wavelengths may involve photodynamic reactions mediated by endogenous sensitizers. Support for this view has come from a number of studies which have demonstrated a dependence upon oxygen for the lethality of wavelengths longer than 320 nm (Webb, 1977). Eisenstark and coworkers (Eisenstark, 1970, 1971; Ferron et al., 1972) have investigated the UV-B and UV-A sensitivity of recombination-deficient strains of B. subtilis and Salmonella typhimurium. During the exponential phase these strains showed sensitivity to UV-A in the presence of oxygen, but they were much less sensitive under anoxic conditions. Several strains of E. coli also show a strong oxygen dependence of inactivation by 365 nm light (Webb and Lorenz, 1970; Tyrell, 1976; Peak et al., 1983). In contrast, inactivation by UV-C radiation has been shown to be independent of the presence of oxygen (Zetterberg, 1964; Webb and Lorenz, 1970; Webb, 1977). An exception to this is Bacteriodes fragilis which is much more sensitive to UV-C radiation under aerobic conditions than under anaerobic conditions (Jones et al., 1980; Slade et al., 1981).

Compared with the investigations of the nature of the deleterious effects of UV-C radiation, relatively little work has been done on elucidating the exact nature of the lethal and mutagenic lesions imposed by UV-B and UV-A radiation. Tyrell (1973) has shown that cyclobutane-type pyrimidine dimers are induced in vivo by UV-A irradiation (365 nm) in E. coli DNA, although with significantly reduced efficiency compared with UV-C radiation. Since 365 nm is within the efficient part of the action spectrum for photoreactivation it could be argued that concomitant photoreactivation is responsible for the reduced efficiency of UV-A radiation in dimer production. However, dimer yields were found to be identical in E. coli B phr, a strain which lacks the photoreactivation enzyme, and E. coli B/r Hcr, indicating that concomitant photoreactivation was not significant for these measurements (Tyrell, 1973). Removal of the UV-A-induced dimers in E. coli DNA by photoreactivation enzyme from yeast and blue light in vitro has confirmed that they are of the same type as the cyclobutane-type pyrimidine dimers induced by UV-C radiation (Tyrell et al., 1973).

The role of pyrimidine dimers in the lethal effects of UV-A radiation will be determined by the repair capability of the cells (Webb, 1977). Since an enzymic photoreactivation system exists which is specific for cyclobutane-type pyrmidine dimers (Setlow, 1967), the absence of photorepair of lethal lesions induced by UV-A radiation, under conditions where functional photoreactivation enzymes could be demonstrated, has led to the conclusion that pyrimidine dimers are probably not significant for UV-A-induced lethality (Webb, 1977). Furthermore, in cells with norma! DNA repair capabilities, concomitant photoreactivation can effectively remove induced dimers during irradiation with lower fluence rates of UV-A (Tyrell et al.,

1973). Therefore, lesions other than dimers would seem to be important in the lethal and mutagenic effects caused by UV-A radiation.

In addition to pyrimidine dimers, UV-A radiation, in common with UV-C and UV-B radiation, also induces single-strand breaks in bacterial DNA at rates approximately equal to the rate of induction of dimers (Fig. 1) (Tyrell *et al.*, 1974; Peak and Peak, 1982). The repair of single-strand breaks induced in *E. coli* DNA by aerobic UV-A radiation has been observed after holding the irradiated cells in phosphate buffer (Ley *et al.*, 1978). However, no repair of DNA breaks was observed in a strain deficient in DNA polymerase I. Similar results have been obtained for *B. subtilis*, where a mutant deficient in DNA

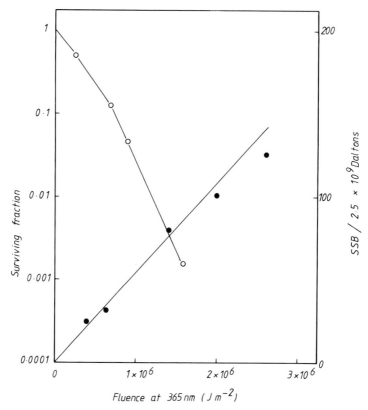

Fig. 1. Fluence response curves for the induction of single-strand breaks (●) and lethality (○) in *E. coli* RT1 by 365 nm radiation. For the determination of viability cells were plated on supplemented minimal medium following irradiation. The total number of single-strand breaks per genome (2.5×10^9 daltons) was calculated from number-average molecular weights of DNA. (Modified from Tuveson *et al.* (1983).)

polymerase I accumulates breaks in response to UV-A radiation significantly faster than the wild-type strain proficient in polymerase I (Peak and Peak, 1982). Ley *et al.* (1978) have provided convincing evidence that the single-strand breaks induced by 365 nm radiation do not result from the enzymic nicking of DNA, as would occur during the excision of a dimer or other DNA lesion. In particular, this conclusion is supported by the observations that isolated DNA is broken more efficiently than cellular DNA under 365 nm radiation and that 405 nm radiation induces single-strand breaks at almost half the rate of 365 nm radiation, but does not cause detectable dimers, even at very high fluence rates.

The efficient photoreactivation of lethality and mutation following UV-C radiation is strong evidence for the predominant role of dimer formation in the effects of this radiation. However, at wavelengths longer than 320 nm, single-strand breaks may be a more important mechanism of cellular damage; this may particularly be the case for wavelengths of 405 nm and longer, where the induction of dimers is not detectable. Several investigations have provided circumstantial evidence that single-strand breaks are the predominant lesions for UV-A-induced lethality. For instance, oxygen is known to sensitize cells (Webb and Lorenz, 1970) and isolated, transforming DNA (Peak *et al.*, 1981) to inactivation by UV-A; oxygen also sensitizes single-strand break induction (Tyrell *et al.*, 1974; Peak and Peak, 1982). A number of agents, including the singlet oxygen quencher diazobicyc-lo[2,2,2]octane (DABCO), protect against single-strand break induction as well as reducing the inactivation of whole cells or isolated transforming DNA by UV-A radiation (Peak *et al.*, 1981; Peak and Peak, 1982).

It seems likely, then, that unrepaired single-strand breaks make a significant contribution to cellular lethality induced by UV-A irradiation. The protective effects of DABCO and the sensitization by oxygen are consistent with a mechanism involving the photodynamic production of singlet oxygen.

In addition to DNA, other biologically important molecules are targets for the damaging effects of light. For example, Ramabhadran (1975) has shown that the fluence rates of UV-A which induce growth delay in continuous cultures of *E. coli* B/r also lead to a complete cessation of net RNA synthesis, with relatively little effect on DNA and protein synthesis. An action spectrum for the inhibition of net RNA synthesis was found to correspond to the action spectrum for growth delay and to the absorbance spectrum of valyl tRNA. Ramabhadran (1975) suggests that inactivation of tRNA through sensitization of 4-thiouridine is the probable first event in the inhibition both of growth and of net RNA synthesis. Inactivation of tRNA induced by UV-A irradiation has also been demonstrated by Thomas and Favre (1975). They suggest that the absorption of light by 4-thiouridine in position 8 of a number of *E. coli* tRNAs and the subsequent formation of 4-thiouridine–cytidine adducts inactivates the tRNA and hence leads to a cessation of macromolecular synthesis.

In some cases, the damaging effects of light-induced DNA lesions are amplified because UV-A radiation can inhibit the full functioning of DNA repair systems (Tyrell, 1973, 1976; Tyrell *et al.*, 1973). The capacity of the photoreactivation enzyme from *E. coli* to monomerize pyrimidine dimers can be inactivated by 365 nm radiation both *in vivo* and *in vitro* (Tyrell *et al.*, 1973). Similarly, the photoreactivation enzyme from yeast can be inactivated *in vitro*, the effect being sensitized by oxygen (Tyrell, 1976). The nature of the damage to the photoreactivation enzyme is not known, but the similarity of inactivation rates *in vivo* and *in vitro* suggests that the photoreactivation enzyme complex may be, or may contain, the photosensitizing chromophore.

Membranes and membrane processes. Hollaender (1943) first suggested that, unlike UV-C radiation, the deleterious effects of UV-A radiation on *E. coli* were, at least partly, due to damage to the cell membrane. Hollaender (1943) observed that irradiating cells with 350-405 nm light induces them to become sensitive to the toxic effect of physiological saline. This saline toxicity was interpreted as being the result of cell membrane damage. Perhaps because of interest in DNA lesions induced by UV-A and visible light, it was not until fairly recently that further evidence for cell membrane damage was presented. Moss and Smith (1981) observed that a DNA-repair-competent strain of *E. coli* K-12 displayed sensitivity to inorganic salts, at concentrations used in minimal media, after irradiation with broad-band UV-A light. The fluences inducing this sensitivity to inorganic salts caused little inactivation of cells plated on complex growth media. Moss and Smith (1981) have proposed that salt sensitivity results from membrane damage formed by UV-A radiation under aerobic conditions, and to a lesser extent under anaerobic conditions. The photosensitizing chromophore responsible for induction of salt sensitivity is unknown, although it does not appear to be 4-thiouridine, since mutants lacking this base show similar light-induced sensitivity. These findings could explain the large variation observed in survival data following UV-A irradiation of cells and suggest that, at least on minimal media plates, membrane damage could contribute towards cell killing. In an extension of these studies, Kelland *et al.* (1983) found that the recovery of *E. coli* from this salt sensitivity induced by UV-A radiation was not affected by chloramphenicol or penicillin but was inhibited by the addition of low concentrations of bacitracin, an antibiotic which acts in part by inhibiting membrane synthesis. Further evidence that membrane damage plays a significant role in lethality induced by UV-A radiation in *E. coli* has come from a study of membrane fatty acid composition. Klamen and Tuveson (1982) have shown that, in the fatty acid auxotroph *E. coli* K1060, the lethality of UV-A radiation is directly related to the number of carbon–carbon double bonds in the fatty acids used to supplement growth (Fig. 2). More direct evidence that UV-A radiation leads to membrane damage has come from a study of light-induced leakage of ^{86}Rb$^+$ from *E. coli* K-12 cells (Kelland *et al.*, 1984). In this organism there is a strong parallelism between the action spectra for cell inactivation and the induction of ^{86}Rb$^+$ leakage. Similarly, Ito and Ito (1983) have observed a correlation between lethality and the permeation of *p*-nitrophenyl-D-glucopyranoside in cells of *Saccharomyces cerevisiae* following UV-A irradiation.

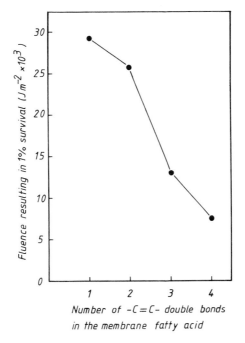

Fig. 2. Dependence of the fluence of 313–425 nm radiation to give 1% survival of *E. coli* K 1060 on the number of double bonds in the supplemented fatty acids. (Redrawn from Klamen and Tuveson (1982).)

Inhibitory effects of UV-A and visible light on several membrane transport systems have also been reported (Barran *et al.*, 1974; Sprott *et al.*, 1975, 1976). Inhibition of transport processes usually occurs at fluences that cause little inactivation in repair-proficient strains. From the differential wavelength dependence of a range of amino acid uptake systems in *E. coli*, Sprott *et al.* (1976) have concluded that the inhibitory effects of UV-A and visible light are not merely due to a non-specific increase in membrane permeability. A similar conclusion was reached by Robb *et al.* (1978) for the inactivation of two separate leucine transport systems in *E. coli*. The action spectra and kinetics of inhibition of both systems were found to be similar. A generalized non-specific membrane damage effect was discounted since cells were able to retain leucine during irradiation by light which reduces the uptake to very low levels but does not cause detectable killing of the cells.

There is a correlation between the inhibitory effects of light on glycine uptake in *E. coli* and the inhibition of respiration (D'Aoust *et al.*, 1974). Transient inhibitions of respiration have been observed previously in response to UV-A radiation at fluences that cause little or no cell death (Jagger, 1964, 1972). Indeed, Jagger (1972) has proposed that the inhibition of respiration forms the basis of light-induced growth delay. Subsequently, however, Doyle and Kubitschek (1976) have failed to observe a parallelism between inhibition of respiration and growth delay, and these workers have suggested that inhibition of carbon source transport may be the primary mechanism of UV-A-induced growth delay in *E. coli* (Kubitschek and Doyle, 1981).

Recently, evidence has accumulated which suggests that menaquinone (vitamin K2) may be both the chromophore and the target for the UV-A effects on membrane transport and other processes in microorganisms (Madden *et al.*, 1981). Menaquinone is the primary electron transport quinone in Gram-positive bacteria, but functions in Gram-negative bacteria primarily under anaerobic conditions, and has also been shown to be involved in the active transport of amino acids. In *Mycobacterium phlei* membrane vesicles, 360 nm irradiation inactivates the natural menaquinone and results in a loss of oxidation, active transport of calcium and proline and phosphorylation (Brodie *et al.*, 1979). Indirect evidence, based on comparative action and absorption spectra, that inactivation of menaquinone may be involved in growth delay in *E. coli* following UV-A irradiation has been summarized by Madden *et al.* (1981). Menaquinone also appears to be one of the targets for the blue light inactivation of the respiratory chain in *Sarcina lutea* (Anwar and Prebble, 1977).

In addition to components of the respiratory electron transport chain a number of enzymes of respiratory metabolism have also been identified as targets for photoinactivation. The flavoenzyme malate dehydrogenase from *S. lutea* has been shown to be inactivated *in vivo* and *in vitro* by blue light (Anwar and Prebble, 1977). Additionally, several agents known to react with flavins and flavin derivatives, such as cysteine and anthranilic acid, have been shown to protect the flavoenzymes L-α-glycerophosphate, succinate and D-lactate dehydrogenase from inactivation by visible light in *E. coli* (D'Aoust *et al.*, 1980). In several prokaryotic and eukaryotic cells a cytochrome exidase has been shown to be photolabile to visible light (see Epel (1973) for a review).

Processes and Components Affected in Phototrophic Organisms

Nucleic acids and viability. Comparatively few studies have been performed on the induction of DNA lesions and lethality by solar visible and UV light in photosynthetic microorganisms. In a small number of studies, it has been established that cyanobacteria contain highly effective photoreactivation and excision repair systems for the removal of UV-irradiation-induced pyrimidine dimers (Singh, 1975; O'Brien and Houghton, 1982a,b); however, in these investigations the DNA lesions were induced by 254 nm UV-C radiation. Van Baalen (1968) studied the effects of 254 nm radiation on the survival and photosynthetic activity of the coccoid cyanobacterium *Agmenellum quadruplicatum*. He observed a very efficient photoreactivation (action maximum at 430 nm) for both survival and photosynthetic activity. Efficient repair mechanisms, in particular a very effective photorecovery, have been proposed as the basis to explain the high tolerance to broad-band UV-A radiation by the cyanobacterium *Anacystis nidulans* (Asato, 1972). The damaging effects of UV-A in this organism were assumed to result from DNA damage because caffeine, which inhibits DNA dark repair mechanisms in *E. coli*, was found to depress the survival of mutant and wild-type *A. nidulans* cells (Asato, 1972). Another unicellular cyanobacterium, *Gleocapsa alpicola*, has been shown to be killed by exposure to UV-A radiation, although a mutant with an elevated carotenoid content appears to be UV-resistant (Buckley and Houghton, 1976).

There have been a few reports of the photooxidative killings of phototrophic microorganisms by visible light, these effects being sensitized by the photosynthetic pigments. The bleaching of bacteriochlorophyll and the subsequent killing of pigment mutants of the purple sulphur bacterium *Rhodopseudomonas sphaeroides* by visible light in air provided some of the earliest evidence on the protective function of carotenoids (Griffiths *et al.*, 1955; Sistrom *et al.*, 1956). Lethality was due to a photodynamic effect initiated by bacteriochlorophyll itself. The photooxidative destruction of the cyanobacterium *A. nidulans* by visible light is one of the few well-documented examples of lethal photodynamic action in photoautotrophic microorganisms (Abeliovich and Shilo, 1972; Abeliovich *et al.*, 1974). Under conditions of elevated oxygen and reduced carbon dioxide levels, *A. nidulans* is rapidly killed by fluence rates of visible light frequently encountered in natural surface waters. Photooxidative death is preceded by an impairment of photosynthesis and the bleaching of photosynthetic pigments (Abeliovich and Shilo, 1972). Furthermore, Shilo and coworkers have accumulated substantial indirect evidence that superoxide functions as the destructive oxygen species for these effects (Abeliovich *et al.*, 1974; Steinitz *et al.*, 1979).

Resistance of cyanobacteria to photooxidation appears to be correlated with their ability to maintain superoxide dismutase activity, providing further indirect evidence that superoxide may be the important oxygen species (Eloff *et al.*, 1976).

Photosynthetic apparatus.   It has been known for more than a century that the photosynthesis of aquatic microalgae is depressed by high fluences of solar radiation and this phenomenon is well documented in the literature (see Harris, 1978). This photoinhibition of the capacity and efficiency of photosynthesis has also been observed in studies with photosynthetic bacteria (Asami and Akazawa, 1978), with isolated chloroplast thylakoids (Kok *et al.*, 1965; Satoh, 1970) and with intact leaves of numerous higher plants (Powles *et al.*, 1979, 1980; Critchley, 1981). The occurrence and assessment of photoinhibition in microalgae and cyanobacteria in natural waters has been reviewed recently by Harris (1978) and Codd (1981). Photoinhibition of higher plant photosynthesis has also been recently reviewed (Osmond, 1981; Powles, 1984).

Photoinhibition in natural populations of phototrophs and in laboratory cultures is a fluence-dependent process, with both the fluence rate and the exposure time being important in determining the extent of photoinhibitory damage (e.g. Takahashi *et al.*, 1971; Whitelam and Codd, 1983a). Additionally, photoinhibition will also clearly be a wavelength-dependent phenomenon. In general, the photosynthetic pigments themselves act as sensitizers for photoinhibition, although in a few cases other sensitizers may be involved, e.g. in some instances blue-light photoreceptors seem to be important (Harnischfeger and Gaffron, 1969; Bauer and Wild, 1976). In field experiments only a few attempts have been made to distinguish between the damaging effects of UV and visible light. The early studies which revealed that broad-band UV radiation could have a significant deleterious effect on aquatic primary productivity have been supported by more recent investigations using spectrally selective containers (Smith *et al.*, 1980; Worrest *et al.*, 1980). Based upon the increase in ^{14}C incorporation by natural phytoplankton during surface layer incubations following the exclusion of UV-B radiation, Lorenzen (1979) has concluded that a significant fraction of the euphotic zone experiences photoinhibition due to these wavelengths. Furthermore, the quantitative analysis of the influence of UV wavelengths on phytoplankton photosynthesis by Smith *et al.* (1980) indicates that 25% of the biologically effective dose for photoinhibition is due to wavelengths less than 390 nm. The chromophores responsible for sensitizing the photosynthetic apparatus to UV radiation are not known, although it may be reasonable to suppose that any component of either photosynthetic electron transport or CO_2 fixation, which has an appreciable extinction in this spectral region, could function as a sensitizer.

Despite the extensive literature, largely from field experiments, on photo-inhibition in the aquatic environment and its significance for phytoplankton productivity, there is a paucity of information on the nature and mechanism of damage to photosynthesis. An early insight into the possible target of visible light damage to algal photosynthesis was provided by the work of Kok (1956). Upon exposure to visible light at photon fluence rates in excess of full sunlight (approximately 2000 μmol m^{-2} second^{-1}) *Chorella* cells showed an immediate, exponential decline in light-limited photosynthesis and a concurrent decline in light-saturated photosynthesis, which proceeded after a short lag (Kok, 1956). From this, Kok concluded that the primary event in photoinhibition was an inactivation of the photochemical process, the reduction in the capacity of this process to pass on light energy then leading to the diminution of light-saturated photosynthesis. This trend has been confirmed in a recent study of photoinhibition of photosynthesis in the cyanobacterium *Microcystis aerguinosa* (Whitelam and Codd, 1983a). Following exposure of the cyanobacterial cells to half full sunlight, there is a time-dependent inactivation of electron transport capacity with a concomitant decline in CO_2 fixation (Fig. 3).

In several other laboratory investigations a lesion in photosynthetic electron transport has also been observed to be a prime consequence of exposure of phototrophic microorganisms to excess visible light (e.g. Harvey and Bishop, 1978; Gerber and Burris, 1981; Samuelsson and Richardson, 1982). In those organisms with higher plant-type photosynthetic electron transport, inhibitions of both photosystem I and photosystem II have been reported. However, the processes by which the two photosystems are inhibited appear to be different. The photoinactivation of photosystem II, which accounts for photoinhibition in the cyanobacterium *M. aeruginosa*, does not seem to be dependent upon oxygen (Whitelam and Codd, 1983a; Tytler *et al.*, 1984a). Photoinhibition of whole-cell oxygen evolution in this organism occurs at the same rate in cells sparged continuously with either oxygen or argon. This is comparable with the photoinhibition which occurs in several higher plants, where exposure to high photon fluence rates leads to an inactivation of photosystem II electron transport which is oxygen independent (Powles, 1984). Therefore, it seems unlikely that the generation of toxic oxygen radicals in photosystem II is responsible for the inhibition.

In contrast, where exposure to excess visible light leads to an inactivation of photosystem I electron transport, this process appears to be strongly dependent upon the presence of oxygen. In the marine diatom *Amphora* and

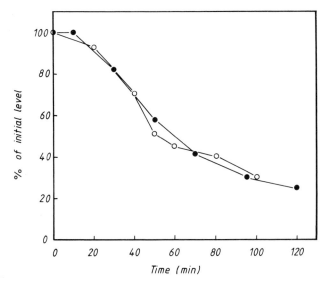

Fig. 3. Concomitant photoinhibition of whole-cell CO_2 fixation (●) and electron transport (○) in *M. aeruginosa*. Cells were exposed to white light at a photon fluence rate of 1000 μmol m⁻² second⁻¹ for the times indicated prior to determination of whole-cell ¹⁴CO_2 fixation capacity or whole-cell electron transport capacity, assayed as methyl-viologen-dependent O_2 consumption. (From Whitelam and Codd (1983a).)

in the green alga *Scenedesmus obliquus*, the photoinactivation of photosystem I electron transport appears to result from the photooxidation of P700, a component of the reaction centre of photosystem I (Harvey and Bishop, 1978; Gerber and Burris, 1981). The photoinhibition of photosystem I activity which occasionally occurs in higher plant leaves and chloroplasts upon exposure to excess visible light is also oxygen dependent (Powles, 1984). This oxygen dependence indicates a different phenomenon from that of photosystem II inactivation and suggests that interactions which lead to the generation of destructive oxygen radicals are involved.

The inhibition of photosynthetic electron transport capacity is also manifested as a change in chlorophyll fluorescence yield. In many algal cells, photoinhibition leads to a reduction in maximal fluorescence yield of photosystem II at room temperature (e.g. Harris and Piccinin, 1977). Analysis of the kinetics of variable chlorophyll fluorescence at room temperature in the cyanobacterium *M. aeruginosa* and in the green alga *Ankistrodesmus braunii* supports the conclusion that the primary site of light damage in these organisms is within photosystem II (Tytler *et al.*, 1984a; Whitelam and Codd, 1984). A similar conclusion was reached on the basis of low temperature − 196 °C) fluorescence data for the marine dinoflagellate *Amphidinium carterae* (Samuelsson and Richardson, 1982).

The precise nature of the lesion(s) within photosystem II is not known, although the decrease in initial fluorescence rise (F_0) observed in photo-inhibited *M. aeruginosa* cells suggests that some destruction of light-harvesting pigments or loss of photosynthetic unit occurs (Tytler *et al.*, 1984a). Furthermore, it is unlikely that the water-splitting complex is affected in this organism because electron transport activity cannot be restored with electron donors that bypass the complex (Tytler *et al.*, 1984a). Experiments with the green alga *Chlamydomonas reinhardii* have indicated that photoinhibition in this organism is associated with a loss of the 32×10^3 dalton herbicide-binding polypeptide rather than with damage to the reaction centre itself (Kyle *et al.*, 1983).

It has been frequently observed that photoinhibitory damage in phototrophic microorganisms is more severe under conditions of carbon dioxide depletion (Asami and Akazawa, 1978; Kaplan, 1981; Krüger and Eloff, 1983a,b; Whitelam and Codd, 1983a). This also appears to be true for photoinhibition in higher plants (Powles, 1984). Photooxidative death and the photobleaching of pigments in cyanobacteria are also enhanced by carbon dioxide depletion (Abeliovich and Shilo, 1972). It has been suggested that in the absence of carbon dioxide, when photosynthetic electron transport is not coupled to carbon dioxide fixation, oxygen functions as the terminal electron acceptor with the resultant production of superoxide and hydrogen peroxide, and that these species are responsible for the oxidative damage which follows (Asami and Akazawa, 1978; Kaplan, 1981). This explanation, for the protective role of carbon dioxide, assumes that the cells do not contain adequate facilities for the removal of these oxygen species and, of course, cannot be applied to the oxygen-independent inactivation of photosystem II (Whitelam and Codd, 1983a). An alternative view is that in the absence of carbon dioxide, the terminal electron acceptor for photosynthesis, the transfer of excitation energy from the light-harvesting pigment assemblies to the photochemical reaction centres will exceed the rate of transfer from the reaction centres to the electron transport chain. This imbalance could be expected to lead to secondary photoreactions, which may or may not involve oxygen, causing inactivation of the reaction centres or associated processes. This explanation assumes that, in the absence of efficient electron transport, other mechanisms do not exist for the adequate de-excitation of photochemical energy (Fig. 4).

In addition to photosystems I and II, other components of the photosynthetic machinery have been found to be photolabile in phototrophic microorganisms exposed to excess visible light. In the filamentous cyanobacterium *Spirulina platensis* high photon fluence rates of visible light, under carbon-

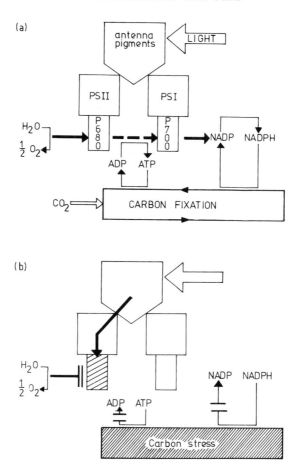

Fig. 4. Schematic representation of the possible processes leading to photoinactivation of photosystem II (PSII) electron transport in *M. aeruginosa*: (a) the normal pathway of electron transport coupled to the fixation of carbon; (b) under conditions of CO_2 stress and high intensity light the inability to dissipate photochemical energy via ATP and NADPH production leads to destructive secondary reactions within photosystem II, with a resultant lesion in electron transport capacity.

dioxide-depleted conditions, lead to inhibition of the ability of the cells to accumulate inorganic carbon from the surrounding medium (Kaplan, 1981). This effect was found to be oxygen dependent and was suggested to involve either a direct effect of oxygen radicals on the bicarbonate transport mechanism of the cells or an indirect effect on the cell membrane, causing a leak of inorganic carbon. Photoinhibition of carbon dioxide fixation in the anaerobic purple sulphur bacterium *Chromatium vinosum* is also an oxygen-dependent process and appears to result from inactivation of the photophos-phorylation capacity of the cells (Asami and Akazawa, 1978). Inactivation of photophosphorylation, observable in preparations of isolated chromato-phores, was found to be significantly reduced by the addition of 1,2-dihydroxybenzene-3,5-disulphonic acid (Tiron), a superoxide scavenger, or α-tocopherol, a singlet oxygen scavenger. Singlet oxygen had previously been shown to lead to inactivation of photophosphorylation in isolated chromato-phores of *Rhodospirullum rubrum* (Slooten and Sybesma, 1976).

   Evidence has been presented that ribulose-1,5-bisphosphate carboxylase, the key enzyme of the Calvin cycle, can also be a target for light damage during the photoinhibition of photosynthesis in some phototrophs (Codd, 1981; Vu *et al.*, 1982, 1984). Blue wavelengths appear to be important for the *in vivo* inactivation of the enzyme in *M. aeruginosa* (Codd, 1981). This observation has been supported by the finding that ribulose-1,5-bisphos-phate carboxylase from this organism is also inactivated by blue light *in vitro* in the absence of added sensitizers (Fig. 5). The identity of the blue-absorbing

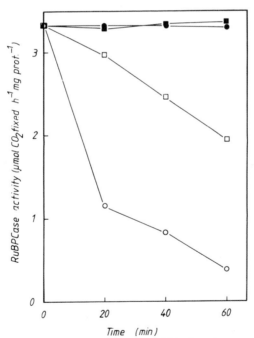

Fig. 5. *In vitro* photoinactivation of ribulose-1,5-bisphosphate carboxylase in crude extracts from *M. aeruginosa*. Enzyme activity was assayed at various times following incubation in blue light (250 μmol m^{-2} second^{-1}) with (○) or without (□) added FMN (0.2 mM) or in darkness with (●) or without (■) sensitizer. (From Whitelam and Codd (1983b).)

photosensitizer is unknown, although added flavins are known to sensitize ribulose-1,5-bisphosphate carboxylase to inactivation by blue light *in vitro* (Codd, 1981; Whitelam and Codd, 1983b) in agreement with the flavin-sensitized photodynamic crosslinking and inactivation of several other multisubunit enzymes (McCarron and Tu, 1983).

In many phototrophs, prolonged exposure to photooxidative conditions leads to bleaching of the photosynthetic pigments; however, this occurs after photoinhibition of photosynthesis and so is not a primary target of light damage (e.g. Abeliovich and Shilo, 1972; Asami and Akazawa, 1978; Whitelam and Codd, 1983a). The bleaching of chlorophyll in the cyanobacterium *M. aeruginosa*, as well as in many other organisms, is a strictly oxygen-dependent process and can be almost completely prevented by anaerobiosis (Whitelam and Codd, 1983a). Thus, this process differs from the initial photoinhibitory lesion in photosynthetic electron transport in this organism, which occurs under high photon fluence rates, even in cells sparged with argon (Whitelam and Codd, 1983a). Photooxidative bleaching of chlorophyll in many organisms is enhanced by low temperatures and conditions which inhibit photosynthetic electron transport (e.g. Sironval and Kandler, 1958; Peschek and Schmetterer, 1978; Schmetterer and Peschek, 1981).

Adaptation–Defence Strategies

Cellular Adaptation and Avoidance

In response to increases in photon fluence rates, microorganisms are capable of displaying a range of responses, or adaptations, enabling the organism to avoid absorbing potentially harmful quantities of radiation. In those microorganisms which are free moving, owing to the possession of cilia or flagellae or as the result of gliding movements, a number of light-induced motor responses are discernible. Häder (1979) describes three classes of photomovement: (1) photokinesis, a response in which the steady state speed of the organism is governed by photon fluence rate; (2) phototaxis, a movement in relation to the light direction; (3) phototrophic reactions, mediated by sudden changes in photon fluence rate. These photomovements have been most intensively studied in photosynthetic microorganisms, but they have also been observed in non-photosynthetic organisms. In principle, any of the three classes of photomovement could function as a mechanism for the avoidance of high light stress (Häder, 1979). For the phototrophic microorganisms, phototaxis appears to be the most effective, since it allows individuals or populations to be guided towards favourable light conditions and away from harmful radiation (see Forward (1976) and Häder (1979) for reviews). Nelson and Castenholz (1982) have observed a vertical diel migration of the filamentous gliding bacterium *Beggiatoa* into and out of sediments in warm spring pools. This response enables the organism to avoid the complete inhibition of growth which would otherwise occur in full sunlight, even when the filaments are aggregated in tufts.

In addition to the relocation of the entire organism in response to unfavourable light conditions, a number of eukaryotic microalgae have been found to orient chloroplasts differently within the cell under such conditions. The uniseriate, filamentous, green alga *Mougeotia* and the cytologically similar unicellular *Mesotaenium* can rotate their single flat chloroplast in such a way as to expose its face or profile to low or high fluence rate white light respectively (Haupt, 1982). This rotation helps to regulate light absorption by the cell and may prevent photooxidative damage under conditions where photosynthesis is saturated.

Microorganisms have adopted several other strategies which regulate the amount of light energy that they receive. These include light shielding by either intracellular or extracellular pigments, association with algal mats or the possession of refractile structures such as calcified sheaths or gas vacuoles. Gas vacuoles provide a means by which aquatic microorganisms can adjust their position in the water column (Walsby, 1972). Thus, photosynthetic organisms can, to some extent, regulate their position with respect to the light environment and obligate aerobic heterotrophs can gain greater access to oxygen by floating to the surface. In addition, studies of the spectroscopic properties of gas vacuoles led to the suggestion that they may act as light-shielding bodies serving to protect individual cells from the destructive effects of high photon fluence rates (Waaland et al., 1970). However, more recent studies of the spectroscopic properties of gas vacuoles have cast doubt on the light-shielding role (Ogawa et al., 1979) and, in two direct studies of this proposed function, it has been found that gas vacuoles do not significantly protect the filamentous cyanobacterium *Anaebena flos-aquae* or the heterotrophic *H. salinarium* from UV inactivation (Shear and Walsby, 1975; Simon, 1980). Nevertheless, a possible role for gas vacuoles in the avoidance of high light stress, based upon buoyancy regulation, has been proposed for *A. flos-aquae* (Dinsdale and Walsby, 1972). At low photon fluence rates, this organism produces large numbers of gas vacuoles, making it buoyant. At high photon fluence rates, the gas vacuoles collapse as a result of the increased cellular turgor pressure which results from increased photosynthesis. In this way, the organism may regulate its position in the water column, avoiding both low and high light stress. This may not be a widespread mechanism for light stress avoidance in the planktonic cyanobacteria, since a strain of *M. aeruginosa* has been described which increases its gas vacuole content in response to increased photon fluence rate (Scott *et al.*, 1981).

Phototrophic microorganisms are capable of altering the amount of their photosynthetic pigments in response to changes in photon fluence rates. These adaptations are most frequently discussed in relation to acclimation to low photon fluence rates, but clearly they may also represent a means for limiting high light stress (see Richardson *et al.*, 1983). Additionally, some phototrophic microbes may have the ability not only to reduce their pigment content but also to alter the morphology of their photosynthetic membranes in response to high light stress. For example, a strain of *M. aeruginosa* resistant to high photon fluence rates is able to reduce the number and length of thylakoid membranes, as well as their arrangement within the cell, in response to high light stress (Scott *et al.*, 1981).

Biochemical Aspects

It is now well established that, at the subcellular level, the major protective strategy for preventing light damage in microorganisms is the possession of coloured carotenoid pigments (see Krinsky, 1978). Since the pioneering studies of Griffiths *et al.* (1955) and Sistrom *et al.* (1956), with carotenoid-free mutants of phototrophic bacteria, there have been numerous investigations demonstrating that the presence of carotenoids either delays the onset of light damage or gives complete protection from high photon fluence rates of visible and UV-A radiation, in both phototrophic and heterotrophic organisms (see Codd, 1981; Paerl, 1984).

The protective effects of carotenoids result from their ability to react with and to quench excited state singlet oxygen $^1O_2$. However, not all carotenoids offer the same degree of protection. There appears to be a distinction between the quenching efficiency of the colourless carotenoids containing less than nine conjugate double bonds and the more unsaturated carotenoids, such as β-carotene which has 11 conjugated double bonds (Mathews-Roth *et al.*, 1974). The significance of the number of conjugated double bonds is due directly to the mechanism of action of carotenoids in quenching singlet oxygen. The quenching reaction involves a direct transfer of energy between singlet oxygen and the carotenoid to yield the excited triplet state carotenoid and ground state oxygen:

$$^1O_2 + car \longrightarrow {}^3O_2 + {}^3car \tag{5}$$

The triplet carotenoid then dissipates its energy to the surrounding solvent and is thus available to continue the reaction in a cyclic fashion. In order for this reaction to proceed, the transition energy from the ground state of the carotenoid to its first excited triplet state must be near or below that of the singlet oxygen to triplet oxygen transition. Theoretical and experimental evidence shows that this is only the case for carotenoids which have nine or more conjugated double bonds, and these are the efficient singlet oxygen quenchers (Krinsky, 1976).

In addition to quenching singlet oxygen, carotenoids also protect against photodynamic inactivation by quenching excited triplet state sensitizers, thereby preventing the generation of singlet oxygen. Direct quenching of a triplet sensitizer involved in photodynamic inactivation has been demonstrated for bacteriochlorophyll in reaction centres isolated from *R. sphaeroides* (Cogdell, 1978). In the reaction centres of a carotenoid-deficient mutant, excited triplet state bacteriochlorophyll decays at room temperature at a rate which is a thousand times slower than in carotenoid-containing reaction centres. This efficient quenching by carotenoids prevents triplet bacteriochlorophyll from interacting with oxygen and the subsequent production of singlet oxygen.

A range of other defence mechanisms exists for the effective removal of the toxic oxygen species generated by photooxidation reactions. The enzyme superoxide dismutase, which catalyses the disproportionation of superoxide radicals (O_2^-) to molecular oxygen and hydrogen peroxide, has been shown to play a fundamental role in the protection of a wide variety of aerobic and anaerobic microorganisms from the toxic effects of oxygen (Fridovich, 1975; Hewitt and Morris, 1975). Abeliovich *et al.* (1974) have shown that the conditions which lead to photooxidative killing of the cyanobacterium *A. nidulans* also lead to a loss of superoxide dismutase activity. We have described the photoinactivation of superoxide dismutase and catalase in whole cells of *M. aeruginosa* during photoinhibitory treatment (Whitelam and Codd, 1982; Tytler *et al.*, 1984b). Furthermore, Eloff *et al.* (1976) and Steinitz *et al.* (1979) have presented evidence that, in cyanobacteria, resistance to photooxidative killing is associated with the ability of the organisms to maintain superoxide dismutase activity under photooxidative conditions (Fig. 6). Photooxidation-resistant cyanobacteria contain a high proportion of superoxide dismutase which has manganese as the metal prosthetic group whereas the sensitive strains contain mainly superoxide dismutase with iron as the prosthetic group. Iron-containing superoxide dismutases are readily inactivated by hydrogen peroxide (Fridovich, 1975).

Other enzymes which are likely to be important defences against photooxidative damage are catalase and peroxidases, responsible for the removal of hydrogen peroxide. Additionally, compounds which scavenge active oxygen species, such as ascorbate, glutathione, α-tocopherol and histidine, may also provide some defence against photooxidative damage.

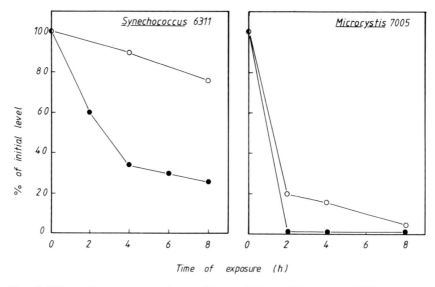

Fig. 6. Effect of exposure to photooxidative field conditions on viability (●)and superoxide dismutase activity (○) in two cyanobacteria. Cells suspensions were immobilized in dialysis tubing to enable exposures to be performed under natural conditions. (Modified from Eloff *et al.* (1976).)

Applied and Economic Aspects

The damaging effects of light on microbes may impose constraints on some applied activities, although these effects are deliberately exploited for useful purposes in other fields. Interest is increasing in the mass cultivation of cyanobacteria and microalgae for the production of food, pigments, lipids, protein, chemical feedstocks and nitrogenous fertilizers and for tertiary waste treatment (Shelef and Soeder, 1980; Ciferri, 1983, Rodriguez-Lopez, 1983). Most of the cyanobacterial–algal mass production schemes using lagoons or shallow tanks are naturally being developed in latitudes receiving high annual irradiance. For example, *Spirulina* spp., which have been harvested from lakes over many centuries as a traditional food in Mexico and North Africa (Ortega, 1972), are being grown on an industrial scale near Mexico City, and mass cultivation of this organism involving modern technology is being developed elsewhere (Ciferri, 1983). Photoinhibition of *S. platensis* has been studied in laboratory cultures and Kaplan (1981) has convincingly shown that the ability of the cells to accumulate inorganic carbon for photosynthesis is reduced. Whether light-induced damage will reduce cell yields on an industrial scale will depend upon reactor design and the relative levels of light, inorganic carbon and oxygen.

The effects of photoinhibition can be seen in several laboratory-scale photobioreactors being developed for useful purposes, e.g. in the photo-chemical fuel cell for hydrogen and electric current production using immobilized *Anabaena* (Kayano et al., 1981). Significantly, photoinhibition has been incorporated recently into a structured, non-segregated computer model for the study of interactions between irradiance, algal physiology and photobioreactor design (Frohlich et al., 1983).

Greater awareness of the lethal effects of sunlight on airborne bacteria has been advocated in view of concern over public health aspects of aerosols produced at wastewater treatment plants and at irrigation schemes using sewage (Fedorak and Westlake, 1978). These workers found that shielding the trapping medium in commercial microbiological air samplers from sunlight during the sampling procedure resulted in threefold to eightfold increases in the bacterial viable counts recovered. Thus, whilst the lethal effects of sunlight may help to reduce the bacterial aerosol hazard, underesti-mations of the numbers of viable microbes being released into such aerosols must arise, unless the trapping media are protected from sunlight.

On the plus side, the lethal action of UV has been used in the food industry for disinfection purposes and the maintenance of sanitary conditions for many years. UV also offers a means of disinfecting drinking water, which is suitable for small-volume or remote rural water supplies with adequate UV transmission properties (Groocock, 1984). A mobile UV-generating unit has also been developed to control microbial growths on stone monuments of historical importance (van der Molen et al., 1980).

Concluding Remarks

Understanding of the damaging effects of light on microorganisms is increasing as a result of observation and research in diverse fields, from the molecular events of DNA damage and repair to the die-off of cyanobacterial blooms in lakes. We have attempted to reflect this breadth of interest in this review. The reduction in the Earth's ozone layer, due to increased human activities and natural events, with the consequent increase in the transmission of solar radiation in the biologically injurious UV-B waveband (see Calkins and Thordardottir, 1980; Worrest et al., 1981) has increased interest in the harmful effects of solar radiation on microbes and we can expect the ecological and applied significance of these effects to increase in the future. Indeed, the approximately equal tolerance and exposure of many aquatic microbes to solar UV (Calkins and Thordardottir, 1980) suggests that there may be no large reservoir of resistance to increased UV, notwithstanding microbial capacity for adaptation. However, the identification of genes involved in bacterial resistance to damage by light and their cloning, amplification and expression (Wakayama *et al.*, 1984) offer exciting prospects for the application of recombinant DNA technology to the mainpulation and control of microbial resistance to the harmful effects of light.

Acknowledgements

Our own work included here was supported by the Natural Environment Research Council. We also acknowledge the valuable contributions made by Dr. E. M. Tytler, whilst at Dundee, Dr. M. F. Hipkins (University of Glasgow) and Professor G. H. Schmid (University of Bielefeld, FRG).

References

Abeliovich, A. and Shilo, M. (1972). Photooxidative death in blue–green algae. *Journal of Bacteriology* **111**, 682–689.

Abeliovich, A., Kellenberg, D. and Shilo, M. (1974). Effect of photooxidative conditions on levels of superoxide dismutase in *Anacystis nidulans*. *Photochemistry and Photobiology* **19**, 379–382.

Ahmad, S. I. (1981). Synergistic action of near ultraviolet radiation and hydrogen peroxide in the killing of coliphage T7: possible role of superoxide radical. *Photobiochemistry and Photobiophysics* **2**, 172–180.

Akazawa, T., Takabe, T., Asami, S. and Kobayashi, H. (1978). Ribulose bisphosphate carboxylase from *Chromatium vinosum* and *Rhodospirillum rubrum* and their role in photosynthetic carbon assimilation. *In* "Photosynthetic Carbon Assimilation" (Eds H. W. Siegelman and G. Hind), pp. 209–226. Plenum, New York.

Albright, L. J. (1980). Photosynthetic activities of phytoneuston and phytoplankton. *Canadian Journal of Microbiology* **26**, 389–392.

Allen, J. F. (1977). Superoxide and photosynthetic reduction of oxygen. *In* "Superoxide and Superoxide Dismutases" (Eds A. M. Michelson, J. M. McCord and I. Fridovich), pp. 417–436. Academic Press, London.

Ananthaswamy, H. N. and Eisenstark, A. (1976). Near-UV-induced breaks in phage DNA sensitization by hydrogen peroxide (a tryptophan photoproduct). *Photochemistry and Photobiology* **24**, 439–442.

Antia, N. J., McAllister, C. D., Parsons, T. R., Stephens, K. and Strickland, K. (1963). Further measurements of primary production using a large-volume plastic sphere. *Limnology and Oceangraphy* **8**, 166–183.

Anwar, M. and Prebble, J. (1977). The photoinactivation of the respiratory chain in *Sarcina lutea* (*Micrococcus luteus*) and protection by endogenous carotenoid. *Photochemistry and Photobiology* **26**, 475–481.

Asada, K. and Kiso, K. (1973). The photooxidation of epinephrime by spinach chloroplasts and its inhibition by superoxide dismutase: evidence for the formation of superoxide radicals in chloroplasts. *Agricultural and Biological Chemistry* **37**, 453–454.

Asami, S. and Akazawa, T. (1978). Photooxidative damage in photosynthetic activities of *Chromatium vinosum*. *Plant Physiology* **62**, 981–986.

Asato, Y. (1972). Isolation and characterization of ultra-violet light sensitive mutants of the blue–green alga *Anacystis nidulans*. *Journal of Bacteriology* **110**, 1058–1064.

Barran, L. R., D'Aoust, J. Y., Labelle, J. L., Martin, W. G. and Schneider, H. (1974). Differential effects of visible light on active transport in *E. coli*. *Biochemical and Biophysical Research Communications* **56**, 522–528.

Bauer, K. and Wild, A. (1976). The effect of blue light on the photosynthetic electron transport in chlorophyll-deficient mutants of *Chlorella fusca*. *Zeitschrift für Pflanzenphysiologie* **80**, 433–454.

Baxter, R. M. and Carey, J. H. (1982). Reactions of singlet oxygen in humic waters. *Freshwater Biology* **12**, 285–292.

Baxter, R. M. and Carey, J. H. (1983). Evidence for photochemical generation of superoxide ion in humic waters. *Nature, London* **306**, 575–576.

Brock, T. D. and Petersen, S. (1976). Some effects of light on the viability of rhodopsin-containing halobacteria. *Archives of Microbiology* **109**, 199–200.

Brodie, A. F., Sutherland, T. O. and Lee, S. H. (1979). The specificity of quinones for restoration of active transport of solutes and oxidative phosphorylation. *In* "Vitamin K Metabolism and Vitamin K-Dependent Proteins" (Ed. J. W. Suttie), pp. 139–202. University Park Press, Baltimore.

Buckley, C. E. and Houghton, J. A. (1976). A study of the effects of near UV radiation on the pigmentation of the blue–green alga *Gloeocapsa alpicola*. *Archives of Microbiology* **107**, 93–97.

Caldwell, M. M. (1979). Plant life and ultraviolet radiation: some perspectives in the history of the earth's UV climate. *Bioscience* **29**, 520–525.

Calkins, J. and Thordardottir, T. (1980). The ecological significance of solar UV radiation on aquatic organisms. *Nature, London* **283**, 563–566.

Carlson, D. J. (1982). Phytoplankton in marine surface microlayers. *Canadian Journal of Microbiology* **28**, 1226–1234.

Carlsson, J. G., Nyberg, G. and Wrethen, J. (1978). Hydrogen peroxide and superoxide radical formation in anaerobic broth media exposed to atmospheric oxygen. *Applied Environmental Microbiology* **36**, 223–229.

Cheng, L., Kellogg, E. W. and Packer, L. (1981). Photoinactivation of catalase. *Photochemistry and Photobiology* **34**, 125–129.

Ciferri, O. (1983). *Spirulina*, the edible microorganism. *Microbiological Reviews* **47**, 551–578.

Codd, G. A. (1972). The photoinhibition of malate dehydrogenase. *FEBS Letters* **20**, 211–214.

Codd, G. A. (1981). Photoinhibition of photosynthesis and photoinactivation of ribulose bisphosphate carboxylase in algae and cyanobacteria. *In* "Plants and the Daylight Spectrum" (Ed. H. Smith), pp. 315–337. Academic Press, London.

Cogdell, R. J. (1978). Carotenoids in photosynthesis. *Philosophical Transactions of the Royal Society of London, Series B* **284**, 569–579.

Cooper, W. J. and Zika, R. G. (1983). Photochemical formation of hydrogen peroxide in surface and ground waters exposed to sunlight. *Science* **220**, 711–712.

Coulombe, A. M. and Robinson, G. G. C. (1981). Collapsing *Aphanizomenon flos-aquae* blooms: possible contributions of photo-oxidation, O_2 toxicity, and cyanophages. *Canadian Journal of Botany* **59**, 1277–1284.

Critchley, C. (1981). Studies on the mechanism of photoinhibition in higher plants. I. Effects of light intensity on chloroplast activities in cucumber adapted to low light. *Plant Physiology* **67**, 1161–1165.

D'Aoust, J. Y., Giroux, J., Barran, L. R., Schneider, H. and Martin, W. G. (1974). Some effects of visible light on *Escherichia coli*. *Journal of Bacteriology* **120**, 799–804.

D'Aoust, J. Y., Martin, W. G., Giroux, J. and Schneider, H. (1980). Protection from visible light damage to enzymes and transport in *Escherichia coli*. *Photochemistry and Photobiology* **31**, 471–474.

Dinsdale, M. T. and Walsby, A. E. (1972). The interrelations of cell turgor pressure, gas-vacuolation, and buoyancy in a blue–green alga. *Journal of Experimental Botany* **23**, 561–570.

Doyle, R. J. and Kubitschek, H. E. (1976). Near ultraviolet light inactivation of an energy-dependent membrane transport system in *Saccharomyces cerevisiae*. *Photochemistry and Photobiology* **24**, 291–293.

Draper, W. M. and Crosby, D. G. (1983). The photochemical generation of hydrogen peroxide in natural waters. *Archives of Environmental Contamination and Toxicology* **12**, 121–126.

Eisenstark, A. (1970). Sensitivity of *Salmonella typhimurium* recombinationless (rec) mutants to visible and near-visible light. *Mutation Research* **10**, 1–6.

Eisenstark, A. (1971). Mutagenic and lethal effects of visible and near-ultraviolet light on bacterial cells. *Advances in Genetics* **16**, 167–198.

Eloff, J. N., Steinitz, Y. and Shilo, M. (1976). Photooxidation of cyanobacteria in natural conditions. *Applied and Environmental Microbiology* **31**, 119–126.

Elstner, E. F. and Oswald, W. (1980). Chlorophyll photobleaching and ethane production in dichlorophenyldimethylurea (DCMU) or paraquat treated *Euglena gracilis* cells. *Zeitschrift für Naturforschung, Teil C* **35**, 129–135.

Epel, B. L. (1973). Inhibition of growth and respiration by visible and near-visible light. *Photophysiology* **8**, 209–229.

Fedorak, P. M. and Westlake, D. W. S. (1978). Effect of sunlight on bacterial survival in transparent air samplers. *Canadian Journal of Microbiology* **24**, 618–619.

Fee, J. A. (1982). Is superoxide important in oxygen poisoning? *Trends in Biochemical Sciences* **7**, 84–86.

Ferron, W. L., Eisenstark, A. and Mackay, D. (1972). Distinction between far- and near-ultraviolet light killing of recombinationless (recA) *Salmonella typhimurium*. *Biochimica et Biophysica Acta* **277**, 651–658.

Forward, R. B. Jr. (1976). Light and diurnal vertical migration: photobehaviour and photophysiology of plankton. *In* "Photochemical and Photobiological Reviews" (Ed. K. C. Smith), Vol I, pp. 157–209. Plenum, New York.

Fridovich, I. (1975). Superoxide dismutases. *Annual Reviews of Biochemistry* **44**, 147–159.

Frohlich, B. T., Webster, A., Ataai, M. M. and Shuler, M. L. (1983). Photobioreactors: models for interaction of light intensity, reactor design and algal physiology. *Biochemical and Bioengineering Symposium* **13**, 331–350.

Gerber, D. W. and Burris, J. E. (1981). Photoinhibiton and P700 in the marine diatom *Amphora* sp. *Plant Physiology* **68**, 699–702.

Giese, A. C. (1968). Ultraviolet action spectra in perspective: with special reference to mutation. *Photochemistry and Photobiology* **8**, 527–546.

Griffiths, M., Sistrom, W. R., Cohen-Bazire, G. and Stanier, R. Y. (1955). Function of carotenoids in photosynthesis. *Nature, London* **176**, 1211–1215.

Groocock, N. H. (1984). Disinfection of drinking water by ultraviolet light. *Journal of the Institution of Water Engineers and Scientists* **38**, 163–172.

Haag, W. R., Hoigne, J., Gassman, E. and Braun, A. M. (1984). Singlet oxygen in surface waters—part I: furfuryl alcohol as a trapping agent. *Chemosphere* **13**, 631–640.

Häder, D.-P. (1979). Photomovement. *In* "Physiology of Movements Encyclopedia of Plant Physiology" (Eds W. Haupt and M. E. Feinleib), Vol. 7, pp. 268–309. Springer, Berlin.

Halliwell, B. (1978). The chloroplast at work. *Progress in Biophysics and Molecular Biology* **33**, 1–54.

Hammer, T. (1983). Limnological studies of the lakes and streams of the upper Qu'Appelle River system, Saskatchewan, Canada. *Hydrobiologia* **19**, 125–144.

Hardy, J. T. and Valett, M. (1981). Natural and microcosm phytoneuston of Sequim Bay, Washington. *Estuarine and Coastal Marine Science* **12**, 3–12.

Harnischfeger, G. and Gaffron, H. (1969). Transient colour sensitivity of the Hill reaction during the disintegration of chloroplasts. *Planta* **89**, 385–388.

Harris, G. P. (1978). Photosynthesis, productivity and growth: the physiological ecology of phytoplankton. *Archiv für Hydrobiologie, Beihilfe Ergebnisse Limnologie* **10**, 1–171.

Harris, G. P. and Piccinin, B. B. (1977). Photosynthesis by natural phytoplankton populations. *Archives of Hydrobiology* **80**, 405–457.

Hartman, P. S. and Eisenstark, A. (1978). Synergistic killing of *Escherichia coli* by near-UV radiation and hydrogen peroxide: distinction between Rec A-repairable and Rec-A-nonrepairable damage. *Journal of Bacteriology* **133**, 769–774.

Harvey, G. W. and Bishop, N. I. (1978). Photolability of photosynthesis in two separate mutants of *Scenedesmus obliquus*. Preferential inactivation of photosystem I. *Plant Physiology* **62**, 330–336.

Haupt, W. (1982). Light-mediated movement of chloroplasts. *Annual Reviews of Plant Physiology* **33**, 205–233.

Hewitt, J. and Morris, J. G. (1975). Superoxide dismutase from some obligately anaerobic bacteria. *FEBS Letters* **50**, 315–318.

Hollaender, A. (1943). Effect of long ultraviolet and short visible radiation (3500–4900 Å) on *Escherichia coli*. *Journal of Bacteriology* **46**, 531–541.

Horrigan, S. G., Carlucci, A. F. and Williams, P. M. (1981). Light inhibition of nitrification in sea-surface films. *Journal of Marine Research* **39**, 557–565.

Ito, A. and Ito, I. (1983). Possible involvement of membrane damage in the inactivation by broad-band near-UV radiation in *Saccharomyces cerevisiae* cells. *Photochemistry and Photobiology* **37**, 395–401.

Jagger, J. (1964). Photoprotection from far ultraviolet effects in cells. *In* "Advances in Chemical Physics" (Ed. J. Duchesne), pp. 584–601. Interscience, New York.

Jagger, J. (1972). Growth delay and photoprotection induced by near-ultraviolet light. *In* "Research Progress in Organic, Biological and Medical Chemistry" (Eds U. Gallo and L. Sautamaria), Vol. 3, pp. 383–401. Elsevier, New York.

Jagger, J. (1981). Near-UV radiation effects on microorganisms. *Photochemistry and Photobiology* **34**, 761–768.

Johnson, F. S., Mo, T. and Green, A. E. S. (1976). Average latitudinal variation in ultraviolet radiation at the earth's surface. *Photochemistry and Photobiology* **23**, 179–188.

Jones, D. T., Robb, F. T. and Woods, D. R. (1980). Effect of oxygen on the survival of *Bacteroides fragilis* after far-UV irradiation. *Journal of Bacteriology* **144**, 1178–1181.

Jørgensen, B. B., Revsbech, N. P. and Cohen, Y. (1983). Photosynthesis and structure of benthic microbial mats: microelectrode and SEM studies of four cyanobacterial communities. *Limnology and Oceanography* **28**, 1075–1093.

Kaplan, A. (1981). Photoinhibiton in *Spirulina platensis*: response of photosynthesis and HCO_3^- uptake capability to CO_2-depleted conditions. *Journal of Experimental Botany* **32**, 669–677.

Kayano, H., Karube, I., Matsunaga, T., Suzuki, S. and Nakayama, O. (1981). A photochemical fuel cell system using *Anabaena* N-7363. *European Journal of Applied Microbiology and Biotechnology* **12**, 1–5.

Keith, G., Rogg, H., Dirheimer, G., Menichi, B. and Heyman, T. (1976). Posttranslational modification of tyrosine tRNA as a function of growth in *Bacillus subtilis*. *FEBS Letters* **61**, 120–123.

Kelland, L. R., Moss, S. H. and Davies, D. J. G. (1983). Recovery of *Escherichia coli* K-12 from near-ultraviolet radiation-induced membrane damage. *Photochemistry and Photobiology* **37**, 617–622.

Kelland, L. R., Moss, S. H. and Davies, D. J. G. (1984). Leakage of $^{86}Rb^+$ after ultraviolet irradiation of *Escherichia coli* K-12. *Photochemistry and Photobiology* **39**, 329–335.

Klamen, D. L. and Tuveson, R. W. (1982). The effect of membrane fatty acid composition on the near-UV (300–400 nm) sensitivity of *Escherichia coli* K 1060. *Photochemistry and Photobiology* **35**, 167–173.

Knowles, A. (1975). The effects of photodynamic action involving oxygen upon biological systems. *In* "Radiation Research: Biochemical, Chemical and Physical Perspectives" (Eds O. F. Nygaard, H. I. Adler and W. K. Sinclair), pp. 612–622. Academic Press, New York.

Kok, B. (1956). On the inhibition of photosynthesis by intense light. *Biochimica et Biophysica Acta* **21**, 234–244.

Kok, B., Gasner, E. and Rurainski, H. J. (1965). Photoinhibiton of chloroplast reactions. *Photochemistry and Photobiology* **4**, 215–227.

Krinsky, N. I. (1976). Cellular damage initiated by visible light. *In* "The Survival of Vegetative Microbes" (Eds T. R. G. Gray and J. R. Postgate), pp. 209–230. Cambridge University Press, Cambridge.

Krinsky, N. I. (1978). Non-photosynthetic function of carotenoids. *Philosophical Transactions of the Royal Society of London, Series B* **284**, 581–590.

Krüger, G. H. J. and Eloff, J. N. (1983a). Effect of CO_2 and HCO_3^- on photosynthetic oxygen evolution by *Microcystis aeruginosa*. *Zeitschrift für Pflanzenphysiologie* **112**, 231–236.

Krüger, G. H. J. and Eloff, J. N. (1983b). Prevention by carbon dioxide of photoinhibition of *Microcystis aeruginosa*. *Zeitschrift für Pflanzenphysiologie* **112**, 237–245.

Krumbein, W. E. and Jens, K. (1981). Biogenic rock varnishes of the Negev Desert (Israel), an ecological study of iron and manganese transformation by cyanobacteria and fungi. *Oecologia* **50**, 25–38.

Kubitschek, H. E. and Doyle, R. J. (1981). Growth delay induced in *Escherichia coli* by near-ultraviolet radiation: relationship to membrane transport functions. *Photochemistry and Photobiology* **33**, 695–702.

Kushner, D. J. (Ed.) (1978). "Microbial Life in Extreme Environments". Academic Press, London, New York and San Francisco, California.

Kyle, D. J., Arntzen, C. J. and Ohad, I. (1983). The herbicide-binding 32KD polypeptide is the primary site of photoinhibition damage. *Plant Physiology Supplements* **72**, 52.

Ley, R. D., Sedita, B. A. and Boye, E. (1978). DNA polymerase I-mediated repair of 365 nm-induced single-strand breaks in the DNA of *Escherichia coli*. *Photochemistry and Photobiology* **27**, 323–327.

van Liere, L. and Walsby, A. E. (1982). Interactions of cyanobacteria with light. *In* "The Biology of Cyanobacteria" (Eds N. G. Carr and B. A. Whitton), pp. 9–45. Blackwell Scientific, Oxford.

Lorenzen, C. J. (1979). Ultraviolet radiation and phytoplankton photosynthesis. *Limnology and Oceanography* **24**, 1117–1120.

Madden, J. J., Boatwright, D. T. and Jagger, J. (1981). Action spectra for modification of *Escherichia coli* B/r menaquinone by near-ultraviolet and visible radiations (313–578 nm). *Photochemistry and Photobiology* **33**, 305–311.

Manzi, J. J., Stefan, P. E. and Dupuy, J. L. (1977). Spatial heterogeneity of phytoplankton populations in estuarine surface microlayers. *Marine Biology* **41**, 29–38.

Marler, J. E. and Van Baalen, C. (1965). Role of H_2O_2 in single-cell growth of the blue–green algae *Anacystis nidulans*. *Journal of Phycology* **1**, 180–184.

Marumo, R., Taga, N. and Nakai, T. (1971). Neustonic bacteria and phytoplankton in surface microlayers of the equatorial waters. *Bulletin of the Plankton Society of Japan* **18**, 36–41.

Mathews-Roth, M. M., Wilson, T., Fujimori, E. and Krinsky, N. I. (1974). Carotenoid chromophore length and protection against photosensitization. *Photochemistry and Photobiology* **19**, 217–222.

McCarron, S. H. and Tu, S.-C. (1983). Flavin-sensitized photodynamic modification of multisubunit proteins. *Photochemistry and Photobiology* **38**, 131–136.

McCord, J. M. and Day, E. D., Jr. (1978). Superoxide-dependent production of hydroxyl radical catalyzed by iron–EDTA complex. *FEBS Letters* **86**, 139–142.

McCormick, J. P., Fischer, J. R., Pochlatko, J. P. and Eisenstark, A. (1976). Characterization of a cell-lethal product from the photo-oxidation of tryptophan: hydrogen peroxide. *Science* **191**, 468–469.

Mitchell, R. L. and Anderson, I. C. (1965). Photoinactivation of catalase in carotenoidless tissues. *Crop Science* **5**, 588–591.

van der Molen, J., Garty, J., Aardema, B. W. and Krumbein, W. E. (1980). Growth control of algae and cyanobacteria on historical monuments by a mobile UV unit (MUVU). *Studies in Conservation* **25**, 71–77.

Moss, S. H. and Smith, K. C. (1981). Membrane damage can be a significant factor in the inactivation of *Escherichia coli* by near-ultraviolet radiation. *Photochemistry and Photobiology* **33**, 203–210.

Nasim, A. and James, A. P. (1978). Life under conditions of high irradiation. *In* "Microbial Life in Extreme Environments" (Ed. D. J. Kushner), pp. 409–439. Academic Press, London, New York and San Francisco, California.

Nelson, D. C. and Castenholz, R. W. (1982). Light responses of *Beggiatoa*. *Archives of Microbiology* **131**, 146–155.

O'Brien, P. A. and Houghton, J. A. (1982a). UV-induced DNA degradation in the cyanobacterium *Synechocystis* PCC 6308. *Photochemistry and Photobiology* **36**, 417–422.

O'Brien, P. A. and Houghton, J. A. (1982b). Photoreaction and excision repair of UV induced pyrimidine dimers in the unicellular cyanobacterium *Gloeocapsa alpicola* (*Synechocystis* PCC 6308). *Photochemistry and Photobiology* **35**, 359–364.

Ogawa, T., Sekine, T. and Aiba, S. (1979). Reappraisal of the so-called light shielding of gas vacuoles in *Microcystis aeruginosa*. *Archives for Microbiology* **122**, 57–60.

Ortega, M. (1972). Study of the edible algae of the Valley of Mexico. *Botanica Marina* **15**, 162–166.

Osmond, C. B. (1981). Photorespiration and photoinhibition: some implications for the energetics of photosynthesis. *Biochimica et Biophysica Acta* **639**, 77–98.

Paerl, H. W. (1984). Cyanobacterial carotenoids: their roles in maintaining optimal photosynthetic production among aquatic bloom forming genera. *Oecologia* **61**, 143–149.

Parkin, T. B. and Brock, T. D. (1980a). The effects of light quality on the growth of phototrophic bacteria in lakes. *Archives of Microbiology* **125**, 19–27.

Parkin, T. B. and Brock, T. D. (1980b). Photosynthetic bacteria production in lakes: the effects of light intensity. *Limnology and Oceanography* **25**, 711–718.

Peak, M. J. and Peak, J. G. (1982). Single-strand breaks induced in *Bacillus subtilis* DNA by ultraviolet light: action spectrum and properties. *Photochemistry and Photobiology* **35**, 675–680.

Peak, J. G., Foote, C. S. and Peak, M. J. (1981). Protection by DABCO against inactivation of transforming DNA by near-ultraviolet light: action spectra and implications for involvement of singlet oxygen. *Photochemistry and Photobiology* **34**, 45–49.

Peak, J. G., Peak, M. J. and Tuveson, R. W. (1983). Ultraviolet action spectra for aerobic and anaerobic inactivation of *Escherichia coli* strains specifically sensitive and resistant to near ultraviolet radiations. *Photochemistry and Photobiology* **38**, 541–543.

Peak, J. G., Peak, M. J. and MacCoss, M. (1984). DNA breakage caused by 334 nm ultraviolet light is enhanced by naturally occurring nucleic acid components and nucleotide coenzymes. *Photochemistry and Photobiology* **39**, 713–716.

Peschek, G. A. and Schmetterer, G. (1978). Reversible photooxidative loss of pigments and of intracytoplasmic membranes in the blue–green alga *Anacystis nidulans*. *FEMS Microbiology Letters* **3**, 295–297.

Peterson, R. B., Friberg, E. E. and Burris, R. H. (1977). Diurnal variation in N_2 fixation and photosynthesis by aquatic blue–green algae. *Plant Physiology* **59**, 74–80.

Powles, S. B. (1984). Photoinhibition of photosynthesis induced by visible light. *Annual Reviews of Plant Physiology* **35**, 14–44.

Powles, S. B., Osmond, C. B. and Thorne, S. W. (1979). Photoinhibition of intact attached leaves of C_3 plants illuminated in the absence of both carbon dioxide and photorespiration. *Plant Physiology* **64**, 982–988.

Powles, S. B., Chapman, K. S. R. and Osmond, C. B. (1980). Photoinhibition of intact attached leaves of C_4 plants: dependence on CO_2 and O_2 partial pressures. *Australian Journal of Plant Physiology* **7**, 737–747.

Ramabhadran, T. V. (1975). Effects of near-ultraviolet and violet radiations (313–405 nm) on DNA, RNA and protein synthesis in *E. coli* B/r: implications for growth delay. *Photochemistry and Photobiology* **22**, 117–123.

Richardson, K., Beardall, J. and Raven, J. A. (1983). Adaptation of unicellular algae to irradiance: an analysis of strategies. *New Phytologist* **93**, 157–191.

Robb, F. T., Hauman, J. H. and Peak, M. J. (1978). Similar spectra for the inactivation by monochromatic light of two distinct leucine transport systems in *Escherichia coli*. *Photochemistry and Photobiology* **27**, 465–469.

Rodriguez-Lopez, M. (1983). Microalgae: their structural organization, uses and applications. *Process Biochemistry* May–June issue, 21–27.

Rowley, D. A. and Halliwell, B. (1982). Superoxide-dependent formation of hydroxyl radicals in the presence of thiol compounds. *FEBS Letters* **138**, 33–36.

Samuelsson, G. and Richardson, K. (1982). Photoinhibition at low quantum flux densities in a marine dinoflagellate *Amphidinium carterae*. *Marine Biology* **70**, 21–26.

Satoh, K. (1970). Mechanism of photoinactivation in photosynthetic systems II. The occurrence and properties of two different types of photoinactivation. *Plant and Cell Physiology* **11**, 29–38.

Schmetterer, G. and Peschek, G. A. (1981). Treatments effecting reversible photo-bleaching and thylakoid degradation in the blue–green alga *Anacystis nidulans*. *Biochemie und Physiologie der Pflanzen* **176**, 90–100.

Scott, W. E., Barlow, D. J. and Hauman, J. H. (1981). Studies on the ecology, growth and physiology of toxic *Microcystis aeruginosa* in South Africa. *In* "The Water Environment: Algal Toxins and Health" (Ed. W. W. Carmichael), pp. 49–69. Plenum, New York.

Setlow, J. K. (1967). The effects of ultraviolet radiation and photoreactivation. *In* "Comprehensive Biochemistry, Photobiology, Ionizing Radiations" (Eds M. Florkin and E. H. Stotz,), pp. 157–203. Elsevier, New York.

Sharma, R. C., Wingo, R. J. and Jagger, J. (1981). Roles of the rel A$^+$ gene and of 4-thiouridine in near-ultraviolet (334 nm) radiation inhibition of induced synthesis of tryptophanase in *Escherichia coli* B/r. *Photochemistry and Photobiology* **34**, 529–533.

Shear, H. and Walsby, A. E. (1975). An investigation into the possible light-shielding role of gas vacuoles in a planktonic blue–green alga. *British Phycological Journal* **10**, 241–251.

Shelef, G. and Soeder, C. J. (Eds) (1980). "Algae Biomass Production and Use." Elsevier–North-Holland, Amsterdam.

Simon, R. D. (1980). Interactions between light and gas vacuoles in *Halobacterium salinarium* strain 5: effect of ultraviolet light. *Applied Environmental Microbiology* **40**, 984–987.

Singh, P. K. (1975). Photoreactivation of UV-irradiated blue–green algae and algal virus LPP-1. *Archives of Microbiology* **103**, 297–302.

Sironval, C. and Kandler, O. (1958). Photooxidation processes in normal green *Chlorella* cells. I. The bleaching process. *Biochimica et Biophysica Acta* **19**, 359–368.

Sistrom, W. R., Griffiths, M. and Stanier, R. Y. (1956). The biology of a photosynthetic bacterium which lacks coloured carotenoids. *Journal of Cellular and Comparative Physiology* **48**, 473–515.

Slade, H. J. K., Jones, D. T. and Woods, D. R. (1981). Effect of oxygen radicals and peroxide on survival after ultraviolet irradiations and liquid holding recovery of *Bacteriodes fragilis*. *Journal of Bacteriology* **147**, 685–687.

Slooten, L. and Sybesma, C. (1976). Photoinactivation of photophosphorylation and dark ATPase in *Rhodospirillum rubrum* chromatophores. *Biochimica et Biophysica Acta* **449**, 565–580.

Smith, R. C., Baker, K. S., Holm-Hansen, O. and Olson, R. (1980). Photoinhibition of photosynthesis in natural waters. *Photochemistry and Photobiology* **31**, 585–592.

Spikes, J. D. (1977). Photosensitization. *In* "The Science of Photobiology" (Ed. K. C. Smith), pp. 87–112. Plenum, New York.

Spikes, J. D. and Livingstone, R. (1969). The molecular biology of photodynamic action: sensitized photooxidation in biological systems. *Advances in Radiation Biology* **3**, 29–121.

Sprott, G. D., Dimock, K., Martin, W. G. and Schneider, H. (1975). Coupling of glycine and alanine transport to respiration in cells of *Escherichia coli*. *Canadian Journal of Biochemistry* **53**, 262–268.

Sprott, G. D., Martin, W. G. and Schneider, J. (1976). Differential effects of near-UV and visible light on active transport and other membrane processes in *Escherichia coli*. *Photochemistry and Photobiology* **24**, 21–27.

Steinitz, Y., Mazor, Z. and Shilo, M. (1979). A mutant of the cyanobacterium *Plectonema boryanum* resistant to photooxidation. *Plant Science Letters* **16**, 327–335.

Sewnson, P. A. and Setlow, R. B. (1970). Inhibition of the induced formation of tryptophanase in *Escherichia coli* by near-ultraviolet radiation. *Journal of Bacteriology* **102**, 815–819.

Taber, H., Pomerantz, J. and Halfenger, G. N. (1978). Near ultraviolet induction of growth delay studied in a menaquinone-deficient mutant of *Bacillus subtilis*. *Photochemistry and Photobiology* **28**, 191–196.

Takahashi, M., Shimura, S., Yamaguchi, Y. and Fujita, Y. (1971). Photo-inhibition of phytoplankton photosynthesis as a function of exposure time. *Journal of the Oceanographical Society of Japan* **27**, 43–50.

Talling, J. F. (1971). The underwater light climate as a controlling factor in the production ecology of freshwater phytoplankton. *Mitteilungen Internationale Vereinigung für Theoretische und Angewandte Limnologie* **19**, 214–243.

Thomas, G. and Favre, A. (1975). 4-thiouridine as the target for near-ultraviolet light induced growth delay in *Escherichia coli*. *Biochemistry and Biophysics Research Communications* **66**, 1454–1461.

Thomson, B. E., Worrest, R. C. and van Dyke, H. (1980). The growth response of an estuarine diatom (*Melosira nummuloides* [DillW.] AG.) to UV-B (290–320 nm) radiation. *Estuaries* **3**, 69–72.

Thorington, I. (1980). Actinic effects of light and biological implications. *Photochemistry and Photobiology* **32**, 117–129.

Tuveson, R. W., Peak, J. G. and Peak, M. J. (1983). Single-strand breaks induced by 365 nm radiation in *Escherichia coli* strains differing in sensitivity to near and far UV. *Photochemistry and Photobiology* **37**, 109–112.

Tyrell, R. M. (1973). Induction of pyrimidine dimers in bacterial DNA by 365 nm radiation. *Photochemistry and Photobiology* **17**, 69–73.

Tyrell, R. M. (1976). RecA$^+$-dependent synergism between 365 nm and ionizing radiation in log-phase *Escherichia coli*: a model for oxygen-dependent near-UV inactivation by disruption of DNA repair. *Photochemistry and Photobiology* **23**, 13–20.

Tyrell, R. M., Webb, R. B. and Brown, M. S. (1973). Destruction of photoreactivating enzyme by 365 nm radiation. *Photochemistry and Photobiology* **18**, 249–254.

Tyrell, R. M., Ley, R. D. and Webb, R. B. (1974). Induction of single-strand breaks (alkali-liable bonds) in bacterial and phage DNA by near-UV (365 nm) radiation. *Photochemistry and Photobiology* **20**, 395–398.

Tytler, E. M., Whitelam, G. C., Hipkins, M. F. and Codd, G. A. (1984a). Photoinactivation of photosystem II during photoinhibition in the cyanobacterium *Microcystis aeruginosa*. *Planta* **160**, 229–234.

Tytler, E. M., Wong, T. and Codd, G. A. (1984b). Photoinactivation *in vivo* of superoxide dismutase and catalase in the cyanobacterium *Microcystis aeruginosa*. *FEMS Microbiology Letters* **23**, 239–242.

Van Baalen, V. (1968). The effects of ultra-violet irradiation on a coccoid blue–green alga: survival, photosynthesis and photoreactivation. *Plant Physiology* **43**, 1689–1695.

Vu, C. V., Allen, L. H., Jr., and Garrard, L. A. (1982). Effects of supplemental UV-B radiation on primary photosynthetic carboxylating enzymes and soluble proteins in leaves of C_3 and C_4 crop plants. *Physiologica Plantarum* **55**, 11–16.

Vu, C. V., Allen, L. H. Jr. and Garrard, L. A. (1984). Effects of enhanced UV-B radiation (280–320 nm) on ribulose-1,5-bisphosphate carboxylase in pea and soybean. *Environmental and Experimental Botany* **24**, 131–143.

Waaland, J. R., Waaland, S. D. and Branton, D. (1970). Gas vacuoles—light shielding in blue–green algae. *Journal of Cell Biology* **48**, 212–215.

Wakayama, Y., Takagi, M. and Yano, K. (1984). Gene responsible for protecting *Escherichia coli* from sodium dodecyl sulfate and toluidine blue plus light. *Journal of Bacteriology* **159**, 527–532.

Walsby, A. E. (1972). Structure and function of gas vacuoles. *Bacteriological Reviews* **36**, 1–32.

Webb, R. B. (1977). Lethal and mutagenic effects of near-ultraviolet radiation. *In* "Photochemical and Photobiological Reviews" (Ed. K. C. Smith), No 2, pp. 169–261. Plenum, New York.

Webb, R. B. and Lorenz, J. R. (1970). Oxygen dependence and repair of lethal effects of near ultraviolet and visible light. *Photochemistry and Photobiology* **12**, 283–289.

Whitelam, G. C. and Codd, G. A. (1982). A rapid whole-cell assay for superoxide dismutase. *Analytical Biochemistry* **121**, 207–212.

Whitelam, G. C. and Codd, G. A. (1983a). Photoinhibition of photosynthesis in the cyanobacterium *Microcystis aeruginosa*. *Planta* **157**, 561–566.

Whitelam, G. C. and Codd, G. A. (1983b). Photoinactivation of *Microcystis aeruginosa* ribulose-1,5-bisphosphate carboxylase: effects of endogenous and added sensitizers and the role of oxygen. *FEMS Microbiology Letters* **16**, 269–272.

Whitelam, G. C. and Codd, G. A. (1984). Photoinhibition of photosynthesis and *in vivo* chlorophyll fluorescence in the green alga *Ankistrodesmus braunii*. *Plant and Cell Physiology* **25**, 465–471.

Wickstrom, C. E. (1984). Depression of *Mastigocladus laminosus* (Cyanophyta) nitrogenase activitiy under normal sunlight intensities. *Journal of Phycology* **20**, 137–141.

Wickstrom, C. E. and Castenholz, R. W. (1978). Association of *Pleurocapsa* and *Calothrix* (Cyanophyta) in a thermal stream. *Journal of Phycology* **14**, 84–88.

Worrest, R. C., Brooker, D. L. and Van Dyke, H. (1980). Results of a primary productivity study as affected by the type of glass in the culture bottles. *Limnology and Oceanography* **25**, 360–364.

Worrest, R. C., Thomson, B. E. and Van Dyke, H. (1981). Impact of UV-B radiation upon estuarine microcosms. *Photochemistry and Photobiology* **33**, 861–867.

Yoakum, G. and Eisenstark, A. (1972). Toxicity of L-tryptophan photoproduct on recombinationless (rec) mutants of *Salmonella typhimurium*. *Journal of Bacteriology* **112**, 653–655.

Zepp, R. G., Wolfe, N. L., Baughman, G. L. and Hollis, R. C. (1977). Singlet oxygen in natural waters. *Nature, London* **267**, 421–423.

Zetterberg, G. (1964). Mutagenic effects of UV and visible light. *In* "Photophysiology", Vol. 3 (Ed. A. C. Giese), pp. 247–281. Academic Press, London and Orlando.

7

Pressure as an Extreme Environment

RICHARD Y. MORITA

Department of Microbiology, College of Science and College of Oceanography, Oregon State University, Corvallis, Oregon 97331-3804, USA

Introduction

Pressure as an environmental factor is expressed in all ecological systems but differs in the degree in which it influences each ecosystem. It can be expressed as hydrostatic pressure, atmospheric pressure, compression or osmotic pressure. The ecosystem that will be addressed in this chapter is mainly the deep sea where the pressure reaches approximately 1100 atm. It is the most studied ecosystem in relation to the pressure factor. For information concerning the other ecosystems where pressure is an environmental factor, Morita (1980a) should be consulted.

The Deep-sea Environment

The rule of thumb is that the hydrostatic pressure in the marine environment increases 1 atm for every 10 m depth. However, in the deepest portions of the ocean, it actually increases more than 1 atm for every 10 m since the sea water is compressed by the water above it, and hence its density increases with depth. The deep sea has been considered a "constant environment" since the salinity does not change to any extent, the oxygen value remains rather steady, the temperature is close to 3–5 °C, the residence time of the water mass is long (measured in terms of hundreds of years) and the organic content is very low. The main variable is the hydrostatic pressure which naturally depends on the depth involved.

The average depth of the oceans is around 3800 m (380 atm) and the deepest known areas are referred to as the hadal region (deeps and trenches).

The trenches represent approximately 1.7% of the oceans. Nevertheless, much of the ocean is dominated by the hydrostatic pressure and it can be a factor in biology just below the water surface (Knight-Jones and Morgan, 1966).

The history of the exploration of the abyssal environment (excluding microbiology) is given by Menzies *et al.* (1973), who provide an excellent review of the fauna that inhabits the deep sea, whereas Zimmerman (1970), Brauer (1972), Kinne (1972), Sleigh and MacDonald (1972), MacDonald (1975) and Rowe (1983) provide the study of pressure from a more physiological viewpoint as well as some of the complication factors in the study of the deep-sea ecosystem such as pressure effects on pH, ionization of salts, etc.

The discovery of the thermal rift vents now adds another dimension to deep-sea microbiology where pressure also plays a dominant role. Thus, in ecosystems that are influenced by hydrostatic pressure, we must deal with the factors of low and high temperature environments as well as the energy availability in each of these environments. In dealing with temperature and pressure, one should always keep in mind the ideal gas law ($PV = nRT$).

Energy Availability

The lack of energy is also an extreme environmental factor and without it organisms will eventually expire. Since the prokaryotes have evolved for a longer time than all other forms of life, the evolutionary processes of bacteria must have taken into consideration the lack of available energy for microbes at one time or another. Thus, microbes must have evolved mechanism(s) for the survival of the species when there is a lack of available energy. One of these mechanisms is the formation of spores or analogous structures. However, there are many bacteria that do not form spores and the survival of these vegetative cells must also be addressed. Available energy is generally lacking in most ecosystems, especially when one addresses the various physiological types of microorganisms present in each ecosystem. As a result of this situation, Morita (1985) states in general that most microbes in nature are in various stages of starvation and starvation is the normal mode of the microbes in most ecosystems.

Organic matter (energy) in the marine environment comes mainly from two sources: land drainage and photosynthesis by the phytoplankton in the surface waters. Land drainage does add quite a bit of organic matter to the near-shore environment, including the trenches which are located near land masses (island arcs). Evidence of organic matter in trenches from terrestrial sources is well documented by Wolff (1970, 1976) and George and Huggins

(1979). Since some of the terrestrial material in the trenches has been identified as coconut husks, branches from trees, etc., the quality (nutrient value) of the energy source must be addressed. For example, in the Pacific Northwest region of the USA, many logs remain in the forest for long periods of time without decomposing to any great extent. These logs may be in streams. Here the energy source is plentiful but the nitrogen and phosphorus sources are lacking for the metabolic activities of the cellulose digestors (Aumen *et al.*, 1983, 1985a,b). An analogous situation occurs in some marine environments where one can measure the organic carbon but this analysis does not give any insight into the other factors in nutrition that microbes must have.

The proteins in the marine environment are probably complexed with lignins and polyphenols; hence they undergo slow degradation (Odum *et al.*, 1979; Martin *et al.*, 1980; Rice, 1982). This would be especially true in near-shore environments since lignins and polyphenols come in with terrestrial run-off. According to Odum *et al.* (1979), as much as 30% of the nitrogen content of aged detritus may exist in the form of non-protein nitrogen compounds as evidenced by their litter bag experiments. They suggest that nitrogen-containing compounds such as amino sugars (glucosamine and chitin), phenol–protein, protein–lignin, protein–chitin, complexes of inorganic clays and amino groups, and nitrogen-containing humic acids are resistant to chemical degradation and therefore are not readily digested and assimilated by many detritivores. Thus, they raise the question as to the usefulness of the detrital C:N ratios as food value indicators. If the nitrogen complexes are not readily chemically degraded, then it holds that the bacterial degradation processes are also inefficient in degrading these protein complexes. During the decomposition, detritus becomes richer in phenolic and carbohydrate groups which may form condensation products with amino acids, yielding precursors to complex nitrogenous humic geopolymers, and biological availability of the humic nitrogen probably depends upon the extent to which proteinoid subunits are retained in the humic macromolecular structure (Rice, 1982).

Many marine biologists consider proteins to be rather resistant to bacterial decomposition. When proteins may suddenly appear (dead animals, etc.), the surrounding immediate environment does not have a readily available energy source. To produce the necessary exoenzymes to degrade the proteins takes much energy since it takes $14\,000\ \text{cal mol}^{-1}$ to acylate one amino acid to the tRNA. It takes a very large number of amino acids to make an exoenzyme. Without the presence of methionine, one does not get translation from the DNA to synthesize the exoenzyme (protein).

There is much literature concerning the organic content of sea water and all reports state that the concentration is low. For instance, Menzel and

Ryther (1970) report the values for particulate organic carbon (POC) and dissolved organic carbon (DOC) to be between 3 and 10 µg carbon l^{-1} and between 0.35 and 70 mg carbon l^{-1} respectively. It should also be noted that some bacteria are present in both the POC and the DOC since some bacteria can pass through or be retained by the use of the Whatman GF/C filter. This filter has a retention of 1.2 µm and is arbitrarily used to make the distinction between POC and DOC. No one has yet determined the DOC content of sea water making certain to exclude the microbes since some of them are ultramicrocells (see below). Furthermore, there is considerable literature concerning the recalcitrant nature of the DOC in deep water. Sheldon *et al.* (1967) determined that less than 5% of the DOC in a seawater sample is utilized by bacterial growth.

It is recognized that particles in the ocean can sink rather rapidly (hundreds to thousands of metres per day) (McCave, 1984). However, it should be pointed out that most of the readily utilizable compounds have already been used by the associated bacteria in and/or on the particles and therefore the more recalcitrant material is left. When starved cells of ANT-300 (a marine psychrophilic bacterium) are exposed to radioactive glutamic acid (a readily utilizable organic compound), the organism can metabolize it when it is in a concentration of 10^{-12} M (Morita, 1984). However, in order to utilize any polymeric compounds, microbes must be capable of degrading the compound first before its subunits can be transported into the cell to be used as an energy source. Thus, there must be sufficient energy to allow the cell to biosynthesize the exoenzyme and the cell must probably wait until it accumulates sufficient energy to synthesize the exoenzyme. If the polymer degrades to subunits that can be passively transported into the cell, then the cell does not need extra energy for the transport process, whereas, if active transport is involved, then extra energy must be expended for the cell to utilize it. Approximately 20% of the energy the cell utilizes is for transport of the substrate (Stouthamer, 1973).

Although there is not a large quantity of organic matter in the deep sea, there are probably mechanisms which the bacteria employ in order to utilize it. The concept of substrate capture has been put forth by Morita (1984), which is based on the work in this laboratory (Griffiths *et al.*, 1974; Geesey and Morita, 1979, 1981). Essentially, the concept is based on the function of the binding proteins of the cell, which bind the substrate to the cell so that it can be utilized.

The quality of the organic matter in the deep sea was initially questioned by Morita (1979), especially in terms of the amino acids present. Lee and Bada (1975) did not report the presence of methionine in deep water. Yet methionine is essential for the biosynthesis of protein in any cell since it is needed for translation from the genome of the cell (via transcription) to synthesize proteins (including enzymes). The role of the essential amino acids in any organism is also important (Morita, 1980b).

In addition to the quality (essential amino acids, vitamins, readily utilizable substrates, etc.) and quantity of the organic matter in the deep sea, one must take into consideration the fact that the residence time of the deep water mass can approach 1000 years (Broecker, 1963). The organic matter is also old and Williams et al. (1969) indicate that it is approximately 3400 years old in water samples taken from the North Pacific from a depth of approximately 1800 m. The turnover time of the organic matter was calculated to be 3300 years (Menzel, 1974) whereas the oxygen utilization was calculated to be 0.004 ml l^{-1} year^{-1} (Craig, 1971). Because of the scarce amount of organic matter, the average size of ascidians decreased from more than 1 cm in the neritic zone to 2.5 mm in the abyssal plains (Monniot, 1979), which is equivalent to a weight decrease of 25 to 1. Dwarfism also occurs among the higher organisms in the deep sea (Allen, 1979). To illustrate further the lack of energy, it was found that a bivalve (*Astarte borealis*) required 50–60 years for gonad development and 100 years to reach 8 mm in size.

Tabor et al. (1981) found that many of the deep-sea isolates were ultramicrocells. These ultramicrocells developed into "normal" size cells when placed on media. Previously, Carlucci et al. (1976) described the existence of ultramicrocells in deep waters. Ultramicrocells are the result of nutrient deprivation (Novitsky and Morita, 1976). Many investigators have reported the presence of ultramicrocells in marine water and soils. They can be found in freshwater environments also. The presence of ultramicrobacteria is a good indication of the starved condition of the organisms but not all bacteria form ultramicrocells when starved. Starved cells also appear to have a better ability to adhere to surfaces which may aid their ability to obtain substrates in a dilute concentration (Dawson et al., 1981). The concept of starvation survival is reviewed by Morita (1982).

Barophilic Bacteria

The Mid-Pacific Expedition offered the first chance after World War II to investigate the deep sea, mainly the abyssal plains of the ocean. Morita and ZoBell (1955) reported the occurrence of bacteria in various types of oozes and red clay but the numbers of bacteria were very low, probably as a result of the lack of organic matter to support the growth of the organisms present. Another indication of the low organic content of the sediment is reflected in the oxidized nature of the sediment (redox reading of + 200 to + 300 mV). However, the finding of bacteria in the lower varves of the sediment cores suggest that these bacteria were dormant for many years. The microbiology initiated during this expedition was the first to address the biology of the deep sea after World War II. During the Danish Galathea Deep-sea Expedition of 1950–1952, ZoBell and Morita (1957, 1959) reported the existence of barophilic bacteria. The majority of barophiles were associated with the gut and surfaces of deep-sea animals (ZoBell and Morita, 1957). Hessler et al. (1978) found that there was an unusually large number of bacteria in the gut of the abyssal scavenging amphipod obtained from the bottom of the Philippine Trench. The gut samples were formalin fixed so there was no isolation of the bacteria. Unfortunately, no barophilic bacteria were isolated in pure culture by the above investigators. However, an enrichment culture of sulphate reducers was cultured for approximately 2 years (ZoBell and Morita, 1959). A heterotrophic diatom was also reported from sediment taken from the Weber Trench (ZoBell and Morita, 1959).

There have been many studies dealing with non-barophilic bacteria in relation to the inhibitory pressure on protein synthesis, DNA and RNA synthesis, enzyme reactions and growth (Johnson et al., 1954; Zimmerman, 1970; Brauer, 1972; Morita, 1972; Marquis, 1976; Landau and Pope, 1980). However, one of the more important aspects of pressure effects on non-barophiles is the ability of ANT-300 cells to survive hydrostatic pressure only when in the starved state. This probably helps to explain why one can isolate non-barophiles in upwelled waters, since the time period between the convergence (down welled) and divergence (up welled) of a water mass may be considerable. Hence, survival of the species is ensured. In contrast, Matsumoto and Marquis (1977) reported that Streptococcus faecalis has the ability to grow at 900 atm when the composition of the medium is changed but, in marine waters, the energy just is not available to the organism to combat the perturbing force.

Renewed interest in the microbiology of the deep sea occurred when Jannasch et al. (1971) reported that the various foods aboard the sunken submersible Alvin, when retrieved approximately 1 year later, were in good condition. This lead Jannasch and Wirsen (1973) to investigate the in situ pressure effects on the ability of the indigenous microflora to degrade various substrates. In all cases, the degradation rate was very slow, which lead Morita (1979a,b) to describe the event as a "blessing in disguise". A fast rate of microbial activity would utilize the indigenous concentration of oxygen as well as taking from the system the organic matter needed by the higher organisms in the deep sea. All the organisms present in the deep sea were in tune with the various factors present and any further depletion of their energy source could result in their death.

The isolation of bacteria from the deep sea only indicates that many can be classified as non-barophiles since their ability to grow at elevated pressures is less than that at 1 atm. This only indicates that these non-barophiles have the ability to survive the rigours of the deep-sea environment. However, there are degrees of barophilism ranging from barophiles that can tolerate pressure to those that absolutely require pressure to grow. Schwartz *et al.* (1976) isolated several barophilic bacteria from the intestinal tract of an amphipod obtained from the bottom of the Aleutian Trench. One of the isolates had the ability to grow just as well at elevated pressures (710 atm) as at 1 atm but there was a short lag in the growth at 710 atm. This suggests that some adaptation to pressure was needed before growth could occur at the same rate as growth at 1 atm. Ultramicrocells, owing to the lack of available energy, were also observed in the deep sea by Tabor *et al.* (1981) and were identified as *Alcaligenes, Flavobacterium, Pseudomonas* and *Vibrio*. Bianchi *et al.* (1979) estimated that the abundance of bacteria in the intestinal tract was 50 times greater than in the sediment. This finding only helps to support the idea that the sediment is lacking in available energy whereas the gut of the animals has a greater abundance of energy for the growth and multiplication of the bacteria. Barophiles associated with the gut of deep-sea animals are reviewed by Deming and Colwell (1981).

In contrast, bacteria which could only grow under elevated pressures but not at 1 atm were discovered by Yayanos *et al.* (1981). The organism was isolated from a decomposing amphipod held under elevated pressure. Silica gel was employed (Dietz and Yayanos, 1978) in the isolation of the barophile since barophiles are now known to be psychrophiles according to the definition of Morita (1975) and short exposures to elevated temperature can kill psychrophiles. Yayanos *et al.* (1981) consider barophiles to be extreme psychrophiles as well, on the basis of the equivalence between temperature and pressure. The temperature for barophiles would then be $-40\,°C$ when correcting for the equivalent pressure. Decompression of barophiles will also bring about death of barophiles that require pressure for growth (Yayanos and Dietz, 1983). Survival due to starvation of a barophile takes place better at elevated pressures than at 1 atm (Fig. 1). Both of the foregoing situations indicate that pressure is preferred by the barophiles. However, we have yet to probe the physiological and biochemical properties of the barophiles to determine why a barophile is a barophile.

Microbiology of the Hydrothermal Vent

An expedition led by Corliss (Corliss *et al.*, 1979) of Oregon State University discovered the existence of thermal vents in the Galapagos Rift area. In some of these thermal vent areas, black smokers were present that extruded quantities of hydrogen sulphide as well as hot sea water. The surprising item,

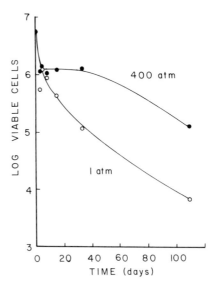

Fig. 1. Starvation survival of barophile JD-4 (UM-40 B) at 1 and 400 atm. Culture provided by J. W. Deming (Pledger and Morita, unpublished data).

aside from the geological aspect, was the large number of macroorganisms present. The Galapagos thermal vents occur at a depth of about 2800 m (280 atm). Recognizing that the energy from the surface was not sufficient to support such a large quantity of life forms and that there was not enough energy in the POC and DOC, it was postulated that the energy for these higher forms was the bacteria present. Bacteria, being extremely nutritious (vitamins, phosphate from their nucleic acids, proteins, fatty acids, etc.), became the only logical choice. Thus, when energy is available the bacteria in the thermal vent area evolve into that ecosystem. In this case, it is the chemolithotrophic bacteria that can utilize the hydrogen sulphide and other reduced compounds present in the thermal vent ecosystem. Photosynthetically derived fallout from the surface is much too low to support the high biomass of the vent animals (Enright *et al.*, 1981; Rau, 1981; Williams *et al.*, 1981; Desbruyeres and Laubier, 1983) and dead animals were found around extinct vents (Corliss *et al.*, 1979). Therefore, bacteria utilizing energy from reduced sulphur compounds, and not the photosynthetic organisms, constitute the first trophic level in this unique ecosystem. As Morita (1980c) concluded, all eukaryotic organisms evolved in a prokaryotic world and therefore the bacteria probably provided the food for many of the higher forms during evolutionary processes on the Earth.

The vesicomyid clams (Turekian *et al.*, 1979; Turekian and Cochran, 1981) and the mytilid mussel (Rhoads *et al.*, 1981) were judged to have a growth rate of 4 cm year^{-1} and 1 cm year^{-1} respectively, probably as a result of the supply of bacterial cells. These growth rates are anywhere from 120 to 500 times faster than that of Tindaria and faster than those of some shallow-water bivalves (Jones, 1983). *Riftia pachyptila* was found to have a chemo-lithotrophic bacterial symbiont in its tissue (Cavanaugh *et al.*, 1981). This plume tissue of the tube worm is insensitive to sulphide (Arp and Childress, 1983) and the blood of the worm contains a sulphide-binding protein that appears to concentrate sulphide from the environment and may function to transport the sulphide to the internal endosymbiotic bacteria contained in the coelomic organ (Powell and Somero, 1983). Apparently the sulphide protein prevents sulphide poisoning as well as preventing sulphide from inhibiting the cytochrome c oxidase. Both *R. pachyptila* and *Calyptogena magnifica* appear to represent 90% of the macroorganisms in the vents and both appear to depend on symbiotic bacteria (Jannasch, 1983). Gastropods graze on microbial mats (Desbruyeres and Laubier, 1983) and suspension-feeding bivalves and planktonic crustacea may filter bacteria from the water (Jannasch, 1983).

Ruby *et al.* (1981) found three distinct physiological types of sulphur-oxidizing bacteria which were *Thiomicrospira*- and *Thiobacillus*-like organisms and *Pseudomonas*-like heterotrophs. They also observed microbial mats which resembled *Beggiatoa* and *Thiothrix*. An anaerobic *Spirochaeta* sp. was later identified by Harwood *et al.* (1982).

A controversy developed concerning the growth of bacteria from the hydrothermal vent at elevated temperatures. Using material taken from a temperature of 350 °C from a black smoker chimney, Baross and Deming (1983) were able to culture microorganisms at 250 °C and 265 atm. In this enrichment culture there appear to be two different morphological types of bacteria. The doubling times for the bacteria were 8 hours at 150 °C, 1.5 hours at 200 °C and 40 minutes at 250 °C under pressure. At 250 °C and 265 atm, there was an orderly increase in protein as well as in the number of bacteria. Trent *et al.* (1984), using the same medium, etc. but no bacteria, reported that bacterial growth at 250 °C may be due to artefacts produced in the medium and to contaminants introduced during the sample processing. However, the amount of protein, etc. does not approach the values reported by Baross *et al.* (1984). They demonstrated an orderly increase in all the various amino acids with incubation time in concentrations greater than that shown by Trent *et al.* (1984). The present author, after reviewing the data obtained by Baross and Becker (colleagues), concludes that Trent *et al.* (1984) may be in error. Artefacts probably do exist but they cannot account for the results obtained by Baross and Deming (1983). Baross *et al.* (1982)

demonstrated that some organisms from the hydrothermal vents could also produce methane, hydrogen, carbon monoxide and dinitrogen oxide when incubated at 1 atm. Further research will resolve this controversy.

Quite some time ago, enzyme reactions were reported to take place at above 100 °C under pressure (Morita and Haight, 1962; Morita and Mathemeier, 1964). In these experiments, it was shown that the molecular volume increases owing to temperature increases; where offset by the pressure, the latter decreases the molecular volume. The presence of the substrate as well as the metal cofactors was also important in helping to offset the increased temperature effects. Growth at elevated temperatures also helps to confirm Brock's (1967) hypothesis that "life is possible at any temperature at which there is liquid water" and the pressure in the hydrothermal vent area keeps the sea water in a liquid phase.

The most interesting hypothesis coming out of the thermal vent studies deals with the site of the origin of life (Corliss et al., 1981). This hypothesis is based on the fact that the hydrothermal vents have the reducing conditions, the thermal energy to synthesize organic compounds abiologically, the presence of carbon monoxide, hydrogen, nitrogen, methane, appropriate catalytic surfaces (Fe–Mg clay minerals) and metals, a mixing gradient of temperature and composition, and continuous flow that removes products from the sites of reactions. This idea has appeal to the author because it demonstrates that bacteria can serve as the first trophic level (which may have occurred during evolution since all eukaryotic organisms evolved in a prokaryotic world). It also demonstrates the interaction of bacteria with higher forms during their evolutionary development.

Conclusion

The research on hydrostatic pressure as an environmental extreme has only begun. We are faced not only with instrumentation problems but also with the problem of hydrostatic pressure–low temperature interactions and hydrostatic pressure–high temperature interactions in the study of the deep-sea environment. This is coupled with the low concentration of available energy in the pressure–low temperature system and sufficient energy (mainly hydrogen sulphide) in the pressure–high temperature system. The latter only indicates that energy is the main environmental factor that dictates the rate at which microbes function in any ecosystem of the biosphere.

References

Allen, J. A. (1979). The adaptations and radiation of deep-sea bivalves. *Sarsia* **64**, 19–27.

Arp, A. A. and Childress, J. J. (1983). Sulfide binding by the blood of the hydrothermal vent tube worm, *Riftia pachyptila. Science* **219**, 295–297.

Aumen, N. G., Bottomley, P. J., Ward, G. M. and Gregory, S. V. (1983). Microbial decomposition of wood in streams: distribution of microflora and factors affecting [^{14}C]lignocellulose mineralization. *Applied and Environmental Microbiology* **46**, 1409–1416.

Aumen, N. G., Bottomley, P. J. and Gregory, S. V. (1985a). Impact of nitrogen and phosphorus on [^{14}C]lignocellulose decomposition by stream wood microflora. *Applied and Environmental Microbiology* **49**, 1113–1118.

Aumen, N. G., Bottomley, P. J. and Gregory, S. V. (1985b). Nitrogen dynamics in stream wood samples incubated with [^{14}C]lignocellulose and [^{15}N]potassium nitrate. *Applied and Environmental Microbiology* **49**, 1119–1123.

Baross, J. A. and Deming, J. W. (1983). Growth of 'black smoker' bacteria at a temperature of at least 250 °C. *Nature, London* **303**, 423–426.

Baross, J. A., Lilley, M. D. and Gordon, L. I. (1982). Is the CH_4, H_2 and CO venting from submarine hydrothermal systems produced by thermophilic bacteria? *Nature, London* **298**, 366–368.

Baross, J. A., Deming, J. W. and Becker, R. R. (1984). Evidence for microbial growth in high-pressure, high-temperature environments. *In* "Current Perspectives in Microbial Ecology" (Eds M. J. Klug and C. A. Reddy), pp. 186–195. American Society for Microbiology, Washington D.C.

Bianchi, A., Scoditti, P.-M. and Bensoussan, M. G. (1979). Distribution des populations bactériennes hétérotrophes dans les sédiments et les tractus digestifs d'animaux benthiques recueillis dans la famille Verma et les planes abyssales du Demerara et de Gambie. *In* "Vie Mer", Vol. I, pp. 7–12. Fondation Oceanographique Richard, Marseilles.

Borecker, W. (1963). Radioisotopes and large-scale organic mixing. *In* "The Sea" (Ed. M. N. Hill), Vol. 2, pp. 88–108. Interscience, New York, London and Sydney.

Brauer, R. W. (1972). "Barobiology and the Experimental Biology of the Deep Sea". University of North Carolina, Chapel Hill, North Carolina.

Brock, T. D. (1967). Life at high temperatures. *Science* **158**, 1012–1019.

Carlucci, A. F., Shimp, S. L., Jumars, P. A. and Paerl, H. W. (1976). *In situ* morphologies of deep-sea and sediment bacteria. *Canadian Journal of Microbiology* **22**, 1667–1671.

Cavanaugh, C. M., Gardiner, S. L., Jones, M. L., Jannasch, H. W. and Waterbury, J. B. (1981). Prokaryotic cell in the hydrothermal vent tube worm *Riftia pachyptila* Jones: possible chemoautotrophic symbionts. *Science* **213**, 340–342.

Corliss, J. B., Dymond, J., Gordon, L. I., Edmond, J. M., von Herzen, R. P., Ballard, R. D., Green, K., Williams, D., Bainbridge, A., Crane, K. and van Andel, T. H. (1979). Submarine thermal springs on the Galapagos Rift. *Science* **203**, 1073–1083.

Corliss, J. B., Baross, J. A. and Hoffman, S. E. (1981). An hypothesis concerning the relationship between submarine hot springs and the origin of life on Earth. *Oceanologia Acta Special Report* 56–69.

Craig, H. (1971). The deep metabolism: oxygen consumption in abyssal ocean water. *Journal of Geophysical Research* **72**, 5078–5086.

Dawson, M. P., Humphrey, B. and Marshall, K. C. (1981). Adhesion: a tactic in survival strategy of a marine vibrio during starvation. *Current Microbiology* **6**, 195–201.

Deming, J. W. and Colwell, R. R. (1981). Barophilic bacteria associated with deep-sea animals. *Bioscience* **13**, 507–511.

Desbruyeres, D. and Laubier, L. (1983). Primary consumers from hydrothermal vents and animal communities. *In* "Hydrothermal Processes at Seafloor Spreading Centers" (Eds P. A. Rona, K. Bostrom, L. Laubier and K. L. Smith, Jr.), pp. 711–733. Plenum, New York.

Dietz, A. S. and Yayanos, A. A. (1978). Silica gel media for isolating and studying bacteria under hydrostatic pressure. *Applied and Environmental Microbiology* **36**, 966–968.

Enright, J. T., Newman, W. A., Hessler, R. R. and McGowan, J. A. (1981). Deep-ocean hydrothermal vent communities. *Nature, London* **289**, 219–221.

Geesey, G. G. and Morita, R. Y. (1979). Capture and uptake of arginine at low concentrations by a marine psychrophilic bacterium. *Applied and Environmental Microbiology* **38**, 1092–1097.

Geesey, C. G. and Morita, R. Y. (1981). Relationship of cell envelope stability to substrate capture in a marine psychophilic bacterium. *Applied and Environmental Microbiology* **42**, 533–540.

George, R. Y. and Huggins, R. P. (1979). Eutrophic hadal benthic community in the Puerto Rico Trench. *Ambio Special Report* **6**, 51–58.

Griffiths, R. P., Baross, J. A., Hanus, F. J. and Morita, R. Y. (1974). Some physical and chemical parameters affecting the formation and retention of glutamate pools in a marine psychrophilic bacterium. *Zeitschrift für Allgemeine Mikrobiologie* **14**, 359–369.

Harwood, C. S., Jannasch, H. W. and Canale-Parola, E. (1982). Anaerobic spirochaete from a deep-sea hydrothermal vent. *Applied and Environmental Microbiology* **44**, 234–237.

Hessler, R. R., Ingram, C. L., Yayanos, A. A. and Burnett, B. R. (1978). Scavenging amphipods from the floor of the Philippine Trench. *Deep Sea Research* **25**, 1029–1048.

Jannasch, H. W. (1983). Microbial processes at deep sea hydrothermal vents. *In* "Hydrothermal Processes at Seafloor Spreading Centers" (Eds P. A. Rona, K. Bostrom, L. Laubier and K. L. Smith, Jr.), pp. 677–709. Plenum, New York.

Jannasch, H. W. and Wirsen, C. O. (1973). Deep-sea microorganisms: *in situ* response to nutrient enrichment. *Science* **180**, 641–643.

Jannasch, H. W., Eimhjellen, K., Wirsen, C. O. and Farmanfarmaian, A. (1971). Microbial degradation of organic matter in the deep sea. *Science* **171**, 672–675.

Johnson, F. H., Eyring, H. and Polissar, M. J. (1954). "The Kinetic Basis of Molecular Biology". Wiley, New York.

Jones, D. S. (1983). Schlerochronology: reading the record of the molluscan shell. *American Scientist* **71**, 384–391.

Kinne, O. (1972). "Marine Ecology", Vol. 1, Part 3. Wiley–Interscience, London.

Knight-Jones, E. W. and Morgan, E. (1966). Responses of marine animals to changes in hydrostatic pressure. *Oceanography Marine Biology Annual Review* **4**, 267–999.

Landau, J. V. and Pope, D. H. (1980). Recent advances in the area of barotolerant protein synthesis in bacteria and implications concerning barotolerant and barophilic growth. *Advances in Aquatic Microbiology* **2**, 49–76.

Lee, C. and Bada, J. L. (1975). Amino acids in equatorial Pacific Ocean warer. *Earth and Planetary Science Letters* **26**, 61–68.

MacDonald, A. G. (1975). "Physiological Aspects of Deep Sea Biology". Cambridge University Press, Cambridge.

Marquis, R. E. (1976). High-pressure microbial physiology. *Advances in Microbial Physiology* **14**, 159–241.

Martin, M. M., Martin, J. S., Kukor, J. J. and Merritt, R. W. (1980). The digestion of protein and carbohydrate by the steam detritivore, *Tipula abdominalis* (Diptera, Tipulidae). *Oecologia* **46**, 360–364.

Matsumoto, P. and Marquis, R. E. (1977). Energetics of streptococcal growth inhibition by hydrostatic pressure. *Applied and Environmental Microbiology* **33**, 885–892.

McCave, I. N. (1984). Size spectra and aggregation of suspended particles in the deep ocean. *Deep Sea Research* **31**, 329–352.

Menzel, D. W. (1974). Primary productivity, dissolved and particulate organic matter, and the sites of oxidation of organic matter. *In* "The Sea" (Ed. E. D. Goldberg), Vol. 5, pp. 659–678. Wiley–Interscience, New York.

Menzel, D. W. and Ryther, J. H. (1980). Distribution and cycling of organic matter in the oceans. *In* "Organic Matter in Natural Waters" (Ed. D. W. Hood), pp. 31–54. Institute of Marine Sciences Publications, College, Alaska.

Menzies, R. J., George, R. Y. and Rowe, G. T. (1973). "Abyssal Environment and Ecology of the World Oceans". Wiley, New York.

Monniot, C. (1979). Adaptation of benthic filtering animals to the scarcity of suspended particles in deep water. *Ambio Special Report* **6**, 73–74.

Morita, R. Y. (1972). Pressure—bacteria, fungi, and blue green algae. *In* "Marine Ecology—Environmental Factors" (Ed. O. Kinne), Vol. 1, Part 3. Interscience, London.

Morita, R. Y. (1975). Psychrophilic bacteria. *Bacteriological Reviews* **39**, 144–167.

Morita, R. Y. (1979a). The role of microbes in the bioenergetics of the deep-sea. Proceedings of the Centenary Symposium of the Kristineberg Marine Biological Laboratory. *Sarsia* **64**, 9–12.

Morita, R. Y. (1979b). Current status of the microbiology of the deep sea. *Ambio Special Report* **6**, 33–36.

Morita, R. Y. (1980a). Biological limits of temperature and pressure. *In* "Limits of Life" (Eds C. Pannamperuma and L. Margulis), pp. 25–32. Reidel, Dordrecht.

Morita, R. Y. (1980b). Microbial life in the deep sea. *Canadian Journal of Microbiology* **26**, 1375–1385.

Morita, R. Y. (1980c). Microbial contributions to the various trophic levels. *In* "Ponecias del Simposio International en Resisitancia a los Antibioticos y Microbiologia Marina, Celebardo en el VI Congreso Nacional Microbiologia", pp. 159–174. Santiago de Compostella.

Morita, R. Y. (1982). Starvation-survival of heterotrophs in the marine environment. *Advances in Microbial Ecology* **6**, 171–198.

Morita, R. Y. (1984). Substrate capture by marine heterotrophic bacteria in low nutrient waters. *In* "Heterotrophic Activitiy in the Sea" (Eds J. E. Hobbie and P. J. le B. Williams), pp. 83–100. Plenum, New York.

Morita, R. Y. (1985). Starvation and miniaturisation of heterotrophs, with special emphasis on maintenance of the starved viable state. *In* "Bacteria in Natural

Environments; the Effect of Nutrient Conditions" (Ed. M. Fletcher), pp. 111–130. Academic Press, London.

Morita, R. Y. and Haight, R. D. (1962). Malic dehydrogenase activity at 101 °C under hydrostatic pressure. *Journal of Bacteriology* **83**, 1341–1346.

Morita, R. Y. and Mathemeier, P. F. (1964). Temperature–hydrostatic pressure studies on partially purified inorganic pyrophosphatase. *Journal of Bacteriology* **88**, 1667–1671.

Morita, R. Y. and ZoBell, C. E. (1955). Occurrence of bacteria in pelagic sediments collected during the Mid-Pacific Expedition. *Deep Sea Research* **3**, 66–73.

Novitsky, J. A. and Morita, R. Y. (1976). Morphological characterization of small cells resulting from nutrient starvation of a psychrophilic marine vibrio. *Applied and Environmental Microbiology* **32**, 619–622.

Odum, W. E., Kirk, P. W. and Zieman, J. C. (1979). Non-protein nitrogen compounds associated with particles of vascular plant detritus. *Oikos* **32**, 363–367.

Powell, M. A. and Somero, G. N. (1983). Blood components prevent sulfide poisoning of respiration of the hydrothermal vent tube worm *Riftia pachyptila*. *Science* **219**, 297–299.

Rau, G. H. (1981). Low $^{15}N/^{14}N$ in hydrothermal vent animals: ecological implications. *Nature, London* **279**, 484–485.

Rhoads, D. C., Lutz, R. A., Revelas, E. and Cerrato, R. M. (1981). Growth of bivalves in the deep-sea hydrothermal vents along the Galapagos Rift. *Science* **214**, 911–913.

Rice, D. L. (1982). The detritus nitrogen problem: new observations and perspectives from organic geochemistry. *Marine Ecology Progress Series* **9**, 153–162.

Rowe, G. T. (1983). "Deep-sea Biology". John Wiley and Sons, New York.

Ruby, E. G., Wirsen, C. O. and Jannasch, H. W. (1981). Chemolithotrophic sulfur-oxidizing bacteria from the Galapagos Rift hydrothermal vents. *Applied and Environmental Microbiology* **42**, 317–324.

Schwartz, J. R., Yayanos, A. A. and Colwell, R. R. (1976). Metabolic activities of the intestinal microflora of a deep-sea invertebrate. *Applied and Environmental Microbiology* **31**, 46–48.

Sheldon, R. W., Evelyn, T. O. T. and Parsons, T. R. (1967). On the occurrence and formation of small particles in seawater. *Limnology and Oceanography* **12**, 367–375.

Sleigh, M. A. and MacDonald, A. G. (1972). "The Effects of Pressure on Organisms". Academic Press, New York.

Southamer, A. H. (1973). A theoretical study of the amount of ATP required for synthesis of microbial cell material. *Antonie van Leeuwenhoek; Journal of Microbiology and Serology* **39**, 545–565.

Tabor, P. S., Ohwada, K. and Colwell, R. R. (1981). Filterable marine bacteria found in the deep sea: distribution, taxonomy and response to starvation. *Microbial Ecology* **7**, 67–83.

Trent, J. D., Chastain, R. A. and Yayanos, A. A. (1984). Possible artefactual basis for apparent growth at 250 °C. *Nature, London* **307**, 737–740.

Turekian, K. K. and Cochran, J. K. (1981). Growth rates of a vesicomyid clam from the Galapagos Apreada center. *Science* **211**, 909–911.

Turekian, K. K., Cochran, J. K., Kharkar, D. P., Cerrato, R. M., Vaisnys, J. R., Sanders, H. L., Grassle, J. F. and Allen, J. A. (1979). Slow growth rate of a deep-sea clam determined by ^{228}Ra chronology. *Proceedings of the National Academy of Sciences of the United States of America* **72**, 2829–2832.

Williams, P. M., Oeschger, H. and Kinney, P. (1969). Natural radiocarbon activity of dissolved organic carbon in the Northeast Pacific Ocean. *Nature, London* **224**, 256–258.

Williams, P. M., Smith, K. L., Druffel, E. M. and Linick, T. W. (1981). Dietary carbon sources of mussels and tubeworms from Galapagos hydrothermal vents determined by tissue ^{14}C activity. *Nature, London* **292**, 448–449.

Wolff, T. (1970). The concept of the hadal or ultra-abyssal fauna. *Deep Sea Resources* **17**, 983–1003.

Wolff, T. (1976). Utilization of seagrass in the deep sea. *Aquatic Botany* **2**, 161–174.

Yayanos, A. A. and Dietz, A. S. (1983). Death of a hadal deep-sea bacterium after decompression. *Science* **220**, 497–498.

Yayanos, A. A., Dietz, A. S. and Van Boxtel, R. (1981). Obligately bariophilic bacterium from the Mariana Trench. *Proceedings of the National Academy of Sciences of the United States of America* **78**, 5212–5215.

Zimmerman, A. M. (1970). "High Pressure Effects on Cellular Processes". Academic Press, New York.

ZoBell, C. E. and Morita, R. Y. (1957). Barophilic bacteria in some deep-sea sediments. *Journal of Bacteriology* **73**, 563–568.

ZoBell, C. E. and Morita, R. Y. (1959). Deep-sea bacteria. *Galathea Reports* **1**, 139–154.

8

Bacterial Adaptations for Growth in Low Nutrient Environments

PHILIP MORGAN* and CRAWFORD S. DOW

*Department of Biological Sciences, University of Warwick, Coventry
CV4 7AL, UK*

Introduction

When considering extreme environments the physiology and ecology of microorganisms growing in low nutrient (oligotrophic) situations are more often than not excluded. In several respects such ecosystems constitute environmental extremes, and yet the low nutrient environment constitutes the bulk of the biosphere, including as it does the open oceans and the majority of unpolluted inland waters, and is most important in terms of the global cycling of nutrients. Consequently such ubiquity must constitute a norm. Whether of either persuasion this is a topic which has attracted relatively little research and our understanding of the physiological nature of the microbes found *in situ* ranges from incomplete to non-existent. In this article it is our intention to review a range of these environments and to examine briefly the presumed and demonstrated physiological adaptations that enable bacteria to grow exclusively and/or competitively in natural low nutrient environments.

Considering a low nutrient environment as being extreme has depended primarily on the use of the term oligotrophic and the definition applied to this term. The definition currently accepted by most microbiologists is that of Kuznetsov *et al.* (1979) and the term is taken to describe a bacterial type rather than an environmental condition. In the former context, oligotrophic bacteria are those able to grow, on first cultivation, at organic carbon levels of $1–15$ mg l^{-1}. If unable to grow at higher concentrations they are termed obligate oligotrophs; otherwise they are referred to as being facultative.

*Present address: Shell Research Ltd, Sittingbourne Research Centre, Sittingbourne, Kent ME9 8AG, UK.

Bacteria growing optimally at organic carbon concentrations higher than this have been termed copiotrophic (Poindexter, 1981) or eutrophic. These terms, oligotrophic and copiotrophic, in this article are taken as indications of the type of environment and the organisms therein, and not as distinct and exclusive microbial types.

The correlation of nutrient status with carbon alone precludes oscillating multiple limitations of carbon, phosphate, sulphate, nitrogen and/or trace elements operating in conjunction with other environmental factors such as pH, temperature and oxygen tension. In low nutrient environments it is far more probable that growth is constrained by several of these parameters and that the physiology of the inherent microbes reflects this. As a consequence of the limited amount of data available, many of the organisms we shall discuss are by no means uniquely characteristic of low nutrient environments; however, the response(s) they show to "starvation" and extreme nutrient limitation, the norm of many natural populations, provide insights into the basic physiology and adaptive response(s) of so-called oligotrophic bacteria.

Low Nutrient Environments

Sea Water

The sheer volume of the oceanic environment makes it difficult to build up a cohesive picture of its ecology. Jannasch (1984) has calculated that, by defining the pelagic deep ocean as being that of depths in excess of 1000 m, over 90% of the volume of the entire biosphere can be accounted for—by including all oceanic regions the figure is obviously higher. Further, for our purposes, we must disregard other factors limiting growth in the open ocean, in particular hydrostatic pressure and temperature.

The average level of organic carbon in sea water is exceedingly low. Disregarding the relatively small volume of the comparatively nutrient-rich photic zone and considering the oceanic volume below 200–300 m in depth, the dissolved organic carbon (DOC) concentration ranges from 0.35 to 70 mg l^{-1} and the particulate organic carbon (POC) concentration from 3 to 10 μg l^{-1} (Menzel and Ryther, 1970). Such figures oversimplify the situation, since both the POC and the DOC are somewhat resistant to bacterial attack. Indeed, Barber (1968) observed no significant alteration in the levels of deep ocean DOC over a 2 month incubation period, despite the presence of viable bacteria. It has also been demonstrated (Gordon, 1970) that hydrolysed deep-ocean POC is a suitable substrate for bacterial growth. It therefore remains unclear how much of the DOC and POC content of oceanic water is available for bacterial growth. Present data indicate that the available percentage is low: calculations of deep-ocean oxygen consumption (Craig, 1971), *in vitro* studies (Carlucci and Williams, 1978) and radiodating of DOC (Williams *et al.*, 1969) all suggest (Morita, 1982) a very low level of organic carbon utilization *in situ*. However, such a view represents gross oversimplification in that these calculations ignore local niches of activity, the importance of interfaces (Wardell *et al.*, 1983) and localized activity in the gut and on the surface of deep-sea animals (Deming *et al.*, 1981; Deming and Colwell, 1982).

Bearing these factors in mind, it is no great surprise that our knowledge of marine bacteria that utilize low levels of organic carbon is limited. However, there are a number of taxonomic studies on assumed "oligotrophic" marine microbes. Austin *et al.* (1979) characterized a variety of taxa attached to wood surfaces in a tropical harbour and obtained a variety of isolates belonging to the somewhat broad genera *Pseudomonas*, *Vibrio* and *Bacillus* plus the prosthecate forms *Hyphomicrobium* and *Hyphomonas*. Mallory *et al.* (1977), in a study of low nutrient estuarine water, isolated and characterized forms belonging to the taxa *Alcaligenes*, *Corynebacterium*, *Listeria*, *Nocardia*, *Planococcus*, *Sphaerotilus*, *Streptothrix* and *Streptomyces* in addition to the above genera. These observations, and the isolation of other prosthecate bacteria from sea water such as *Pedomicrobium* (Mallory *et al.*, 1977) and *Prosthecomicrobium* (Bauld *et al.*, 1983), despite the criticisms that can be levelled at the selection pressures imposed by agar plate cultivation, serve to illustrate the wide taxonomic and morphological range of marine bacteria. In addition, a large number of such isolates do not immediately correlate with the "model oligotroph" concepts of, for example, Poindexter (1981) and Shilo (1980).

Of particular importance in low nutrient oceanic ecosystems are the large numbers of so-called filterable bacteria or ultramicrobacteria (Anderson and Heffernan, 1965; Torrella and Morita, 1981; MacDonell and Hood, 1982). These will be considered later. However, at this point emphasis must be placed on the conclusions of a large number of investigators (e.g. Novitsky and Morita, 1977; Stevenson, 1978; Jones and Rhodes-Roberts, 1981; Tabor *et al.*, 1981) who believe these organisms to be the true autochthonous marine bacteria.

Fresh Water

The spectrum of low nutrient freshwater environments ranges from fast flowing rivers through deep, comparatively static lacustrine systems to a range of man-made environments, most notably distilled water storage systems. Our knowledge of the bacterial flora of rivers is largely limited to polluted systems, the natural low nutrient situation having received comparatively little attention, not least owing to the difficulties in monitoring and sampling such environments. One notable feature of the literature, if one that is not surprising, is the importance to microbial activity of surface attachment to both biotic and abiotic surfaces (McFeters *et al.*, 1978).

The comparatively static oligotrophic lakes have received more detailed study. Although a large number of surveys have emphasized the occurrence of the morphologically distinct prosthecate and stalked bacteria both *in situ* and in enrichment cultures (e.g. Nikitin and Kuznetsov, 1967; Hirsch and Rheinheimer, 1968; Schmidt and Starr, 1978), it is important to emphasize the wide morphological variety of bacteria that occur (Fig. 1). In addition, studies on isolates from Lake Biwa, Japan (Ishida *et al.*, 1980; Ishida and Kadota, 1981) have led to the proposal that bacteria unable to grow in "high" nutrient media, which are hence the classical oligotrophs of Kuznetsov *et al.* (1979), are unable to grow at solid–water interfaces. This supposition is questionable, however, as a consequence of the experimental procedures used to isolate and study these microbes.

The potential of distilled water as a suitable "culture medium" for oligotrophic bacteria became evident in the mid-1950s when the morphologically distinct prosthecate bacteria were observed as contaminants of electron microscope preparations (Houwink, 1952; Bystricky, 1954; Kandler *et al.*, 1954). Reports of such observations are still deemed interesting enough to merit publication (Callerio *et al.*, 1983) and indeed we have frequently observed a variety of morphological types in double-distilled water storage vessels (Fig. 2). Hirsch (1964) designated the bacteria growing under such conditions as being oligocarbophiles—bacteria capable of utilizing solubilized volatile organic compounds for growth. By using the simple expedient of closing vessels containing sterile minimal medium with various types of stopper Geller (1983) has elegantly demonstrated that even uninoculated medium exhibits a steady increase in DOC content and that the observed increase is a direct result of the permeability of the stopper to air. It was further demonstrated that *Pseudomonas fluorescens*, an organism certainly not noted for its oligotrophic characteristics, could successfully grow as an oligocarbophilic bacterium under such conditions. Whilst this type of observation is by no means new, it is worth noting the potential and actual medical importance of the growth of pseudomonads and enteric bacteria in both tap and distilled water (Favero *et al.*, 1971; Botzenhart and Kufferath, 1976; van der Kooij *et al.*, 1980, 1982).

Soil

Soil has been widely presumed to be a potentially oligotrophic habitat (Hirsch, 1979). However, the physical nature of soil and its concomitant

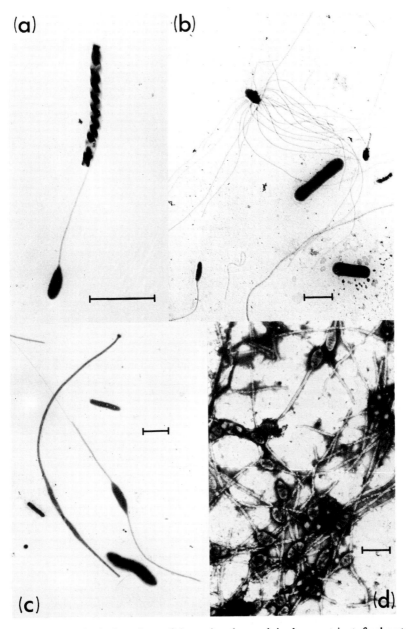

Fig. 1. Morphological variety of bacteria observed in low nutrient freshwater environments. (a) *Caulobacter*-like prosthecate bacterium and a *Seliberia* cell. (b) Variety of forms including multiprosthecate cell. Electron micrographs of gold–palladium-shadowed specimens. (c) Two varieties of *Prosthecobacter*-like cells plus rod-shaped bacteria. (d) Multicellular array of *Hyphomicrobium*-like cells. Electron micrograph of phosphotungstic-acid-stained specimens. (Scale bars, 3.0 μm.)

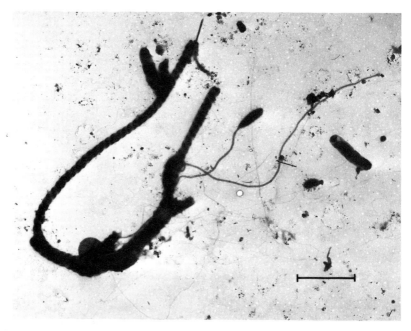

Fig. 2. Morphological variety of bacteria occurring in double-distilled water. A range of cell types can be seen, including a long *Seliberia*-like cell and a *Caulobacter*-like prosthecate bacterium possessing a very long prostheca. Electron micrograph of gold–palladium-shadowed specimen. (Scale bar, 4.0 μm.)

heterogeneity would, in most situations, render nutrient limitation alone, over a comparatively large area, an unlikely occurrence. It is undeniable, however, that such limitations will frequently occur within individual microcosms. In the light of such a probability, the observations that the prosthecate bacteria characteristic of low nutrient aquatic environments are present in soil (Aristovskaya, 1963) and that many soil bacteria are inhibited by relatively high concentrations of organic nutrients (Hattori and Hattori, 1980) are significant. One particularly interesting observation is the ability of unidentified soil and aquatic bacteria to utilize carbon monoxide (CO) as a source of carbon at concentrations down to 1 nM (Conrad and Seiler, 1982). These "oligotrophic carboxydobacteria" have affinities for CO that are 1–2 orders of magnitude higher than those observed in currently cultivated microbes. It is therefore a distinct possibility that oligotrophic bacteria in both soil and water are physiologically competent and capable of growth with the observed affinity for the extremely low atmospheric levels of CO (Conrad and Seiler, 1980).

Potential Physiological Adaptations for Growth in Low Nutrient Environments

From the above somewhat selective review of bacteria from low nutrient environments, it is clear that a range of physiological and morphological types can apparently compete successfully. The obvious question which arises relates to the adaptations necessary for an organism to grow and compete successfully in such an environment. This section highlights the observed and probable adaptations necessary for such a mode of existence, although it is essential to point out that the actual importance *in situ* of a number of the presumed adaptations has yet to be determined.

Miniaturization

In both soil and aquatic environments, under low nutrient conditions, the predominant bacterial population consists of ultramicrobacteria (filterable bacteria: those able to pass through a 0.45 μm pore size filter) and it is clear that these are the true indigenous forms. The physiology and formation of these cells has largely been elucidated from studies on an Antarctic marine *Vibrio* ANT-300.

Under laboratory conditions in a nutrient-rich medium this organism produces morphologically "typical" rod-to-comma-shaped cells which, following harvesting and resuspension in a carbon-free salts solution, respond to starvation by fragmenting to produce large numbers of ultramicrobacteria (Novitsky and Morita, 1976). The viability of ANT-300 under such conditions, expressed as total viable count, increases by up to 10 orders of magnitude within a week but subsequently declines to, and stabilizes at, a level similar to that observed in the initial, unstarved culture (Novitsky and Morita, 1977). Viable ultramicrobacteria exhibit a marked reduction in endogenous respiration (Novitsky and Morita, 1978), have chemotactic capabilities (Geesey and Morita, 1979; Morita, 1982) and, following an initial dramatic decline, show a steady increase in levels of cellular RNA, DNA and protein (Amy et al., 1983a). ATP levels in the starved cells also exhibit a rapid decline during the initial stages of starvation but subsequently show a steady increase. the metabolic energy source for ATP synthesis has yet to be determined but is presumably an endogenous storage compound, available evidence suggesting lipids (Amy et al., 1983a). Similar effects have been demonstrated in the marine *Vibrio* DW1 (Kjelleberg and Dahlback, 1984). It has also been shown that starved ANT-300 cells possess a number of proteins not found in actively growing cells (Amy and Morita, 1983a) but these are of unknown function. On inoculation of starved ultramicrobacteria into fresh medium, recovery occurred after an initial lag period proportional to the starvation period—owing to a possible relationship between the period of starvation and the time taken for the ultramicrobacteria to attain a size suitable for cell division (Amy *et al.*, 1983b). Synthesis of RNA, DNA and protein during recovery was similarly affected, suggesting a relationship between the starvation period and the extent of dormancy (Amy *et al.*, 1983b). The nature of the ANT-300 ultramicrobacteria as a survival stage and the recovery of the cells following starvation indicates their functional role *in situ*. Their physiological activity in the natural environment is still unclear, particularly with respect to their response(s) to natural fluctuations in nutrient levels.

Analogous behaviour to that observed in ANT-300 has been observed in a range of marine isolates under identical conditions (Amy and Morita, 1983b). In the natural environment ultramicrobacteria are common in sea water and soil (for a review see Morita (1982)) and we have observed analogous effects with bacteria isolated from distilled water reservoirs (McQuaid, Morgan and Dow, unpublished observations). The significance of such cell types has been demonstrated *in vitro* and the weight of available data points to their being the predominant active forms in natural low nutrient environments. The significance of their apparent physiological adaptation, as we shall discuss later, extends beyond the observations made to date and is potentially applicable to a wide range of bacteria.

Surface Attachment

The statement of Wardell *et al.* (1983) that it was the paper of ZoBell and Anderson (1936) which first stimulated widespread interest in bacterial growth at surfaces is indicative of the considerable biological importance of solid–liquid interfaces in the low nutrient environment. ZoBell and Anderson (1936) observed that in stored sea water both the numbers of bacteria and their metabolic activity were directly proportional to the ratio of the surface area to the volume of the container in which the water was stored. Subsequent investigations demonstrated that there was a threshold level of nutrients below which the presence of surfaces increased metabolic activity and that this was due to the association of bacteria with these surfaces (Heukelakian and Heller, 1940; Jannasch, 1958). The physical nature of a solid–liquid interface results in the concentration of nutrients and ions (Marshall, 1979, 1980) at this site and initially accounts for attraction and attachment of bacterial cells. Subsequent population development, however, leads to a far more complex situation in which intraspecific cross-feeding, cometabolism, proton transfer and competition (Wardell *et al.*, 1983) become important.

It is not our intention to consider the literature relating to surface attachment and subsequent population development since this has been reviewed elsewhere (Fletcher and Marshall, 1982; Wardell *et al.*, 1983). However, since the evidence relating to increased functional activity of bacteria at surfaces is clear (Jannasch and Pritchard, 1972; Hendricks, 1974; Goulder, 1977), it is necessary to consider the adaptations exhibited by bacteria in low nutrient environments for surface attachment.

The nature of bacterial attraction to surfaces is initially physical (Wardell *et al.*, 1983). However, subsequent attachment appears to depend both on the hydrophobicity and electrostatic nature of the cell envelope and on the production of adhesive polymers. Bacteria characteristic of low nutrient environments exhibit both. Indeed, certain, although not all, marine vibrios examined by Kjelleberg and Hermansson (1984) exhibited increased surface hydrophobicity and/or increased prevalence for surface binding during the starvation–miniaturization process. Large numbers of bacteria also produce variable amounts of adhesive polysaccharide for attachment and this is clearly evident in the prosthecate, stalked and budding bacteria—thought of as model oligotrophs. Most possess "holdfast" material at the distal end of the prostheca (e.g. *Caulobacter, Prosthecobacter* and *Planctomyces*—(Dow and Lawrence, 1980; Fig. 3) although *Asticcacaulis* (Umbreit and Pate, 1978) has the holdfast on the main body of the cell and the prosthecae are not involved in attachment, a situation somewhat similar to that in *Rhodopseudomonas palustris* and certain *Hyphomicrobium* species.

It has been postulated (Ishida *et al.*, 1980; Ishida and Kadota, 1981; Kjelleberg *et al.*, 1982) that true oligotrophic bacteria are those unable to attach to surfaces, and hence to take advantage of the local increase in nutrients, and that those at the surface are, most typically, resting stages of copiotrophs. From both an ecological and an evolutionary point of view it is difficult to accept such arguments as presented—it would appear to be inconceivable that an organism adapted for growth in the "extreme" low nutrient environment should fail to exploit a local increase in nutrients at a surface. Most bacteria conceived of as oligotrophs exhibit surface attachment and even those exhibiting inhibition at high substrate concentrations are unlikely to be inhibited by the actual concentration of nutrients at an interface, however large it may be relatively. In terms of competitive potential, the advantages to be gained at biotic and abiotic surfaces are too great, in an environment as challenging as that with very low nutrient concentrations, to be ignored.

Endogenous Metabolism

The subject matter that can be covered under the heading of endogenous

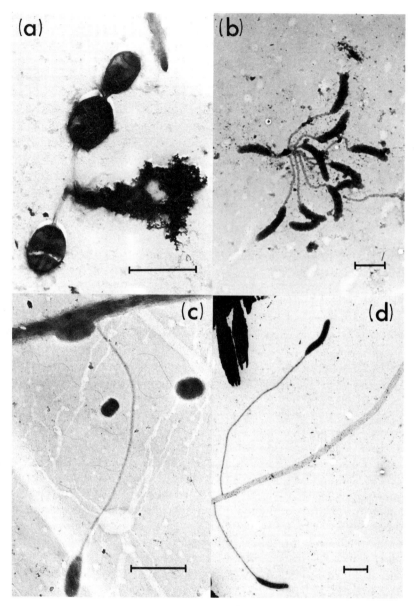

Fig. 3. Attachment by prosthecate and stalked bacteria in the low nutrient fresh-water environment. (a) Two cells of the stalked genus *Planctomyces*. (b) Rosette of *Caulobacter*-like cells. (c), (d) *Caulobacter*-like cells attached to microbial filaments. (Scale bars, 2.0 μm; electron micrographs of gold–palladium-shadowed specimens.)

metabolism is considerable but, unfortunately, little has been published regarding the metabolic processes of bacteria from low nutrient environments. However, studies on a range of bacteria in continuous culture have yielded several insights into what may be occurring in such organisms and in the "common laboratory forms" growing under more natural conditions. As an indication of how important such adaptations are, it is only necessary to consider the competitive advantage of a freshwater *Spirillum* sp. over a *Pseudomonas* sp. as observed by Matin and Veldkamp (1978). Among the metabolic features conferring an advantage on the former in a mixed continuous culture, at low dilution rates, were a more active and/or efficient respiratory chain, a lower energy of maintenance (0.016 compared with 0.066 g L-lactate (g dry wt)$^{-1}$ hour^{-1}) and a lower minimal growth rate.

Storage Polymers

A wide range of bacteria are known to accumulate excess nutrients and to store them as polymers. Both phosphate (as polyphosphate) and carbon (as polyhydroxybutyrate, carbohydrate polymers or lipids) are stored by bacteria from low nutrient environments. *Vibrio* ANT-300 is believed to accumulate lipid (Amy *et al.*, 1983a) and the model oligotrophs *Caulobacter* and *Hyphomicrobium* are known to store both polyphosphate and polyhydroxybutyrate (Poindexter and Eley, 1983). It is important to note that the benefits of intracellular storage polymers are linked to the rate of their utilization (Dawes, 1976); they will only be of benefit in a low nutrient situation if they are utilized relatively slowly.

Metabolic Enzymes

Before considering the nature and control of metabolic pathways in bacteria adapted to, or surviving in, oligotrophic environments it is necessary to emphasize the following point. In the natural environment the range of potential substrates available to a bacterium is likely to be large; yet, unfortunately, comparatively few data are available concerning mixed substrate utilization by bacteria and even fewer regarding the metabolic interactions involved. As an illustration of how important mixed substrate utilization is likely to be in the low nutrient environment the behaviour of *Pseudomonas oxalaticus* when grown on a mixture of oxalate and acetate is revealing. In batch culture the organism exhibits diauxic growth with the acetate utilized first. In carbon-limited continuous culture at dilution rates below 0.15 hour^{-1} both acids are utilized equally, and between 0.15 and 0.30

hour^{-1} acetate is used preferentially, excess oxalate remaining unutilized at a level that is proportional to the dilution rate (Harder and Dijkhuizen, 1983). The situation in the natural environment is likely to be analogous but significantly more complex and a degree of caution must be exercised in interpreting much of the published data.

The response of catabolic enzymes to alterations in growth rate in continuous culture has been widely studied and recently reviewed by Harder and Dijkhuizen (1983). It is a frequent observation that decreasing substrate concentration leads to an increase in the biosynthesis of the enzyme(s) involved in the catabolism of that substrate with the consequence that utilization of the substrate is more effective. Similarly, a number of bacteria have been shown to synthesize different substrate-utilizing enzyme(s) when a substrate is limiting, resulting in a net increase in the organism's affinity for the said substrate (Neijssel et al., 1975, for example). Koch has observed (Koch and Coffman, 1970; Koch, 1979) that of a population of Escherichia coli growing at low dilution rates in continuous culture (doubling time 16 hours) only two-thirds were actively synthesizing protein at a given time. However, all cells could actively induce enzyme synthesis within a 3 hour period of nutrient pulsing. It is unfortunate that little is known regarding enzyme affinities and metabolic pathways employed by bacteria characteristic of oligotrophic ecosystems. In particular, the distinct cell type expression of prosthecate bacteria (see below) offers an opportunity for studies of differential, cell-type-specific enzymological control which is unparalleled elsewhere. However, preliminary studies have shown there to be no appreciable difference in the activities of tricarboxylic acid cycle enzymes in two distinct cell types of Rhodomicrobium vannielii, nor any difference in inhibitor effects on the organism's citrate synthase (Morgan, Kelly and Dow, in press). More detailed studies on enzymes nearer the "first utilization" steps of a limiting substrate are more likely, however, to prove fruitful.

In addition to the specific metabolic pathways, it has long been known that competitiveness under conditions of nutrient limitation is related to a maintenance energy requirement. The debate over the existence or otherwise of a need for true maintenance energy over a long period of time (see, for example, Dawes, 1976; Morita, 1982) need not concern us here for in chemostat competition studies the background basal level of metabolism is clearly important (Matin and Veldkamp, 1978) and in the ANT-300 starvation survival studies of Novitsky and Morita (1977) it was demonstrated that endogenous respiration declined to 0.0071% of its peak level within 7 days. It is therefore clear that the basal metabolic rate of bacteria growing in a low nutrient situation will indeed play a highly significant role in competition — energy wasted on routine tasks, whatever they may be, is not available for uptake, tactic movements, biosynthesis or growth.

The observations that *E. coli, Bacillus polymyxa* and *Paracoccus denitrificans* grown in continuous culture exhibit three discontinuous modes of growth (Chesbro *et al.*, 1979; Arbige and Chesbro, 1982; van Verseveld *et al.*, 1984) is of considerable interest. In these organisms the three modes occur at almost identical dilution rates with identical results (Table 1). The importance of these responses in ensuring survival is as yet unclear. It has been suggested (van Verseveld *et al.*, 1984) that the most profound effect of the reduction in protein synthesis associated with the full stringent response is to decrease the inherent likelihood of mistranslation of mRNA in slow-growing organisms. This is achieved by decreasing the speed of protein chain elongation and hence increasing the time available for discrimination between amino acid-tRNA species. Wider physiological effects are as yet undocumented and nothing is known with regard to different growth modes in other bacterial species.

Protonmotive Force

The importance of the transmembrane electrochemical proton gradient to bacteria in low nutrient environments lies primarily in its roles in driving substrate uptake systems and flagellar rotation. It has been suggested that the ability of bacteria to survive in low nutrient environments depends on the ability of the cells to maintain a functional gradient across the membrane (Harder *et al.*, 1984). The ability of bacteria characteristic of low nutrient environments to regulate their electrochemical gradients has yet to be investigated, as with those organisms capable of long-term survival in low nutrient environments. In the light of the ability of certain bacteria to block their transport systems below critical protonmotive force levels (Konings and Veldkamp, 1983) and hence to prevent efflux of accumulated nutrients, such information should prove enlightening.

Nutrient Transport and Binding Proteins

The existence in Gram-negative bacteria of periplasmically located substrate binding proteins which can be released from the periplasmic space by osmotic shock has been clearly documented. A wide range of such proteins has been described, principally in *E. coli*, that bind a range of compounds including sulphate (Pardee, 1968), phosphate (Medveczky and Rosenberg, 1969), galactose (Anraku, 1967) and a whole range of amino acids (see, for example, Rosen, 1971; Morita, 1982). It is important to note that the sole function of such proteins is to bind a substrate; the proteins possess no

Table 1. *Features of the three growth-rate-dependent growth modes of* E. coli, P. denitrificans *and* B. polymyxa *(Chesbro et al., 1979; Arbige and Chesbro, 1982; van Verseveld et al., 1984)*

Characteristic	Mode 1[a]	Mode 2[a]	Mode 3[a]
Approximate doubling time (T_D)	Minutes to 15 hours	33–50 hours	$\geqslant$ 100 hours
Molar growth yield (Y) observed (as percentage of maximum)	100	50–75	30–40
Level of HPGN[b]	Basal	Increasing; maximal at about 80 hours	Maximal
Stringent response	Basal	Increasing[c]	Full
Endogenous nutrient reserves	Accumulated	Variable response; may be utilized	Utilized

[a] van Verseveld et al. (1984) designated mode 1 exploitation, mode 2 adaptation and mode 3 stringent control.

[b] HPGN, highly phosphorylated guanosine nucleotides (ppGpp and pppGpp).

[c] The effects are variable and increase with increasing doubling time. However, the characteristic suppression of RNA synthesis observed in the stringent response does not appear to occur (van Verseveld et al., 1984).

demonstrated enzymological activity. The physiological roles of such proteins appear to be threefold: they function in transport systems; they are involved in chemotactic responses (discussed below) and they are sensors of the environment with a potential role in metabolic regulation and cell type expression. Unfortunately no specific physiological information on such proteins in oligotrophic bacteria is available.

The importance of substrate transport systems under starvation conditions to *E. coli* is such that they are constitutive (Koch, 1979). The advantages of such behaviour are obvious; in order to be competitive when nutrients become available the cell must be able to assimilate the nutrients without delay. This is even more critical for bacteria growing in low nutrient environments.

Bacteria growing under nutrient limitation can effectively increase their substrate uptake potential in two ways. Firstly they can increase the concentration of a given system in the cell envelope and hence increase the net uptake velocity (V_{max}). Secondly a system with a higher affinity (K_m) for a substrate can be synthesized. Although there is extensive evidence for the latter in a range of bacteria and using a range of substrates, e.g. phosphate (Rosenberg *et al.*, 1977), certain amino acids in *E. coli* (Rosen, 1971) and glucose in *Pseudomonas aeruginosa* (Harder and Dijkhuizen, 1983), and evidence of the involvement of binding proteins in high affinity uptake systems (Kalckar, 1971; Rosen, 1971), there are comparatively few data on how general growth conditions affect the synthesis and activity of substrate transport systems. Nor are there any data specifically relating to the nature of the protonmotive force under such conditions.

Data for bacteria typical of low nutrient environments are similarly lacking, although it is clear that the K_m and V_{max} values of transport systems play an important part in ensuring an organism's competitive ability (Matin and Veldkamp, 1978). Geesey and Morita (1979) demonstrated the presence of high and low affinity arginine transport systems in the vibrio ANT-300 and Akagi and Taga (1980) demonstrated that a natural marine "oligotrophic" isolate possessed a higher affinity proline transport system than did its "copiotrophic" counterpart. Interestingly, the latter workers also observed a coupling between transport and metabolism (i.e. "first utilization") of glucose in the "oligotrophic" organism and that the affinity for glucose was lower in this organism than in its "copiotrophic" counterpart, suggesting that the relationship between K_m and adaptation to low nutrient environments is not as clear cut as it may at first appear. However, because of this coupling of utilization and uptake, glucose utilization at low substrate concentrations was much more efficient in the "oligotroph". Certain amino acid transport systems in two bacteria from low nutrient environments have been shown to have far higher affinities than those of analogous "copiotrophs" (Ishida *et al.*, 1982). Studies on glucose transport in natural assemblages of marine bacteria have demonstrated that the population exhibits a diversity in individual K_m and V_{max} values (Azam and Hodson, 1981). The data available to date therefore indicate that high affinity nutrient transport systems are competitively important *in situ* but that this is only part of the story and that a coupling between uptake and utilization is important.

The situation in the prosthecate bacteria is even more complicated. With the exception of the demonstration of high and low affinity glucose uptake systems in the prosthecae of *Asticcacaulis biprosthecum* (Larson and Pate, 1976) little is known regarding the kinetics of transport systems in these organisms. It has long been clear (Schmidt and Stanier, 1966; Whittenbury and Dow, 1977; Harder and Attwood, 1978) that the length of prosthecae is inversely proportional to the level of nutrients in the environment and it appears reasonable to assume that such behaviour results in a higher V_{max} for substrate uptake as a consequence of the increased surface area available. The belief of Poindexter (1984a, b) that the prosthecae of both *Caulobacter* and *Hyphomicrobium* are the principal sites of phosphate uptake in these organisms would appear to be questionable. Carbon-limited chemostat cultures of *Hyphomicrobium* exhibit prostheca lengths in inverse proportion to the dilution rate (Harder and Attwood, 1978) and we find it unlikely that a nutrient as essential to the cells as phosphate would be of limited availability to stalk-free cells (be they mutants or cells grown in a high nutrient medium). The whole topic of nutrient transport in bacteria from low nutrient environments at present has raised more questions than answers and, in view of its importance, undoubtedly holds a number of important clues as to the nature of these bacteria.

Tactic Responses

The importance of chemotactic responses in bacterial competitiveness in low nutrient environments has been demonstrated (Smith and Doetsch, 1969; Pilgram and Williams, 1976); and similar importance can be attached to aerotaxis and phototaxis in appropriate organisms. It has been suggested (Rowbury *et al.*, 1983) that a rapidly respiring bacterium with a single flagellum expends approximately 0.1% of its available energy on flagellar rotation. In a peritrichous organism or one undergoing limited metabolic activity the relative cost must therefore be higher. In the absence of clearly defined solute gradients such expenditure of energy, if the estimates are correct, is a competitive disadvantage (Pilgram and Williams, 1976; Rowbury *et al.*, 1983). In the natural low nutrient environment, with few exceptions, nutrient gradients would be expected to be widespread and the organisms therein would therefore be at an advantage if they were chemotactic, despite the energy drain.

Sensory transduction in chemotactic, phototactic and aerotactic bacteria is mediated by specific binding proteins for a number of chemoattractants and by changes in the protonmotive force for other attractants (Taylor, 1983). As we discussed above, our knowledge of such systems in bacteria characteristic of low nutrient environments is virtually non-existent.

It is interesting to note that the *Vibrio* ANT-300 displays chemotaxis towards arginine (Geesey and Morita, 1979) and a range of amino acids, carbohydrates and other organic compounds (Morita, 1982) and that cells freshly harvested from nutrient-rich media are very poorly chemotactic. It is only after 2–3 days of nutrient deprivation that the response is optimal (Morita, 1982). The logic behind such a situation is clear: when growing in high nutrient niches chemotaxis is superfluous and wasteful; it is only when limitation occurs that nutrient gradients, and hence chemotaxis, become ecologically significant. Chemotaxis has also been shown to be important in the "model oligotroph" *Caulobacter* and has been partially characterized at the molecular level (e.g. Shaw *et al.*, 1983; Gomes and Shapiro, 1984) but ecological data are lacking.

Cell Cycles and Cell Type Expression

Although most clearly visible in the cell cycles of the prosthecate and stalked bacteria, a large number of bacteria in low nutrient environments exhibit growth cycles of varying degrees of complexity and nature (Morgan and Dow, 1985). Associated with such a growth cycle are a characteristic cell polarity (Whittenbury and Dow, 1977; Shapiro *et al.*, 1982; Gomes and Shapiro, 1984; Huguenel and Newton, 1984; Kelly and Dow, 1984) and the potential for cell type differentiation (Figs. 4 and 5). The result of such a potential for differentiation in bacteria growing in low nutrient environments appears to be the production of "swarmer", "shut down" or "growth precursor" cells (Dow and Whittenbury, 1980; Dow *et al.*, 1983) which may or may not be morphologically distinct from the mother cells but are physiologically distinct in that they are metabolically quiescent and that cellular differentiation, growth and reproduction only proceed under suitable environmental conditions. The physiological nature (Table 2) and potential ubiquity of swarmer-type cells in organisms growing in low nutrient environments have recently been reviewed (Dow *et al.*, 1983; Morgan and Dow, 1985). From these the following should be emphasized: the range of organisms performing such growth cycles; the evidence for metabolically limited growth precursor cells in a range of organisms (Morgan and Dow,

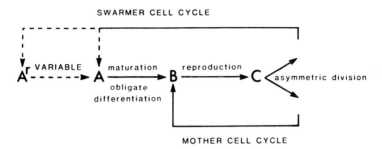

Fig. 4. Generalized life cycle of budding and prosthecate bacteria. A^r designates the metabolically restricted swarmer (growth precursor) cell whose development is controlled by environmental conditions. Following the onset of differentiation (A), a sequence of steps is followed to produce a mother (prosthecate, if appropriate) cell (B) which grows polarly to produce a daughter (swarmer) cell which separates by asymmetric division (C). The mother cell, under suitable environmental conditions, can initiate reproduction immediately whereas the swarmer cell is again under environmental control. (From Morgan and Dow (1985).)

1985); and the probability that such cell types play a significant role in the natural low nutrient environment.

Other Potential Adaptations

Outlined above are the main physiological and morphological adaptations for bacterial growth and survival in low nutrient environments. Bacterial sporulation and related responses have not been discussed although their potential as survival propagules is undeniable. It is also possible that flotation in the natural environment may play a role, either by ensuring aerobiosis or by retaining cells in the nutrient-rich photic zone in otherwise barren waters. Acellular stalks, prosthecae and gas vesicles may all play a part in this process. Finally the question of rosette formation and cell–cell aggregation in the stalked and prosthecate bacteria, a widespread phenomenon in laboratory culture and the natural environment, must not be ignored. In view of the observed specificity of the process (Starr and Schmidt, 1984) it is unlikely that such behaviour is functionless. It may be involved in flotation, in the capture of leaked nutrients and/or in the physical concentration of substrates at or around cell surfaces.

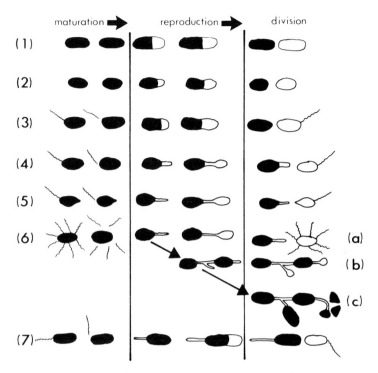

maturation ➡   reproduction ➡   division

Fig. 5. Morphogenetic gradient comparing the cell cycle of *E. coli* growing at a generation time in excess of 60 minutes and exhibiting unidirectional polar growth (Donachie and Begg, 1970; Begg and Donachie, 1977) with those of the budding and/ or prosthecate bacteria. The latter exhibit two distinct phases: an inhibition phase where development of a swarmer (growth precursor) cell is under time-independent environmental control and a maturation–reproduction phase involving a definite series of physiological and morphological steps ending in the production of a daughter swarmer cell whose subsequent development is again under environmental control. The mother cell (shaded) can continue reproduction immediately without the need for a second maturation period. The organisms illustrated are as follows: (1) *E. coli*; (2) *Rhodopseudomonas blastica*; (3) *Rhodopseudomonas acidophila*; (4) *R. palustris*; (5) *Hyphomicrobium*; (6) *R. vannielii* undergoing (a) the simplified cell cycle, (b) chain formation and (c) exospore production (Whittenbury and Dow, 1977); (7) *Caulobacter*. (Figure modified from Kelly and Dow (1984).)

Oligotrophy and Oligotrophic Bacteria

As stated at the beginning of this article, the authors' dislike of the current use of the term oligotrophic and its application to "types" of bacteria is strong. Growth on 1–15 mg organic carbon 1^{-1} *in vitro* appears to bear little

Table 2.   *Adaptations of the growth precursor (swarmer) cell to survival and dispersal in the low nutrient environment*

Physiological Parameter	References
Metabolic repression	
Cell-type specific proteins	Dow *et al.*, 1983
Suppression of DNA and RNA synthesis	Potts and Dow, 1979; Dow *et al.*, 1983
Suppression of protein synthesis	Dow *et al.*, 1983
Alteration in RNA polymerase subunits	Dow *et al.*, 1983
Alteration in nucleoid configuration	Swoboda *et al.*, 1982a,b
Response to environment	
Motility and chemotaxis	Shaw *et al.*, 1983; Gomes and Shapiro, 1984
Differentiation on environmental "trigger"	Whittenbury and Dow, 1977; Dow *et al.*, 1983

Data relate mainly to *R. vannielii* and *C. crescentus* although environmentally triggered cell types have been observed in a range of physiological and morphological forms (Morgan and Dow, 1985). Not all parameters are applicable to all species.

resemblance to *in situ* organic nutrient fluxes of less than 0.1 mg l^{-1} day^{-1} in the lacustrine environment and the very low levels of available organic carbon in the deep ocean (Poindexter, 1981), and in any case presupposes no overgrowth by "copiotrophic" or "facultatively oligotrophic" forms. The bulk of the biosphere is indeed oligotrophic, by whatever definition one may choose, and the very nature of global nutrient cycling tells us that bacterial mineralization of DOC and POC is of vital importance.

The current concept of model oligotrophs and their presumed adaptation to such a mode of existence (Hirsch, 1979) is also questionable. Firstly, many of the organisms presented as models will grow optimally in comparatively high concentrations of nutrients. For example, *Hyphomicrobium* X has been grown in the presence of 100 mM phosphate buffer and *Caulobacter crescentus* CB15 in 10 mM phosphate buffer and 0.25% w/v peptone, equivalent to approximately 420 mg organic carbon l^{-1} (Morgan and Dow, unpublished observations). More importantly, however, most of the presumed adaptations of model oligotrophs to their niches have been demonstrated to apply equally effectively to *E. coli* (Koch, 1979), which indeed is the only organism which has been shown to use its energy source preferentially for uptake at low nutrient concentrations (Purdy and Koch, 1976).

If a proportion of the physiological evidence is wanting, what of "oligotrophic" isolates? We are not entirely happy with the isolation techniques used nor with the precautions, or rather lack of them, taken against oligocarbophilic growth (Geller, 1983). More interestingly, the observations of Martin and MacLeod (1984) have demonstrated that natural "oligotrophic" isolates are only inhibited by high concentrations of certain organic substrates and not by others. These observations and those of workers who have grown "model copiotrophs" at "oligotrophic" nutrient concentrations (e.g. van der Kooij et al., 1980, 1982) in laboratory culture must cast extreme doubt on the nature of all "oligotrophic" isolates. Is the toxicity of high nutrient concentrations to bacteria from low nutrient environments merely a manifestation of substrate-accelerated death (Strange and Dark, 1965) or is it due to the rapid influx of nutrients into a cell previously adapted to scavenging them (Koch, 1971, 1979)? Certainly many bacteria characterized as oligotrophic on first isolation rapidly "adapt" to high nutrient concentrations on subculture (Kuznetsov et al., 1979). Consequently, we find ourselves in agreement with Koch (1979) and Martin and MacLeod (1984) in expressing extreme doubts about the existence of strictly oligotrophic bacteria as defined at present.

Bacterial Responses to Low Nutrient Environments

Despite our criticisms of current concepts outlined above there is no doubt that many bacteria are adapted primarily for growth in the vast volume of low nutrient ecosystems. Which of the responses outlined above are important is as yet unclear; many readers must have tired of the repeated use of phrases along the lines of "data are not yet available". The ability of bacteria under conditions of extreme nutrient stress to produce metabolically repressed vegetative cell types has long been proposed (Tuckett and Moore, 1959; Sussman and Halvorson, 1966; Stevenson, 1978). In the light of our knowledge of the ultramicrobacteria (Morita, 1982) and the apparent ubiquity of swarmer-type cells (Dow et al., 1983; Morgan and Dow, 1985) such behaviour appears to be extremely common. What is as yet unclear is what proportion of ultamicrobacterial populations are metabolically active, what thresholds control the production, maintenance and exit from the swarmer cell state and how natural populations react physiologically in the natural environment. In view of the volume of oligotrophic environments on a global scale, the individual answers are undoubtedly local, but ecological insights will enable us to begin to appreciate both the organisms and the scale of their importance in the most ubiquitous "extreme" environment.

References

Akagi, Y. and Taga, N. (1980). Uptake of D-glucose and L-proline by oligotrophic and heterotrophic marine bacteria. *Canadian Journal of Microbiology* **26**, 454–459.

Amy, P. S. and Morita, R. Y. (1983a). Protein patterns of growing and starved cells of a marine *Vibrio* sp. *Applied and Environmental Microbiology* **45**, 1748–1752.

Amy, P. S. and Morita, R. Y. (1983b). Starvation-survival patterns of sixteen freshly isolated open-ocean bacteria. *Applied and Environmental Microbiology* **45**, 1109–1115.

Amy, P. S., Pauling, C. and Morita, R. Y. (1983a). Starvation-survival processes of a marine vibrio. *Applied and Environmental Microbiology* **45**, 1041–1048.

Amy, P. S., Pauling, C. and Morita, R. Y. (1983b). Recovery from nutrient starvation by a marine *Vibrio* sp. *Applied and Environmental Microbiology* **45**, 1685–1690.

Anderson, J. I. W. and Heffernan, W. P. (1965). Isolation and characterization of filterable marine bacteria. *Journal of Bacteriology* **90**, 1713–1718.

Anraku, Y. (1967). The reduction and restoration of galactose transport in osmotically shocked cells of *Escherichia coli*. *Journal of Biological Chemistry* **242**, 793–800.

Arbige, M. and Chesbro, W. R. (1982). Very slow growth of *Bacillus polymyxa*: stringent response and maintenance energy. *Archives of Microbiology* **132**, 338–344.

Aristovskaya, T. V. (1963). Natural forms of existence of soil bacteria. *Mikrobiologiya* **32**, 564–568.

Austin, B., Allen, D. A., Zachary, A., Belas, M. R. and Colwell, R. R. (1979). Ecology and taxonomy of bacteria attaching to wood surfaces in a tropical harbour. *Canadian Journal of Microbiology* **25**, 447–461.

Azam, F. and Hodson, R. E. (1981). Multiphasic kinetics for D-glucose uptake by assemblages of natural marine bacteria. *Marine Ecology Progress Series* **6**, 213–222.

Barber, R. T. (1968). Dissolved organic carbon from deep waters resists microbial oxidation. *Nature, London* **220**, 274–275.

Bauld, J., Bigford, R. and Staley, J. T. (1983). *Prosthecomicrobium litoralum*, a new species from marine habitats. *International Journal of Systematic Bacteriology* **33**, 613–617.

Begg, K. J. and Donachie, W. D. (1977). The growth of the *Escherichia coli* cell surface. *Journal of Bacteriology* **129**, 1524–1535.

Botzenhart, K. and Kufferath, R. (1976). Über die vermehrung verschiedener Enterobacteriaceae sowie *Pseudomonas aeruginosa* und *Alkaligenes* spec. in destilliertem wasser, entionisiertem wasser, leitungswasser und mineralsalzlosung. *Zentralblatt für Bakteriologie, Parasitenkunde, Infektionskrankheiten und Hygiene. Abteilung 1 Originale Reihe B* **163**, 470–485.

Bystricky, V. (1954). *Caulobacter*—an unusual contaminant of preparations for the electron microscope. *Biologiya* **9**, 566–569.

Callerio, D., Gagliardi, R., Chersicla, M. and Callerio, C. (1983). Sulla presenza del genus *Caulobacter* nell'acqua distillata. *Istituto Sieroterapico Milanese Bolletino* **62**, 251–256.

Carlucci, A. F. and Williams, P. M. (1978). Simulated *in situ* growth rates of pelagic marine bacteria. *Naturwissenschaften* **65**, 541–542.

Chesbro, W. R., Evans, T. and Eifert, R. (1979). Very slow growth of *Escherichia coli*. *Journal of Bacteriology* **139**, 625–638.

Conrad, R. and Seiler, W. (1980). Role of microorganisms in the consumption and production of atmospheric carbon monoxide by soil. *Applied and Environmental Microbiology* **40**, 437–445.

Conrad, R. and Seiler, W. (1982). Utilization of traces of carbon monoxide by aerobic, oligotrophic microorganisms in oceans, lakes and soil. *Archives of Microbiology* **132**, 41–46.

Craig, H. (1971). The deep metabolism: oxygen consumption in abyssal ocean water. *Journal of Geophysical Research* **76**, 5078–5086.

Dawes, E. A. (1976). Endogenous metabolism and the survival of starved prokaryotes. *In* "The Survival of Vegetative Microbes" (Eds T. R. G. Gray and J. R. Postgate), pp. 19–53. Cambridge University Press, Cambridge.

Deming, J. W. and Colwell, R. R. (1982). Barophilic bacteria associated with deep sea animals. *BioScience* **31**, 507–511.

Deming, J. W., Tabor, P. S. and Colwell, R. R. (1981). Barophilic growth of bacteria from intestinal tracts of deep-sea invertebrates. *Microbial Ecology* **7**, 85–94.

Donachie, W. D. and Begg, K. J. (1970). Growth of the bacterial cell. *Nature, London* **227**, 1220–1225.

Dow, C. S. and Lawrence, A. (1980). Microbial growth and survival in oligotrophic freshwater environments. *In* "Microbial Growth and Survival in Extremes of Environment" (Eds G. W. Gould and J. E. L. Corry), pp. 1–19. Academic Press, London.

Dow, C. S. and Whittenbury, R. (1980). Prokaryotic form and function. *In* "Contemporary Microbial Ecology" (Eds D. C. Ellwood, J. N. Hedger, M. J. Latham, J. M. Lynch and J. H. Slater), pp. 391–417. Academic Press, London.

Dow, C. S., Whittenbury, R. and Carr, N. G. (1983). The "shut down" or "growth precursor" cell—an adaptation for survival in a potentially hostile environment. *In* "Microbes in Their Natural Environments" (Eds J. H. Slater, R. Whittenbury and J. W. T. Wimpenny), pp. 187–247. Cambridge University Press, Cambridge.

Favero, M. S., Carson, L. A., Bond, W. W. and Petersen, N. J. (1971). *Pseudomonas aeruginosa*: growth in distilled water from hospitals. *Science* **173**, 836–838.

Fletcher, M. and Marshall, K. C. (1982). Are solid interfaces of ecological significance to aquatic bacteria? *Advances in Microbial Ecology* **6**, 199–236.

Geesey, G. G. and Morita, R. Y. (1979). Capture of arginine at low concentrations by a marine psychrophilic bacterium. *Applied and Environmental Microbiology* **38**, 1092–1097.

Geller, A. (1983). Growth of bacteria in inorganic medium at different levels of airborne organic substances. *Applied and Environmental Microbiology* **46**, 1258–1262.

Gomes, S. L. and Shapiro, L. (1984). Differential expression and positioning of chemotaxis methylation proteins in *Caulobacter*. *Journal of Molecular Biology* **178**, 551–568.

Gordon, D. C. (1970). Some studies on the distribution and composition of particulate organic carbon in the North Atlantic Ocean. *Deep Sea Research* **17**, 233–243.

Goulder, R. (1977). Attached and free bacteria in an estuary with abundant suspended solids. *Journal of Applied Bacteriology* **43**, 399–405.

Harder, W. and Attwood, M. M. (1978). Biology, physiology and biochemistry of hyphomicrobia. *Advances in Microbial Physiology* **17**, 303–359.

Harder, W. and Dijkhuizen, L. (1983). Physiological responses to nutrient limitation. *Annual Review of Microbiology* **37**, 1–23.

Harder, W., Dijkhuizen, L. and Veldkamp, H. (1984). Environmental regulation of microbial metabolism. *In* "The Microbe 1984. Part II. Prokaryotes and Eukaryotes" (Eds D. P. Kelly and N. G. Carr), pp. 51–95. Cambridge University Press, Cambridge.

Hattori, R. and Hattori, T. (1980). Sensitivity to salts and organic compounds of soil bacteria isolated on diluted media. *Journal of General and Applied Microbiology* **26**, 1–14.

Hendricks, C. W. (1974). Sorption of heterotrophic and enteric bacteria to glass surfaces in the continuous culture of river water. *Applied Microbiology* **28**, 572–578.

Heukelakian, H. and Heller, A. (1940). Relation between food concentration and surface for bacterial growth. *Journal of Bacteriology* **40**, 547–558.

Hirsch, P. (1964). Oligocarbophilie (wachstum auf kosten von luftverunreinigungen) bei Mycobacterien und einigen ihnen nahestehenden Actinomyceten. *Zentralblatt für Bakteriologie, Parasitenkunde, Infektionskrankheiten und Hygeine, Abteilung 1* **194**, 70–82.

Hirsch, P. (1979). Life under conditions of low nutrient concentrations. *In* "Strategies of Microbial Life in Extreme Environments" (Ed. M. Shilo), pp. 357–372. Verlag Chemie, Weinheim.

Hirsch, P. and Rheinheimer, G. (1968). Biology of budding bacteria. V. Budding bacteria in aquatic habitats: occurrence, enrichment and isolation. *Archiv für Mikrobiologie* **62**, 289–306.

Houwink, A. L. (1952). Contamination of electron microscope preparations. *Experientia* **8**, 385.

Huguenel, E. and Newton, A. (1984). Isolation of flagellated membrane vesicles from *Caulobacter crescentus* cells: evidence for functional differentiation of polar membrane domains. *Proceedings of the National Academy of Sciences of the United States of America* **81**, 3409–3413.

Ishida, Y. and Kadota, H. (1981). Growth patterns and substrate requirements of the first isolated naturally occurring obligate oligotrophs. *Microbial Ecology* **7**, 123–130.

Ishida, Y., Shibihara, K., Uchida, H. and Kadota, H. (1980). Distribution of obligately oligotrophic bacteria in Lake Biwa. *Bulletin of the Japanese Society for the Science of Fisheries* **46**, 1151–1158.

Ishida, Y., Imai, I., Miyagaki, T. and Kadota, H. (1982). Growth and uptake kinetics of a facultatively oligotrophic bacterium at low nutrient concentrations. *Microbial Ecology* **8**, 23–32.

Jannasch, H. W. (1958). Studies on planktonic bacteria by means of a direct membrane filter method. *Journal of General Microbiology* **18**, 609–620.

Jannasch, H. W. (1984). Microbes in the oceanic environment. *In* "The Microbe 1984. Part II. Prokaryotes and Eukaryotes" (Eds D. P. Kelly and N. G. Carr), pp. 97–122. Cambridge University Press, Cambridge.

Jannasch, H. W. and Pritchard, P. H. (1972). The role of inert particulate matter in the activity of aquatic microorganisms. *Memorie dell'Istituto Italiano di Idrobiologia Dott (Suppl.)* **29**, 289–308.

Jones, K. L. and Rhodes-Roberts, M. E. (1981). The survival of marine bacteria under starvation conditions. *Journal of Applied Bacteriology* **50**, 245–258.

Kalckar, H. M. (1971). The periplasmic galactose binding protein of *Escherichia coli*. *Science* **174**, 557–565.

Kandler, O., Zehender, C. and Huber, O. (1954). Uber das vorkommen von *Caulobacter* spec. in destilliertem wasser. *Archiv für Mikrobiologie* **21**, 57–59.

Kelly, D. J. and Dow, C. S. (1984). Microbial differentiation: the role of cellular asymmetry. *Microbiological Sciences* **1**, 214–219.

Kjelleberg, S. and Dahlback, B. (1984). ATP level of a starving surface-bound and free-living marine *Vibrio* sp. *FEMS Letters* **24**, 93–96.

Kjelleberg, S. and Hermansson, M. (1984). Starvation-induced effects on bacterial surface characteristics. *Applied and Environmental Microbiology* **48**, 497–503.

Kjelleberg, S., Humphrey, B. A. and Marshall, K. C. (1982). The effect of interfaces on small starved marine bacteria. *Applied and Environmental Microbiology* **43**, 1166–1172.

Koch, A. L. (1971). The adaptive properties of *Escherichia coli* to a feast and famine existence. *Advances in Microbial Physiology* **6**, 147–217.

Koch, A. L. (1979). Microbial growth in low concentrations of nutrients. *In* "Strategies of Microbial Life in Extreme Environments" (Ed. M. Shilo), pp. 341–356. Verlag Chemie, Weinheim.

Koch, A. L. and Coffman, R. (1970). Diffusion, permeation, or enzyme limitation: a probe for the kinetics of enzyme induction. *Biotechnology and Bioengineering* **12**, 651–677.

Konings, W. N. and Veldkamp, H. (1983). Energy transduction and solute transport mechanisms in relation to environments occupied by micro-organisms. *In* "Microbes in Their Natural Environments" (Eds J. H. Slater, R. Whittenbury and J. W. T. Wimpenny), pp. 153–186. Cambridge University Press, Cambridge.

van der Kooij, D., Visser, A. and Hijnen, W. A. M. (1980). Growth of *Aeromonas hydrophila* at low concentrations of substrates added to tap water. *Applied and Environmental Microbiology* **39**, 1198–1204.

van der Kooij, D., Oranje, J. P. and Hijnen, W. A. M. (1982). Growth of *Pseudomonas aeruginosa* in tap water in relation to utilization of substrates at concentrations of a few micrograms per liter. *Applied and Environmental Microbiology* **44**, 1086–1095.

Kuznetsov, S. I., Dubinina, G. A. and Lapteva, N. A. (1979). Biology of oligotrophic bacteria. *Annual Review of Microbiology* **33**, 377–387.

Larson, R. J. and Pate, J. L. (1976). Glucose transport in isolated prosthecae of *Asticcacaulis biprosthecum*. *Journal of Bacteriology* **126**, 282–293.

MacDonell, M. T. and Hood, M. A. (1982). Isolation and characterization of ultramicrobacteria from a gulf coast estuary. *Applied and Environmental Microbiology* **43**, 566–571.

Mallory, L. M., Austin, B. and Colwell, R. R. (1977). Numerical taxonomy and ecology of oligotrophic bacteria isolated from the estuarine environment. *Canadian Journal of Microbiology* **23**, 733–750.

Marshall, K. C. (1979). Growth at interfaces. *In* "Strategies of Microbial Life in Extreme Environments" (Ed. M. Shilo), pp. 281–290. Verlag Chemie, Weinheim.

Marshall, K. C. (1980). Reactions of microorganisms, ions and macromolecules at interfaces. *In* "Contemporary Microbial Ecology" (Eds. D. C. Ellwood, J. N. Hedges, M. J. Latham, J. M. Lynch and J. H. Slater), pp. 93–106. Academic Press, London.

Martin, P. and MacLeod, R. A. (1984). Observations on the distinction between oligotrophic and eutrophic marine bacteria. *Applied and Environmental Microbiology* **47**, 1017–1022.

Matin, A. and Veldkamp, H. (1978). Physiological basis of the selective advantage of a *Spirillum* sp. in a carbon-limited environment. *Journal of General Microbiology* **105**, 187–197.

McFeters, G. A., Stuart, S. A. and Olson, S. B. (1978). Interactions of algae and heterotrophic bacteria in an oligotrophic stream. *In* "Microbial Ecology" (Eds M. W. Loutit and J. A. R. Miles), pp. 57–61. Springer, Berlin.

Medveczky, N. and Rosenberg, H. (1969). The binding and release of phosphate by a protein isolated from *Escherichia coli*. *Biochimica et Biophysica Acta* **192**, 369–371.

Menzel, D. W. and Ryther, J. H. (1970). Distribution and cycling of organic matter in the oceans. *In* "Organic Matter in Natural Waters" (Ed. D. W. Hood), pp. 31–54. Institute of Marine Science, Alaska.

Morgan, P. and Dow, C. S. (1985). Environmental control of cell type expression in prosthecate bacteria. *In* "Bacteria in Their Natural Environments" (Eds M. Fletcher and G. D. Floodgate), pp. 131–169. Academic Press, London.

Morita, R. Y. (1982). Starvation-survival of heterotrophs in the marine environment. *Advances in Microbial Ecology* **6**, 171–198.

Neijssel, O. M., Hueting, S., Crabbendam, K. J. and Tempest, D. W. (1975). Dual pathways of glycerol assimilation in *Klebsiella aerogenes* NCIB 418. Their role and possible functional significance. *Archives of Microbiology* **104**, 83–87.

Nikitin, D. I. and Kuznetsov, S. I. (1967). Electron microscope study of the microflora of water. *Mikrobiologiya* **36**, 789–794.

Novitsky, J. A. and Morita, R. Y. (1976). Morphological characterization of small cells resulting from nutrient starvation of a psychrophilic marine vibrio. *Applied and Environmental Microbiology* **32**, 617–622.

Novitsky, J. A. and Morita, R. Y. (1977). Survival of a psychrophilic marine vibrio under long-term nutrient starvation. *Applied and Environmental Microbiology* **33**, 635–641.

Novitsky, J. A. and Morita, R. Y. (1978). Possible strategy for the survival of marine bacteria under starvation conditions. *Marine Biology* **48**, 289–295.

Pardee, A. B. (1968). Membrane transport proteins. *Science* **162**, 632–637.

Pilgram, W. K. and Williams, F. D. (1976). Survival value of chemotaxis in mixed cultures. *Canadian Journal of Microbiology* **22**, 1771–1773.

Poindexter, J. S. (1981). Oligotrophy. Fast and famine existence. *Advances in Microbial Ecology* **5**, 63–89.

Poindexter, J. S. (1984a). Role of prostheca development in oligotrophic aquatic bacteria. *In* "Current Perspectives in Microbial Ecology" (Eds M. J. Klug and C. A. Reddy), pp. 33–40. American Society for Microbiology, Washington, District of Columbia.

Poindexter, J. S. (1984b). The role of calcium in stalk development and in phosphate acquisition in *Caulobacter crescentus*. *Archives of Microbiology* **138**, 140–152.

Poindexter, J. S. and Eley, L. F. (1983). Combined procedure for assays of poly-β-hydroxybutyric acid and inorganic polyphosphate. *Journal of Microbiological Methods* **1**, 1–17.

Potts, L. E. and Dow, C. S. (1979). Nucleic acid synthesis during the developmental cycle of the *Rhodomicrobium vannielii* swarm cell. *FEMS Letters* **6**, 393–395.

Purdy, D. R. and Koch, A. L. (1976). Energy cost of galactoside transport to *Escherichia coli*. *Journal of Bacteriology* **127**, 1188–1196.

Rosen, B. P. (1971). Basic amino-acid transport in *Escherichia coli*. *Journal of Biological Chemistry* **246**, 3653–3662.

Rosenberg, H., Gerder, R. G. and Chegwidden, K. (1977). Two systems for the uptake of phosphate in *Escherichia coli*. *Journal of Bacteriology* **131**, 505–511.

Rowbury, R. J., Armitage, J. P. and King, C. (1983). Movement, taxes and cellular interactions in the response of microorganisms to the natural environment. *In*

"Microbes in Their Natural Environments" (Eds J. H. Slater, R. Whittenbury and J. W. T. Wimpenny), pp. 299–350. Cambridge University Press, Cambridge.

Schmidt, J. M. and Stanier, R. Y. (1966). The development of cellular stalks in bacteria. *Journal of Cell Biology* **28**, 423–436.

Schmidt, J. M. and Starr, M. P. (1978). Morphological diversity of freshwater bacteria belonging to the *Blastocaulis–Planctomyces* group as observed in natural populations and enrichments. *Current Microbiology* **1**, 325–330.

Shapiro, L., Mansour, J., Shaw, P. and Henry, S. (1982). Synthesis of specific membrane proteins is a function of DNA replication and phospholipid synthesis in *Caulobacter crescentus*. *Journal of Molecular Biology* **159**, 303–322.

Shaw, P., Gomes, S. L., Sweeney, K., Ely, B. and Shapiro, L. (1983). Methylation involved in chemotaxis is regulated during *Caulobacter* differentiation. *Proceedings of the National Academy of Sciences of the United States of America* **80**, 5261–5265.

Shilo, M. (1980). Strategies of adaptation to extreme conditions in aquatic microorganisms. *Naturwissenschaften* **67**, 384–389.

Smith, J. L. and Doetsch, R. N. (1969). Studies in negative chemotaxis and the survival value of motility in *Pseudomonas fluorescens*. *Journal of General Microbiology* **55**, 379–391.

Starr, M. P. and Schmidt, J. M. (1984). *Planctomyces stranskae* (ex Wawrik 1952) sp. nov., nom. rev. and *Planctomyces guttaeformis* (ex Hortobagyi 1965) sp. nov., nom. rev. *International Journal of Systematic Bacteriology* **34**, 470–477.

Stevenson, L. H. (1978). A case for bacterial dormancy in aquatic systems. *Microbial Ecology* **4**, 127–133.

Strange, R. E. and Dark, F. A. (1965). "Substrate-accelerated death" of *Aerobacter aerogenes*. *Journal of General Microbiology* **39**, 215–228.

Sussman, A. S. and Halvorson, H. O. (1966). 'Spores, Their Dormancy and Germination'. Harper and Row, New York.

Swoboda, U. K., Dow, C. S. and Vitkovic, L. (1982a). Nucleoids of *Caulobacter crescentus* CB15. *Journal of General Microbiology* **128**, 279–289.

Swoboda, U. K., Dow, C. S. and Vitkovic, L. (1982b). *In vitro* transcription and translation directed by *Caulobacter crescentus* CB15 nucleoids. *Journal of General Microbiology* **128**, 291–301.

Tabor, P. S., Ohwada, K. and Colwell, R. R. (1981). Filterable marine bacteria found in the deep sea: distribution, taxonomy and response to starvation. *Microbial Ecology* **7**, 67–83.

Taylor, B. L. (1983). Role of proton motive force in sensory transduction in bacteria. *Annual Review of Microbiology* **37**, 551–573.

Torrella, F. and Morita, R. Y. (1981). Microcultural study of bacterial size changes and microcolony and ultramicrocolony formation by heterotrophic bacteria in sea water. *Applied and Environmental Microbiology* **41**, 518–527.

Tuckett, J. D. and Moore, W. E. C. (1959). Production of filterable particles by *Cellvibrio gilvus*. *Journal of Bacteriology* **77**, 227–229.

Umbreit, T. H. and Pate, J. L. (1978). Characterization of the holdfast region of wild-type cells and holdfast mutants of *Asticcacaulis biprosthecum*. *Archives of Microbiology* **118**, 157–168.

van Verseveld, H. W., Chesbro, W. R., Braster, M. and Stouthamer, A. H. (1984). Eubacteria have 3 growth modes keyed to nutrient flow. Consequences for the concept of maintenance and maximal growth yield. *Archives of Microbiology* **137**, 176–184.

Wardell, J. N., Brown, C. M. and Flannigan, B. (1983). Microbes and surfaces. *In*

"Microbes in Their Natural Environments" (Eds J. H. Slater, R. Whittenbury and J. W. T. Wimpenny), pp. 351–378. Cambridge University Press, Cambridge.

Whittenbury, R. and Dow, C. S. (1977). Morphogenesis and differentiation in *Rhodomicrobium vannielii* and other budding and prosthecate bacteria. *Bacteriological Reviews* **41**, 754–808.

Williams, P. M., Oeschger, H. and Kinney, P. (1969). Natural radiocarbon activity of dissolved organic carbon in the northeast Pacific Ocean. *Nature, London* **224**, 256–258.

ZoBell, C. E. and Anderson, D. Q. (1936). Observations on the multiplication of bacteria in different volumes of stored sea water and the influence of oxygen tension and solid surfaces. *Biological Bulletin* **71**, 324–342.

9

Biotechnological Implications for Microorganisms from Extreme Environments

R. J. SHARP and M. J. MUNSTER

Microbial Technology Laboratory, Public Health Laboratory Service
Centre for Applied Microbiology and Research, Porton Down,
Salisbury SP4 0JG, Wiltshire, UK

Introduction

Microorganisms or their enzymes have a multitude of industrial, medical and environmental applications. These involve a diverse range of physical operating parameters including extremes of temperature, pressure and pH. These processes include the microbial desulphurization of coal, the extraction of metals from naturally occurring ores or from industrial wastes, the treatment of toxic industrial wastes, the enhancement of oil recovery from deep oil wells operating at high barometric pressures, the high temperature production of alcohol from cellulose and the production of enzymes to operate at extremes of temperature and pH in a range of industrial processes.

Microorganisms suited to all these processes are found occurring naturally in a range of environments which in general are considered extreme but to many organisms are essential for their normal growth and development. What may be considered to be an optimum environment for many organisms may be extreme to others.

In this review we have attempted to discuss many of the biotechnological implications for some of these organisms. We have concentrated particularly on the thermophiles, alkalophiles, acidophiles, halophiles, barophiles and psychrophiles, although a number of other groups of organisms from "extreme environments" such as radiation-resistant strains and heavy-metal-resistant strains also have potentially significant roles in biotechnology.

One of the more recent developments is the use of biosensors based on the immobilization of cells or enzymes. These have considerable potential in the fields of medical diagnosis, the monitoring of pollution and the control of industrial processes. Those processes which are carried out under extreme conditions of temperature, pH, pressure or osmolarity will require biosensors which remain functional and stable under these conditions. Many of the organisms and enzymes discussed here will have wide application in these areas.

Thermophilic Microorganisms

Definition and Occurrence

Thermophilic microorganisms have been isolated from many prokaryotic groups of microorganisms including cyanobacteria, photosynthetic bacteria, the spore formers *Bacillus* and *Clostridium*, lactic acid bacteria, methane producers, methane utilizers, sulphur oxidizers and sulphate reducers, Mycoplasma, pseudomonads, Actinomycetes and Gram-negative aerobes. A comprehensive list is presented by Brock (1978).

A number of definitions or classification systems for thermophilic microorganisms have been proposed (Ljungdahl, 1979), but the most commonly accepted definition is that of Williams (1975). Organisms with a maximum growth temperature of more than 60 °C and an optimum of more than 50 °C were described as thermophilic, whilst those with a maximum growth temperature of more than 90 °C and an optimum of more than 65 °C were described as caldoactive. Any classification system is essentially arbitrary, and consideration must now be given to the organisms isolated from superheated deep-sea waters which grow under high pressure at 300 °C (Baross and Deming, 1983).

Thermophilic microorganisms have been isolated from a wide range of environments including volcanic and geothermal regions in the USA, Mexico, Japan, Iceland, the USSR and New Zealand, solar-heated environments such as soil and ground litter, self-heating organic-rich materials such as compost heaps, seaweed piles, hay, straw, sawdust and coal refuse piles, domestic and industrial hot water and cooling systems, and steam lines and steam condensate lines.

The Major Groups of Thermophilic Microorganisms

The aerobic spore formers. *Bacillus stearothermophilus* is the most ubiquitous and widely studied thermophilic microorganism. The type strain of the species was described by Donk (1920) and for the following 50 years almost all studies of thermophilic microorganisms were centred on the genus

Bacillus and to a much lesser extent on *Clostridia*. The sixth edition of *Bergey's Manual* (Breed *et al.*, 1948) characterized 18 species of thermophilic *Bacillus* able to grow at or above 60 °C and listed a further 30 which were not individually characterized. Following the work of Gordon and Smith (1949) the seventh edition of *Bergey's Manual* (Breed *et al.*, 1957) had considerably rationalized the position, listing only *B. stearothermophilus* and *B. coagulans* as thermophilic species of *Bacillus*. Since then other thermophilic *Bacillus* isolates have been described which do not comply with the description of *B. stearothermophilus* and *B. coagulans* and consequently the taxonomy of thermophilic *Bacillus* has been re-examined (Klaushofer and Hollaus, 1970; Walker and Wolf, 1971; Sharp *et al.*, 1980; Wolf and Sharp, 1981; Sharp and Woodrow, 1982). There appears to be a minimum of four clearly defined taxonomic groups of neutrophilic thermophilic *Bacillus* species. The type strain of Donk (1920) (NCA 26) is taxonomically similar to most of the other early strains held by the National Canners Association (NCA) including NCA 1503, NCA 1518 and NCA 1356. *B. stearothermophilus* ATCC 8005 (*B. kausttophilus*) (Prickett, 1928) closely resembles *B. caldotenax, B. caldovelox* and *B. caldolyticus* (Heinen and Heinen, 1972) and differs from the type strain group in having a slightly high percentage of G + C, 62–66% compared with 59%. A third group are the non-amylolytic strains first described in detail by Epstein and Grossowicz (1969) and Daron (1970). The strains first described by Ambroz (1913) as *Denitrobacterium thermophilum* and later renamed *B. thermodenitrificans* by Mishustin (1950) are characterized predominantly by their flat, rough, erose colonies and their ability to reduce nitrate and nitrite under anaerobic conditions (Sharp and Woodrow, personal communication).

Minimal media for the cultivation of *B. stearothermophilus* strains have been described by Baker *et al.* (1960) and Rowe *et al.* (1975). Defined medium for *B. caldotenax* was described by Kuhn *et al.* (1979). Studies by Baker *et al.* (1960) of 68 strains of *B. stearothermophilus* indicated most of them to be auxotrophic for methionine and several required vitamins and other cofactors.

The thermophilic *Bacillus* species generally grows well on rich medium. Atkinson *et al.* (1979a) described the cultivation of *B. stearothermophilus* NCA 1503 in a 3000 l fermenter on sucrose (4% w/v), yeast extract (5% w/v) and various salts, which yielded 3 mg dry wt ml^{-1}.

B. stearothermophilus (NCA 1503) yields up to 9 mg dry wt ml^{-1} in a defined medium on glycerol when grown in continuous culture (Atkinson *et al.*, 1975b), which is higher than the yields obtained in batch culture or in complex medium (Sargeant *et al.*, 1971).

Acidophilic aerobic spore formers. *B. coagulans* (Hammer, 1915) is a facultative thermophile growing well at 45–55 °C. Physiologically strains of *B. coagulans* can be divided into two distinct groups (Wolf and Barker, 1968): type A, which grows at 65 °C if the pH is acidic (pH 6.2), and type B, which is more acidophilic and grows at pH 4.5.

B. coagulans is a saccharolytic, non-proteolytic, microaerophilic bacterium and has a G + C content of 47–48%. It has been implicated as a common cause of overacidification in canned fruits and vegetables, particularly tomato juice. Studies of 26 strains of *B. coagulans* by Baker *et al.* (1960) indicated that most strains were auxotrophic for methionine or glutamic acid.

Uchino and Doi (1967) described extreme thermoacidophilic strains of *Bacillus* isolated from hot springs in Japan. Similar isolations were reported by Darland and Brock (1971) who proposed the name *B. acidocaldarius*. The main characteristics were aerobic growth at 45–70 °C over a pH range of 2.0–6.0. The amount of G + C is 59–66% and the spores are remarkably heat sensitive (Wolf and Sharp, 1981).

Thermophilic clostridia. Anaerobic spore formers have been isolated from soil with a maximum growth temperature in the region of 75 °C. *Clostridium thermocellum*, which ferments cellulose and cellobiose to ethanol, acetate and CO_2, has an optimum growth temperature of 55–60 °C and a maximum in the region of 70 °C (McBee, 1950). *C. thermohydrosulfuricum* was first isolated from the extraction juices in Austrian sugar beet factories (Klaushofer and Parkinen, 1965) and has a maximum growth temperature of 76 °C. *C. thermosaccharolyticum* has an upper growth temperature of 67 °C and ferments a wide range of carbohydrates including glycogen, starch and pectin. *C. thermoaceticum*, which has a growth optimum of 60 °C, ferments glucose, fructose, xylose and pyruvate to acetate and has little DNA homology with the other thermophilic clostridia (Matteuzzi *et al.*, 1978).

Thermophilic clostridia are a problem in the food and canning industry since their spores survive normal heat treatment including boiling. They do, however, have considerable commercial potential in the production of organic acids and alcohols.

Thermus *species.* *Thermus aquaticus*, isolated from Yellowstone National-al Park, was first described by Brock and Freeze (1969) as a Gram-negative, aerobic rod able to grow at up to 75 °C. Similar strains have since been isolated from other geothermal areas and also from hot water systems (Brock and Boylen, 1973; Pask-Hughes and Williams, 1975). Other species described include *T. thermophilus* (Oshima and Imahori, 1971), *T. flavus* (Saiki *et al.*, 1972) and *T. ruber* (Loginova and Egorova, 1975). *T. aquaticus* is grown in a basal salts medium supplemented with a low concentration of organic constituents (0.1% w/v tryptone and 0.1% w/v yeast extract); higher concentrations are inhibitory to most strains. *T. thermophilus* will grow at concentrations of 1% organic material and can be cultivated on standard tryptone soya media. The optimum pH for growth is pH 7.0–8.0. The majority of strains are able to utilize pyruvate, glutamate and acetate as sole carbon source whilst a few utilize succinate, lactose, galactose, maltose, sucrose and trehalose (Munster *et al.*, 1984). Although the majority of strains appear to be strict aerobes some strains appear to grow anaerobically in the presence of nitrite. Preliminary taxonomic studies using numerical taxonomy of 55 strains indicate the occurrence of two major and two minor taxonomic groups (Munster *et al.*, 1984). Several strains produce extracellular amylase and protease.

Sulfolobus *species.*   *Sulfolobus acidocaldarius* has been isolated from hot springs in Yellowstone National Park, Italy, Dominica, El Salvador and Iceland. It is a facultative autotroph growing on elemental sulphur or on a range of amino acids and sugars (Brock, 1978). The pH range for growth varies from 1.0 to 4.0 and the optimum temperature is 60–70 °C. Oxidation of sulphur to sulphuric acid was reported by Shivers and Brock (1973). Reduction of ferric iron using elemental sulphur and glutamic acid as electron donors was demonstrated by Brock and Gustafson (1976).

Two other species have been described as *S. solfataricus* and *S. brierleyi* (Zillig *et al.*, 1980). The cells are small and spherical with a primitive submicroscopic morphology; they lack a true cell wall and are surrounded by a plasma membrane and a very fine extracellular coat. Pili appear to be involved in their attachment to substrates (Weiss, 1973).

Ferrolobus, an obligate autotroph which showed no growth in the absence of either sulphur or iron (Brierley and Brierley, 1973), was considered to be closely related to *Sulfolobus* (Brierley, 1977). Isolates described as *Caldariella acidophila* (De Rosa *et al.* (1975) were considered by Zillig *et al.* (1980) to be also closely related to *Sulfolobus* species.

Chloroflexus *species.*   The isolation of thermophilic, photosynthetic, flexibacteria was first reported by Pierson and Castenholz (1974). *Chloroflexus aurantiacus* has been isolated from hot alkaline springs in Yellowstone National Park, Iceland, Japan and New Zealand (Pierson and Castenholz, 1974). The optimum growth temperature is 52–60 °C at pH 7.8–8.0. It grows photoheterotrophically under anaerobic conditions, whereas in aerobic conditions it grows chemoheterotrophically in darkness or light. Details of isolation, media and cultivation are presented by Castenholz and Pierson (1983). Large-scale production of *C. aurantiacus* for the production of restriction endonucleases CauI and CauII has been carried out in medium D of Castenholz (1969) at pH 8.2 and supplemented with yeast extract (2 g l^{-1}). The cells were grown photoheterotrophically in a 550 l dual-connected tower reaction system equipped with illumination from tungsten filament lamps (Bingham and Darbyshire, 1982). Anaerobiosis was maintained by flushing the culture with oxygen-free nitrogen, the temperature being maintained at 55 °C. Cell yields of 1.2 g wet wt l^{-1} were obtained.

The Stability of Thermophilic Enzymes

The ability of microorganisms not only to survive but to grow and multiply actively at temperatures which normally denature proteins and particularly enzymes has led to considerable interest in the molecular basis of thermophily. A number of reviews discuss the stability of proteins and other cellular constituents at elevated temperatures (Singleton and Amelunxen, 1973; Williams, 1975; Ljungdahl and Sherod, 1976; Singleton, 1976; Welker, 1976; Amelunxen and Murdock, 1978; Zuber, 1978; Shilo, 1979; Zeikus, 1979; Wolf and Sharp, 1981; Hartley and Payton, 1983).

Comparative studies of homologous enzymes from thermophilic and mesophilic strains have been carried out by a number of researchers (Militzer et al., 1949; Amelunxen and Lins, 1968; Devanathan et al., 1969; Tanaka et al., 1971; Zuber, 1978). Generally proteins from thermophilic microorganisms showed greater thermal stability. A number of mechanisms have been proposed to account for this increase in stability. Singleton and Amelunxen (1973) suggested that thermophiles may contain factors which increase thermal stability, or conversely mesophiles may contain factors increasing thermal lability. Attempts to transfer heat stability or lability by mixing cell-free extracts from thermophiles and mesophiles failed to change the properties of the individual enzymes (Koffler, 1957; Koffler and Gale, 1957; Amelunxen and Lins, 1968). Thermal stability is retained following extensive purification and recrystallization of enzymes (Amelunxen and Murdock, 1978). Some enzymes from thermophiles have proved to be unstable at the optimum growth temperature of the organism and are stabilized by cell components such as membranes or cofactors. Welker (1976) demonstrated the thermal stabilization of alkaline phosphatase by the cell membrane. Wedler (1978) demonstrated glutamate synthetase to be stabilized threefold by (i) the binding of L-glutamate, ATP and Mn^{2+}, (ii) changes in protein structure caused by an increase in polar amino acids and probably by formation of disulphide bridges and (iii) enzyme aggregation driven by energy released from hydrophobic interactions.

Calcium has been demonstrated to confer stability on several extracellular enzymes such as α-amylase (Yutani, 1976), thermolysin (Tajima *et al.*, 1976) and proteases (Sidler and Zuber, 1977).

Most studies of thermostable proteins suggest an inherent stability of the protein structure which is dependent upon the amino acid sequence and composition. Several proteins from thermophiles and mesophiles have been demonstrated to differ in only a very few amino acids although they showed considerable differences in thermal stability. These changes affect the stability of the secondary and tertiary structure of the molecule by increases in hydrogen bonding, hydrophobic bonding and ionized group interactions. Most data appear to indicate fewer disulphide bridges in thermophilic enzymes compared with their mesophilic counterparts.

A study of clostridial ferredoxins from mesophilic and thermophilic species (Devanathan *et al.*, 1969; Tanaka *et al.*, 1971) indicated considerable similarity in their molecular weight, iron, inorganic sulphide and cysteine contents, adsorption spectra and number of amino acid residues. The proteins from the thermophilic strains, however, showed greater thermal stability. Ferredoxin (molecular weight 6000 daltons) has no secondary or tertiary structure and hence increased thermal stability is presumably associated with changes in the amino acid composition. Comparative studies of protein sequence data indicated that only the thermophilic ferredoxins contained histidine which replaced either serine or tyrosine (Table 1). These appear to serve as ligands for the tighter binding of atoms of iron, or the differences in charge may increase the opportunities for hydrogen bonding. There also appears to be a slightly greater number of basic amino acids in proteins from thermophiles. Perutz and Raidt (1975) suggested that this might permit the formation of additional ionic bonds. In the thermophilic ferredoxins a maximum number of four ionic bonds are formed by *Clostridium thermosaccharolyticum* and only one in the mesophile *C. acidurici*. Increased stability correlates with an increased number of glutamic acid residues. Six glutamic acid residues are present in ferredoxin from *B. stearothermophilus* (Hase *et al.*, 1976) and Robson and Pain (1971) indicated that glutamic acid residues are the best α-helix-promoting amino acids.

The half-life of rhodanese from *Thiobacillus denitrificans* was found to be 0.5 minutes at 65 °C; from *B. subtilis* the half-life was 4.5 minutes and from *B. stearothermophilus* 36 minutes. The only significant changes in the amino acid composition were in the aspartate:glutamate ratio from 8:2 in *T. denitrificans* to 5:5 in *B. subtilis* and to 1:10 in *B. stearothermophilus* (Atkinson, 1976). A change from aspartate to glutamate would require only a single base change in the third base of the triplet code.

A 3% increase in hydrogen bonding is sufficient to increase the melting temperature of a polypeptide chain (molecular weight 35 000 daltons) from 35 to 45 °C and up to 10 kcal mol^{-1} differences in activation energy can be

Table 1. *Mesophilic and thermophilic ferredoxins: differences in basic and acidic amino acid residues and possible ionic bond formation*

| | Basic residues | | | Acidic residues | | | | Possible salt bridges[a] | | |
	His	Lys	Arg	Glu	Gln	Asp	Asn	(1)	(2)	(3)
C. pasteurianum	0	1	0	2	2	5	3	1	0	0
C. acidurici	0	0	1	2	2	6	1	1	1	1
C. tartarivorum	2	2	0	5	2	4	0	4	2	2
C. thermosaccharolyticum	2	2	0	7	0	4	0	4	4	4

Data from Zuber (1978).
[a] Salt bridges: (1) based on amino acid composition; (2) based on amino acid sequence data; (3) based on three-dimensional structure data.

obtained by changes in a relatively few amino acid residues (Atkinson, 1976). A thirtyfold difference in thermal stability at 60 °C of triosephosphate isomerase from rabbit muscle and *B. stearothermophilus* can be accounted for by a difference in activation energy of only 2.2 kcal mol^{-1} (Hocking and Harris, 1976). The differences in thermal stability of the ferredoxins can similarly be accounted for by differences of only 4.5–8.5 kcal mol^{-1}.

Glyceraldehyde-3-phosphate dehydrogenase from rabbit muscle is 93% inactivated after 10 minutes at 70 °C, whereas the enzyme from *B. stearothermophilus* is relatively stable—only 5% inactivated after 10 minutes at 90 °C (Amelunxen, 1966). The enzyme from *T. aquaticus* is stable for several hours at 90 °C (Hocking and Harris, 1980). The stability of the *B. stearothermophilus* enzyme is attributed to additional hydrophobic interactions and an additional ionic bond. The *Thermus* enzyme is further stabilized by more hydrophobic interactions and increased ionic bond formation (Walker *et al.*, 1980).

The evidence indicates that small and subtle changes in a number of amino acid residues appear to account for significant differences in the thermal stability of enzymes which generally show close structural similarity. These substitutions are responsible for increasing the opportunities for hydrogen bonding, hydrophobic bonding and ionic bonding which appear to account substantially for the increased thermal stability of enzymes from thermophilic microorganisms.

Products, Processes and Applications for Thermophilic Microorganisms

Enzymes for Industry, Medicine and Research

Proteases. The commercial exploitation of protease enzymes was given considerable impetus in 1963 when the first biological washing powder was introduced into Europe. The leather industry utilizes protease enzymes for the removal of unwanted hair and interfibrillar material. The use of a protease and amylase mixture from *B. subtilis* as a mouthwash has been demonstrated to be effective against dental plaque (Shaver and Schiff, 1969). Proteases are widely used in the food industry including brewing, baking, meat tenderization and cheese processing.

Microbial proteases are classified into three main groups: acid proteases (maximum activity pH 2.0–4.0), neutral proteases (maximum activity pH 7.0–8.0), which are inhibited by metal chelators and often prove to be metalloenzymes with similar specificities for the peptide bonds cleaved (Morihara *et al.*, 1968), and alkaline proteases (maximum activity pH 9.0–

11.0) which cleave a wide range of peptide bonds, possess esterase activity and are unaffected by metal chelators.

The most widely studied thermophilic protease is thermolysin produced by *B. thermoproteolyticus* (Endo, 1962), an organism now considered to be a strain of *B. stearothermophilus*. Thermolysin is a neutral protease which differs from the *B. subtilis* enzyme mainly by retaining 86% of its activity after 30 hours at 70 °C.

Comparative studies of neutral protease from *B. thermoproteolyticus* and *B. subtilis* are presented by Keay and Wildi (1970) and Levy et al. (1975). Sidler and Zuber (1972) reported the isolation of two neutral proteases from a strain of *B. stearothermophilus* (NCIB 8924) isolated at two different temperatures. The enzyme produced at 40 °C was denatured between 55 and 60 °C at a rate similar to that for *B. subtilis* neutral protease. The enzyme produced at 50 °C was stable up to 65–70 °C, although it was not as stable as thermolysin which will survive up to 75–80 °C. Like the *B. subtilis* and *B. thermoproteolyticus* enzymes both of these proteases were stabilized by Ca^{2+} ions.

The production of neutral proteinase from *B. stearothermophilus* (NCIB 8924 and NRRL B-3880) at 53 °C is growth associated (Sidler and Zuber, 1977). The formation and stabilization of the enzyme was dependent upon the concentration of free Ca^{2+} ions. Excessive use of phosphate reduced the yields of protease. Procedures for purification of proteinase from these strains are reported by Sidler and Zuber (1980). *Thermoactinomycetes vulgaris* produces a thermostable neutral protease at 55 °C, with maximum activity in the range 60–80 °C. The enzyme is produced in the stationary phase of growth on corn starch and corn steep liquor (Klingenberg et al., 1979). Protease production is suppressed in the presence of glucose (Sonnleitner, 1984) and media based on lactose, starch, grain, fishmeal, cerelose, casein hydrolysate, soya extract or nutrient gelatin are generally used on the industrial scale (Keay et al., 1972). *T. aquaticus* strain YT-1 produces two proteases. Aqualysin I is an alkaline protease and is produced from mid-exponential growth until the cells cease to grow. Aqualysin II is a neutral protease and is produced after the cells have ceased to grow, a period of about 4 days (Matsuzawa et al., 1983). Production from cells is maintained for up to 17 days, at which time the cell density of the culture decreases to half the maximum reached during growth, although cells still remain viable. The optimum pH values for Aqualysin I and II activity were 10.4 and 7.0 respectively.

Caldolysin, a protease produced by *T. aquaticus* T351, is stable between pH 4.0 and pH 12.0 at 75 °C. It has a half-life of 30 hours at 80 °C compared with 1 hour at 80 °C for thermolysin (Morgan et al., 1981; Khoo et al., 1984); it is stabilized by the presence of six Ca^{2+} ions, whereas thermolysin is stabilized by four Ca^{2+} ions. The enzyme is produced on a trypticase yeast extract medium and appears to be the most thermostable protease described.

Cellulases.   Because of the increases in the price of oil and the predicted depletion of fossil energy reserves (Ghose and Ghosh, 1979), research has been directed into the development of renewable sources of energy and chemical feedstocks. The most widely available resource is cellulose which is available as agricultural residues or byproducts and as waste from industries and communities. The utilization of cellulose as an energy and chemical feedstock supply has been reviewed by Wilke (1975). Cellulose consists of chains of D-glucopyranose linked by β-1;4 glycosidic bonds; total hydrolysis yields D-glucose and partial hydrolysis yields cellobiose.

C. *thermocellum* and related strains have been widely studied for their cellulolytic properties (McBee, 1948, 1950; Ng *et al.*, 1977; Weimer and Zeikus, 1977; Ng and Zeikus, 1981). Extracellular cellulase production on cellulose is growth associated; the enzyme is stable at 70 °C for 45 minutes (Ng *et al.*, 1977). Maximum yields of cellulase are obtained using native cellulose and α-cellulose but not on carboxymethylcellulose or other carbohydrate sources. Cellulase activity was inhibited on cellobiose. The optimum pH and temperature for endoglucanase activitiy are pH 5.2 and 65 °C, while optimum exoglucanase activity is observed at pH 5.4 and 64 °C.

The ratio of endoglucanase:exoglucanase activity is higher than that found in cellulase from *Trichoderma reesei* (Ng and Zeikus, 1981). Extracellular cellobiase and β-xylosidase activity are absent; the initial hydrolysis of microcrystalline cellulose produces long chain oligosaccharides, whilst longer hydrolysis (24 hours) produces cellobiose and xylobiose. A total of 39 strains of *Clostridia* hydrolysing cellulose and the hemicellulose xylan were isolated from various anaerobic digesters (Le Ruyet *et al.*, 1984) and were classified into three main groups according to their hydrolytic properties. Strains in group I hydrolysed cellulose and cellobiose and appeared to be strains of C. *thermocellum*. Strains in group II did not hydrolyse cellulose but fermented a range of polysaccharides, including starch, xylan and cellobiose. Strains in group III did not hydrolyse cellulose and hydrolysed cellobiose only slowly, whereas xylan and starch were actively fermented.

The preliminary characterization of seven anaerobic microorganisms isolated from geothermal sites in the USA and able to hydrolyse xylan was reported by Weimer *et al.* (1984). Harkin *et al.* (1974) discussed the use of thermophilic Actinomycetes in the hydrolysis of waste from paper mills and the production of high value protein as animal feed. Thermophilic Actinomycetes produce extracellular cellulase with pH and temperature optima of 6.0 and 65–70 °C respectively. Their use as a source of single-cell protein following digestion of ligocellulose is discussed by Bellamy (1977).

Bacillus species 11–15 produced an extracellular xylanase with optimum activity at pH 4.0 and was stable at 70 °C for 15 minutes. The enzyme was also active on cellulose and carboxymethyl cellulose (Uchino and Nakane, 1981).

Amylases. Amylases are widely used in the textile, confectionery, paper, brewing and alcohol industries. They are utilized where rapid hydrolysis of starch is required or the viscosity of a starch slurry needs to be lowered. α-amylase (α-1,4-glucan-4-glucanohydrolase) is produced by a wide range of microorganisms, and the thermophiles are no exception. The enzyme hydrolyses α-1,4 glucosidic linkages in amylose, amylopectin and glycogen in an endo fashion, the end products of its action being sugars having the α-D configuration. Bechina *et al.* (1983) reported investigations of 237 strains of thermophilic *Bacillus*, Actinomycetes and fungi for the production of α-amylase. The production of α-amylase from *B. stearothermophilus* is generally growth linked such that in batch culture production occurs in the exponential phase of growth (Sidler and Zuber, 1977; Davis *et al.*, 1980). Similar results have been reported for other thermophilic *Bacillus* isolated from sewage sludge (Grueninger *et al.*, 1984) and *B. caldolyticus* (Emanuilova and Toda, 1984) where production is constitutive throughout the exponential and stationary phases. This is unlike α-amylase production from *B. subtilis* and *B. licheniformis* which occurs between the late exponential and the stationary phase, prior to the onset of sporulation (Priest, 1977). Amylases from thermophiles are generally stabilized by Ca^{2+} ions (Oghasahara *et al.*, 1970). The enzyme produced by *B. caldolyticus* divides into non-active subunits upon loss of Ca^{2+} (Heinen and Lauwers, 1975), but activity is restored by the addition of 0.5 mM Ca^{2+} at 70 °C. Similar results were observed with the enzyme from *B. stearothermophilus* NCA 1503 (Pfueller and Elliott, 1969). *B. stearothermophilus* strain NCA 1518 does not produce amylase in the absence of starch (Campbell, 1955). *B. caldolyticus* produces amylase on starch, glucose or brain heart infusion (Grootegoed *et al.*, 1973).

The production of an inducible thermoacidophilic α-amylase from *B. acidocaldarius* was reported (Buonocore *et al.*, 1976) with temperature and pH optima for activity of 75 °C and pH 3.5 respectively. A similar thermoacidophilic α-amylase was described by Uchino (1982) with a temperature optimum of 70 °C and a pH optimum of 2.0. Both these enzymes were non-growth associated, only being produced during the stationary phase.

Other significant thermophilic enzymes. β-galactosidases can be used industrially for hydrolysing lactose syrups and mixtures to glucose and galactose. They may also have clinical applications since there is evidence of a lactase deficiency in the population which is either inherited or the result of ageing. Such deficiencies lead to intestinal disorders when dietary lactose is high.

β-galactosidase has been isolated from a number of thermophiles including *T. aquaticus* (Ulrich *et al.*, 1972; Cowan *et al.*, 1984), *B. stearothermophilus* (Goodman and Pederson, 1976), *B. coagulans* (Long and Lee, 1974) and *Caldariella acidophila* (Buonocore *et al.*, 1980). *B. stearothermophilus* produces the enzyme constitutively on tryptone and yeast extract media. It has pH and temperature optima of pH 6.0–6.4 and 65 °C. The enzyme has a half-life of 20 days at 50°C: however, after 10 minutes at 60 °C, 60% of enzyme activity was lost. *Thermus* strain 41A (Cowan *et al.*, 1984) cultured at 75 °C produced β-galactosidase with a half-life of 8 minutes at 90 °C. At 65 °C the enzyme lost 10% of its activity after 36 days. *B. coagulans* produces a β-galactosidase with optimum activity at 60–65 °C and pH 4.5–6.0.

α-glucosidase hydrolyses α-1,4 and/or α-1,6 linkages in short chain saccharides which arise from the activity of other amylolytic enzymes on starch. The production of an extracellular α-glucosidase from *B. stearothermophilus* KP 1006 was optimal at 60 °C and pH 6.5 in medium containing soluble starch, peptone, yeast extract and meat extract (Suzuki *et al.*, 1976). The enzyme accumulated in the cytoplasm during exponential growth and was secreted into the medium from mid-logarithmic growth to a maximum at the end of the stationary phase. The enzyme was stable at 60 °C. α-glucosidase production from *Thermoactinomycetes* at temperatures of 40–70 °C is described by Horwath *et al.* (1977).

Cyclodextrin glycosyltransferase, which produces a series of non-reducing cyclic dextrins from starch, is produced by *B. stearothermophilus* strains (Shiosaka, 1976). Maximum production on a medium containing soluble starch and mineral salts was obtained after 48 hours of growth at 50 °C.

Glucose isomerase, which is widely used in the food industry to convert glucose to fructose for use as a sweetener, is produced by *B. coagulans* (Outtrup, 1976; Diers, 1977). The organism is cultured at 50 °C with oxygen limitation in a glucose–mineral salts medium. *B. stearothermophilus*, cultured at 55 °C on a xylose, starch, peptone, corn steep liquor, meat extract and yeast extract medium, produced glucose isomerase which was relatively heat stable at 80 °C for 30 minutes (Chen, 1980).

Alcohol dehydrogenase is of commercial interest for its analytical use in the detection of alcohol (Smith and Olson, 1975) and in stereospecific organic synthesis (Jones and Beck, 1976). Alcohol dehydrogenase produced from *Thermoanaerobium brockii* is stable at high solvent concentrations and can be used at 60 °C with 0.4% v/v or less of alcohol or ketones. It has a wide substate specificity including linear and branched alcohols and linear and cyclic ketones (Lamed *et al.*, 1981). The enzyme was immobilized in an enzyme electrode and was stable for 2 weeks at 25 °C and qualitatively detected alcohols and ketones at 0–200 µM in test solutiions. A similar enzyme has been described from *C. thermohydrosulfuricum* (Lamed and Zeikus, 1981). Alcohol dehydrogenase from *Thermoanaerobacter ethanolicus* is NADP dependent, is stable at 70 °C for 2 days and has a maximum activity at 95 °C. Ethanol is a substrate, but secondary alcohols such as 2-propanol, 2-butanol and 2-pentanol are oxidized more quickly (Bryant and Ljungdahl, 1981; Ljungdahl *et al.*, 1981).

Hydrogenases are found in many microorganisms and algae; they reversibly activate molecular hydrogen. In recent years there has been considerable interest in the biophotolysis of water by solar energy to produce hydrogen. Most of the systems examined depend on hydrogenase for the evolution of hydrogen from a photochemically reduced compound (Krasna, 1979). Cell-free extracts of *C. thermoaceticum* have been demonstrated to catalyse the hydrogen-dependent reduction of various artificial electron acceptors. The activity of the hydrogenase was optimal at pH 8.5–9.9 and was stable for at least 32 minutes at 55 °C. *Methanobacterium thermoautotrophicum* produced a nickel-containing hydrogenase; when immobilized the enzyme showed no loss of activity under H_2 over 2 weeks at 25 °C (Graf and Thauer, 1981).

Restriction endonucleases for gene cloning have been isolated from a number of thermophiles including *B. stearothermophilus* Bst II (Roberts, 1980), *B. caldolyticus* Bcl I (Bingham *et al.*, 1978), *T. thermophilus* TtH III (Shinomiya and Sato, 1980), *T. aquaticus* Taq I (Sato *et al.*, 1977) and *C. aurantiacus* Cau I, Cau II (Bingham and Darbyshire, 1982). Modificiation methylases which are used in the preparation of gene libraries have been isolated from *T. aquaticus* and *T. thermophilus* (McClelland, 1981).

L-asparaginase, which is used in the treatment of lymphatic leukaemia, is produced by a range of microorganisms including *Erwinia chrysanthemi* (Elsworth *et al.*, 1973). L-asparaginase produced from *T. aquaticus* strain T351 has a molecular weight of 60 000 daltons and a half-life of 25 minutes at 85 °C (Guy and Daniel, 1982).

The use of enzymes in clinical diagnosis is becoming increasingly significant. The absolute specificity of many enzymes is utilized to measure the concentration of metabolites such as glucose, cholesterol, triglycerides and urea in body fluids. Glycerokinase from *Escherichia coli* and *Candida mycoderma* is commercially available for the estimation of serum triglycerides; this is of use in the diagnosis of medical conditions such as arteriosclerosis. The enzyme, which is widely used in clinical laboratories, has been isolated from constitutive mutants of *B. stearothermophilus* NCA 1503. The half-life for this enzyme at 60 °C is 310 minutes compared with 4.5 minutes and 0.72 minutes for the enzyme from *E. coli* and *C. mycoderma* respectively (Comer *et al.*, 1979). A similar system using glycerol dehydrogenase from *B. stearothermophilus* RS93 was described by Atkinson *et al.* (1979a).

Atkinson *et al* (1979a) described the isolation of over 15 enzymes from 70 kg of *B. stearothermophilus* cell paste, including superoxide dismutase, rhodanese, tyrosyl–tRNA synthetase and tryptophanyl–tRNA synthetase.

Alcohol Production

The increasing costs of petroleum and natural gas have focused attention on the bioconversion of renewable biomass resources such as cellulose, hemicellulose and starch into ethanol via anaerobic fermentation. The production of ethanol from the fermentation of glucose by *Saccharomyces cerevisiae* is well known and is still the most efficient conversion. There are several disadvantages in using yeast, particularly their restricted substrate range. One of the advantages of using a thermophilic process is the possibility of being able to evaporate or distil the ethanol produced (Cysewski and Wilke, 1978). A continuous process could be envisaged where carbohydrate is fed into an active thermophilic fermentation. As ethanol is produced it would continuously evaporate from the vessel into a suitable collecting device. The production of ethanol by thermophilic bacteria has been reviewed by Wiegel (1980), Zeikus *et al.* (1981) and Payton (1984).

C. *thermocellum* degrades cellulose directly to ethanol, acetic acid, hydrogen and carbon dioxide. Continuous fermentation at 60 °C and pH 7.0 using 1.5% w/v and 3% w/v cellulose as a limiting substrate yields maximum ethanol concentrations of 0.3% and 0.9% respectively. The overall yields of ethanol are approximately 0.45 and 0.75 g g^{-1} of cellulose utilized (Zertuche and Zall, 1982). C. *thermocellum* has also been examined in mixed culture with C. *thermosaccharolyticum* for ethanol production from cellulose (Avgerinos *et al.*, 1981). Mixed cultures of C. *thermocellum* and C. *thermohydrosulphuricum* fermented a variety of cellulosic substrates including hexose and pentose components in polymeric wood saccharide, resulting in ethanol concentrations considerably higher than those with C. *thermocellum* alone. C. *thermohydrosulphuricum* is able to utilize pentoses as well as rapidly utilizing glucose and cellobiose. Coupled with the ability of C. *thermocellum* to degrade β-1,4 xylans and glucanes, a twofold increase in ethanol concentrations was obtained compared with the use of monocultures (Ng *et al.*, 1981). The ethanol concentration did not exceed 1% and the selection of strains of C. *thermocellum* which are relatively tolerant to 6% ethanol should be advantageous (Avgerinos *et al.*, 1981). C. *thermohydrosulfuricum* strains tolerant to 8% ethanol at 45 °C have been isolated (Lovitt *et al.*, 1984) and in mixed cultures these may lead to increased yields.

T. *brockii* (Zeikus *et al.*, 1979) is a thermophilic strict anaerobe which produces lactic acid. It can be modified to produce ethanol as a result of the regulatory properties of lactate dehydrogenase which is specifically activated by fructose-1,6-diphosphate (Lamed and Zeikus, 1980). The ethanol:lactate ratio is also affected by the yeast extract concentration in the medium. Compared with glucose, growth on starch produced a twofold increase in ethanol, only half the yield of lactic acid but twice the level of acetic acid. Lamed and Zeikus (1980) carried out comparative studies with *C. thermocellum* and *T. brockii*. Batch and continuous culture studies were reported by Sonnleitner *et al.* (1984). *Thermoanaerobacter ethanolicus*, a non-sporing anaerobic bacterium, grows between 37 and 78 °C and pH 4.4–9.8. Its main fermentation products from hexoses were ethanol and CO_2. It has a wide substrate range, producing ethanol (Wiegel and Ljungdahl, 1981) as the main fermentation product from starch, cellobiose, lactose, xylose, fructose, ribose, maltose, mannose, galactose and pyruvate. Carriera *et al.* (1984) reported studies on the regulation of carbohydrate utilization by *T. ethanolicus*. A mutant strain JW200 Fe(4) was able to produce up to 3% v/v ethanol from 6% w/v soluble corn starch. Wiegel *et al.* (1984) demonstrated that JW200 Fe(4) utilized xylan from the hemicellulose fractions of steam-exploded birch and beechwood.

B. stearothermophilus NCA 1503 grown on complex medium with 3% w/v glucose produced ethanol in continuous culture at 65 °C under anaerobic conditions (Atkinson *et al.*, 1975a). The major fermentation product was lactate; however, selection of strains deficient in lactate dehydrogenase resulted in a 100% increase in the yield of ethanol (Payton, 1984).

Acetic Acid Production

The main organisms currently used in the commercial production of acetic acid are *Acetobacter schuetzenbachii* or *A. curvum* in the quick vinegar process and *A. orleanense* used in the quick vinegar or the Orleans process. Acetate is formed as a byproduct of thermophilic fermentations producing alcohol.

C. thermoaceticum ferments 1 mol of glucose to 3 mol of acetic acid at pH 7.0. Since the production of acetic acid reduces the pH the fermentation stops unless the pH is controlled with sodium hydroxide. This permits fermentation to continue but complicates the recovery of the acetic acid since it must be removed as an acetic acid salt and regenerated to free acetic acid. To overcome this problem Swartz and Keller (1982a,b) adapted an acid-tolerant strain which was capable of growth at pH 2.0–5.0 and between 45 and 65 °C. On a nutrient medium containing tryptone (0.25% w/v) and yeast extract (0.25% w/v), 4.5 g of acetic acid per litre was produced.

Methane Production and Waste Treatment

Thermophilic methane production by non-defined mixed cultures has been a feature of the anaerobic digestion of stabilized sludge in municipal sewage treatment plants. The value of methane as a fuel which can be utilized in existing equipment or stored and transported has promoted considerable research. Anaerobic digestion has not been generally accepted since it suffers from two main disadvantages: (i) the process has poor stability to shock loadings and (ii) it has a long solids residence time (SRT), up to 10–30 days for some plants (Spencer, 1979).

Comparative studies of mesophilic and thermophilic digesters (Cooney and Wise, 1975) indicated several advantages in using a thermophilic process (65 °C). The rate of digestion and the conversion of organic waste to methane were increased, the fluid viscosity and the amount of biomass formed were decreased and the accumulation of bacterial and viral pathogens was inhibited. Studies on the effect of temperature and SRT on methane production from beef cattle waste indicated little difference in the rates of methane production between 40 and 60 °C. The microbial communities in digesters vary with temperature and there is evidence that the population of thermophilic microorganisms selected at 60 °C has restricted catabolic activity. Chen (1983) indicated that enrichment of the fermentation with thermophilic, celluloytic, pectinolytic and acetate-utilizing bacteria was essential for the digestion of municipal sewage at 60 °C.

Anaerobic digestion studies of cattle waste using thermophilic processes have been examined (Varel *et al.*, 1977; Rorick *et al.*, 1980; Shelef *et al.*, 1980). Shelef *et al.* (1980) reported studies using 1000 and 10 000 l vessels for pilot studies in which the digested sludge was used in animal feeding and fertiliser trials and the generated methane was examined in gas utilization studies. Hashimoto (1983) discussed the use of straw–manure mixtures for the production of methane at 35 and 55 °C. Little information is available on the bacterial species present in thermophilic fermenters. Converse *et al.* (1977) reported *C. thermocellum* and *M. thermoautotrophicum* to be the main cellulolytic and methanogenic bacteria in a thermophilic cattle waste digester. Varel (1984) reported preliminary characterization data on anaerobic thermophiles isolated from a methane-producing beef waste digester.

The use of *M. autotrophicus* (Zeikus and Wolfe, 1972) may eventually have potential for the production of natural gas (Zeikus, 1979).

Thermophilic Spores

B. stearothermophilus spores are available as commercial preparations for the monitoring of steam sterilization processes. The use of *B. stearothermophilus* spores has also been investigated as a tracer to follow water movements. Satter *et al.* (1972) considered them to be ideal as airborne tracers.

Spoilage in the Food Industry

Thermophilic species of *Bacillus* were isolated as spoilage organisms growing in hot sucrose solutions at 60 °C in sugar refineries (Laxa, 1900). They were reported by Jacobsen (1918) to be actively growing in milk during pasteurization. Cameron and Esty (1926) at the NCA, Washington, District of Columbia, isolated over 200 strains of *Bacillus* from spoilt cans, 150 of which proved to be facultative or obligate thermophiles. *B. stearothermophilus* and *B. coagulans* were originally named the "flat sour bacteria". The name was derived from the flat ends of the cans of food spoilt by these organisms (there was no gas production and hence no swelling of the cans). Flat sour spoilage of acidic foods such as canned tomatoes or citrus fruit is generally caused by *B. coagulans* strains, whereas *B. stearothermophilus* is responsible for the spoilage of low acid foods such as canned peas, beans, corn and asparagus. Flat sour spoilage may be caused by a number of *Bacillus* species, but generally mesophilic spores are less heat resistant and are usually killed by heat processing. Obligate thermophiles do not cause spoilage unless the product is retained at high temperatures for an extended time or cooling is carried out very slowly; alternatively facultative thermophiles are able to grow at room temperature. The main source of spoilage organisms is the plant equipment; once established the organisms can be isolated from warm holding tanks, blanchers and warm filler bowls (Ito, 1981).

Klaushofer *et al.* (1971) discussed the problems of thermophilic species of *Bacillus* in the refining of sugar beet. The extraction process for beet pulp is generally completed at temperatures between 65 and 73 °C and organisms isolated at this stage included *B. stearothermophilus, B. thermodenitrificans, B. coagulans* and *B. sphaericus.* The consequence of bacterial activity during the extraction of beet pulp was loss of sucrose through conversion into acids, an increase in the production of molasses and a decrease in the extraction yield. Deterioration in the quality of the final product resulted from

(1) contamination with microorganisms resulting in an increase in the ash content,
(2) reduction of nitrate to nitrite resulting in the production of imido sulphonates through the reaction of nitrite and sulphite in factories which sulphurize dilute juices, and
(3) crystallization of the imido sulphonates with the sucrose and a consequent increase in the ash content.

The Advantages of using Thermophilic Microorganisms in
Biotechnological Processes

As a consequence of the thermal stability of their enzymes, as well as of a number of other significant processing advantages, thermophiles have been the subject of intense investigation. Some of the major advantages of thermophilic processes are as follows.

(1) The costs of cooling large-scale thermophilic fermentations are reduced; expensive heat exchange and refrigerated equipment required for mesophilic cultures is not required since cooling water at ambient temperature is sufficient. Most of the heat required for the fermentation is self-generated by exothermic growth of the microorganism.

(2) The reduced viscosity of the medium increases the efficiency of mixing, reduces the power input to stirrers and aids harvesting by centrifugation.

(3) The solubility of reactants is increased and permits the use of higher concentrations of certain relatively insoluble compounds.

(4) Volatile products such as ethanol may be removed from the fermentation broth by applying a mild vacuum or by sparging with CO_2. This reduces the build-up of inhibitory levels of alcohol in the broth.

(5) Operation at elevated temperatures reduces the possibility of contamination by mesophilic microorganisms but does not eliminate contamination by thermophilic spore formers.

(6) The decreased solubility of oxygen is advantageous when culturing strict anaerobes.

(7) Thermostable enzymes are more resistant to the denaturing activities of detergents and organic solvents.

(8) In theory the catalytic activity and hence the growth rate should increase with temperature. *B. caldolyticus* has been shown to have a doubling time of 15 minutes at 60 °C (Emanuilova and Toda, 1984); this is not a general rule, however, and other thermophiles are much slower.

(9) Immobilized enzyme reactors would have a longer useful period of operation, thus reducing costs.

(10) Enzyme isolation and purification can be carried out at room temperature.

(11) Higher enzyme recoveries may be obtained owing to the enhanced enzyme stability.

Process Considerations

The operation of fermenters and laboratory equipment at high temperatures has an effect on the medium, equipment and system design. The thermal lability of medium supplements such as vitamins, cofactors and antibiotics should be considered when operating at extreme temperatures. The solubility of O_2 in water at 70 °C is 5% of the solubility at 20 °C; consequently aerobic cultures require greater mixing and aeration. Increased aeration results in increased evaporation which leads to the concentration of potentially toxic fermentation products and condensation in exit gas lines and filters. Compensation for loss of water can be made and evaporation reduced by increased condenser capacity.

The sensitivity of *T. aquaticus* and *T. thermophilus* to polypropylene or polyacetal tubing connections at 80°C has been observed (Sonnleitner *et al.*, 1982). The use of this material in the feed lines of chemostats of *B. caldotenax* was also inhibitory. Furthermore, silicon tubing sterilized with medium was found to be slightly inhibitory to *Thermus* strains. Similar results were found with the silicon rubber membranes used to seal ports on fermenters. The use of different qualities of stainless steel (AISI 304L and AISI 316L) combined within one reaction vessel resulted in the solubilization of metal ions and resulted in loss of culture reproducibility or total failure of organisms to grow (Sonnleitner *et al.*, 1982).

The common suggestion that contamination is less of a problem with thermophilic cultures is not generally borne out in practice. Contamination of *Thermus* cultures with thermophilic, aerobic spore formers is not uncommon; Sharp *et al.* (1979) reported the contamination of a 400 l culture of *B. stearothermophilus* NCA 1503 with a bacteriocinogenic strain of *B. stearothermophilus*. One point regarding media is significant: batches of gelatin used in our laboratory and obtained from a well-known supplier were found be contaminated with spores of a thermophilic *Bacillus*. The spores were not killed by normal sterilization and only germinated following incubation of the medium at 60 °C. No germination of spores and hence evidence of contamination was observed at 37 °C.

The study and exploitation of microorganisms which are extreme in their environmental requirements requires consideration of the suitability of materials and equipment exposed to these conditions. With a few exceptions, most large-scale fermentation and bioprocessing equipment is manufactured from stainless steel. Stainless steel combines robustness, good machinability and good resistance to corrosion, the latter being derived primarily from the addition of chromium. To provide greater resistance to corrosion in specific environments other elements are added, giving a range of stainless steels each with its own specific characteristics (Table 2).

The addition of nickel improves the ability of stainless steel to form a protective oxide film (passive film) over the surface, especially in the presence of strong oxidizing acids such as HNO_3. This process is known as passivation. When exposed to strong hydrochloric acid, acidity dominates the mildly

Table 2. *The chemical composition of various grades of stainless steel and their relative resistance to corrosion*

Type of steel	AISI[a] No	Cr (%)	Ni (%)	Mo (%)	Resistance to acid salts	Resistance to chemical oxidation	Resistance to chemical reduction
Austenitic	302	18	9	—	—	+	—
Austenitic	303	18	9	—	—	—	—
Austenitic	304	19	10	—	—	+	—
Austenitic	316	17	12	2	+	+	+
Austenitic	Carpenter 20[b]	29	20	2.3	+	+	+
Austenitic	Nitronic 60[c]	17	9	—	—	+	—
Ferritic	410	12.5	—	—	—	—	—
Ferritic	430	16	—	—	—	+	—

[a] AISI, American Iron and Steel Institute (Weast and Selby, 1966; Davies, 1982).
[b] Carpenter 20 also contains 3.3% Cu; it has the best corrosion resistance.
[c] Nitronic 60 also contains 8% Mn and 4% Si.

oxidizing capacity of dissolved oxygen and corrosion is rapid. In dilute phosphoric acid stainless steels are generally passive, particularly in the presence of dissolved oxygen or oxidizing ions such as the ferric ion. Organic acids are considerably less corrosive than inorganic acids. Acetic acid solutions are on the boundary of being passive or active. Austenitic stainless steels become active as the temperature is increased.

Sodium hydroxide at low temperature and low concentration causes negligible corrosion; at higher concentrations corrosion is increased as pressure is applied.

The addition of molybdenum to stainless steels improves resistance to pitting corrosion, particularly in the presence of liquids with high chloride contents. Halide ions are particularly effective in penetrating passive films and causing pitting. Further discussions regarding microbial corrosion can be found in Tiller (1982) and Wakerley (1979).

Alkalophilic Microorganisms

Definition

Alkalophiles may be defined as microoganisms with optimum growth in alkaline environments. Two classes of alkalophiles may be distinguished; those that are alkaline tolerant (alkalotolerant) and those that are obligately alkalophilic. Alkalotolerant microorganisms grow optimally at pH values between 8.5 and 9.0, but they can also grow under neutrophilic conditions. Such strains cannot normally grow above pH 9.5. In contrast, obligate alkalophiles do not grow at neutral pH values and often grow optimally at pH values in excess of pH 10.0.

The criterion which assigns alkalophiles to one or other group is very arbitrary and is often further complicated by differing pH profiles obtained using different substrates. *B. alcalophiles*, for example, grows at higher pH values on L-malate as a substrate than with lactose (Guffanti *et al.*, 1978, 1979a).

Microorganisms from alkaline environments represent a relatively new, unexplored and unexploited area of microbiology. Alkalophilic bacteria and enzymes have been used in a number of industrial applications and have the potential of yielding more new novel products, enzymes and applications.

Occurrence, Isolation and Classification

Alkalophiles are most commonly isolated from alkaline environments and include a wide spectrum of microorganisms, including both photosynthetic and non-photosynthetic bacteria, fungi, yeasts and blue–green algae. Recent studies imply that alkalophilic microorganisms are more ubiquitous than first considered with the isolation of alkalophilic bacteria from both neutral and acidic (pH 4.0) soils (Horikoshi and Akiba, 1982).

Man-made alkaline environments are generally the result of waste from industries such as the food processing and the textile industry. Natural alkaline environments are relatively uncommon but include desert soils (Rupela and Tauro, 1973) and springs (Souza et al., 1974). The most alkaline natural environments are soda lakes and deserts where pH values in excess of pH 10.0 have been recorded. Such areas are listed by Te-Pang (1890), but two of the most well known are Wadi Natrum in Egypt and Lake Magadi in Kenya. Environments such as these usually develop in a closed basin where concentration by evaporation takes place. The alkaline nature of these areas is believed to result from the leaching of metal bicarbonates by CO_2-charged water from surrounding rocks; this is concentrated to produce an $NaHCO_3$–Na_2CO_3 brine (Hardie and Euguster, 1970). The environment may also contain high concentrations of other salts such as NaCl.

Alkalophilic microorganisms can be isolated using alkaline media usually containing Na_2CO_3, although $NaHCO_3$, K_2CO_3, Na_2PO_4 or occasionally $Na_2B_4O_7$ can be used. Sodium carbonate, which commonly occurs in natural alkaline environments, is generally considered to be most successful (Nakamura and Horikoshi, 1976a; Imhoff and Trüper, 1977; Grant et al., 1979). At concentrations between 0.5% and 5% w/v Na_2CO_3, pH values between 8.5 and 11.0 may be achieved, although at high pH values the medium should be checked regularly as absorption of CO_2 from the atmosphere may cause a decrease in pH. Isolation may be performed with solid or liquid media at mesophilic temperatures. A detailed account of isolation procedures is given by Grant and Tindall (1980).

The genus *Bacillus* provides the most abundant source of alkalophilic bacteria. Both alkalotolerant and obligately alkalophilic *Bacillus* strains have been isolated, but the species identification is often uncertain. Vedder (1934) reported the first alkalophilic *Bacillus* with an optimum pH of 8.6–10.0, this organism subsequently being named *B. alcalophilus*. Other alkalotolerant strains isolated later include *B. circulans* (Chislett and Kushner, 1961; Guffanti et al., 1979b) and *B. firmus* (Guffanti et al., 1980). There is also a report of alkalophilic adaptation in a strain of *B. cereus* (Kushner and Lisson, 1959). *B. pasterii* is a novel alkalophile in that it requires ammonium ions as well as a high pH for growth (Gibson, 1934). It has been considered an obligate alkalophile growing well at pH 11.0; however, these values are disputed and may be somewhat lower (Krulwich and Guffanti, 1983).

Obligate *Bacillus* alkalophiles are relatively common (Horikoshi, 1971a,b,

1972; Boyer and Ingle, 1972; Horikoshi and Atsukawa, 1973a,b; Ohta *et al.*, 1975; Ando *et al.*, 1981a,b). However, very few have been assigned a species name, reflecting the inadequate taxonomy of these organisms. *B. alkaliphilus* has been proposed as a new species of obligate alkalophile isolated from fermenting indigo leaves. Its pH optimum is reported to be pH 10.0–11.5 with a temperature optimum of 35–40 °C and it requires a peptide growth factor for normal growth (Takahara *et al.*, 1961; Takahara and Tanabe, 1962). Guffanti *et al.* (1980) isolated an obligately alkalophilic strain of *B. firmus* which grew exclusively at pH values in excess of pH 10.0.

A number of other Gram-positive alkalophiles have also been identified. Akiba and Horikoshi (1976) isolated an alkalophilic strain No 31-2 identified as a *Micrococcus* species. The strain, which grew between pH 9.0 and 11.0 and at 20–37 °C, produced an intacellular β-galactosidase. Souza *et al.* (1974) reported the isolation of alkalophilic *Clostridium* species.

An obligately alkalophilic *Streptomyces* species grew well at pH 9.0 and produced an alkaline protease (Nakanishi, 1975). It was named *S. alkalophilus* on the basis of its alkalophilic growth but resembles *S. griseus* in some of its characteristics.

Souza and Deal (1977) isolated an alkalophilic coryneform bacteria (A-1) from an alkaline spring. Similar strains were isolated from potato-processing effluent, although these strains produced an orange cellular pigment (Gee *et al.*, 1980). *Corynebacterium* species designated 93-1 and 150-1 isolated from soil had a pH range of 7.0–11.0 with an optimum of 9.6 and a temperature range of 20–40 °C (Kobayashi and Horikoshi, 1980; Kobayashi *et al.*, 1980). Gram-negative alkalophiles have also been reported including *Pseudomonas* species isolated from soil and industrial alkaline wastes (Hale, 1977; Watanabe *et al.*, 1977). A *Flavobacterium* species was isolated from a Californian mineral spring (pH 10.8–12.1) by Souza *et al.* (1974). It was obligately alkalophilic, unable to grow at pH 7.6 and had a pH optimum of 9.0–10.0. Strains of *Halobacterium* (Tindall *et al.*, 1980) and the photosynthetic bacterium *Ectothiorhospira* (Grant *et al.*, 1979) have been isolated from the alkaline soda lakes (pH 10.0–11.0) in Kenya. These strains required a high salt concentration for growth.

Cyanobacteria have a preference for alkaline environments (Fogg, 1956), although they are not generally believed to be obligate alkalophiles. Physiological studies of *Microcystis aeruginosa* and *Synechococcus* suggest that they are alkalotolerant (McLachlan and Gorham, 1962; Kallas and Castenholz, 1982a,b).

Microbial Physiology in Alkaline Environments

In order to colonize alkaline environments, alkalophilic microorganisms have evolved highly specialized physiological processes. The physiology of alkalophilic microorganisms has recently been reviewed by Horikoshi and Akiba (1982) and Krulwich and Guffanti (1983).

Intracellular pH may be measured by a number of methods (Mizushima *et al.*, 1964; Mela, 1966; Harold *et al.*, 1970; Hsung and Haug, 1975; Thomas *et al.*, 1976; Ogawa *et al.*, 1978); however, the most widely used technique is the use of weak acids, such as 5,5-dimethyl-2,4-oxazolidiniedione (DMO). DMO passively diffuses across membranes, particularly in cells whose internal pH is more alkaline than the external pH (Waddel and Butler, 1959). The internal pH can then be calculated from the distribution by means of the Henderson–Hasselbach equation. The intracellular pH of a number of bacteria have been calculated this way including *B. acidocaldarius*, pH 6.5 (Thomas *et al.*, 1976), *S. faecalis*, pH 7.7 (Harold *et al.*, 1970), *B. megaterium*, pH 8.03 (Decker and Lang, 1978), and *E. coli* K12, pH 7.7 (Padan *et al.*, 1976). DMO is unsatisfactory where the pH of the external environment is higher than that of the internal environment, as is the case with alkalophiles. Guffanti *et al.* (1978) reported that no DMO uptake could be observed when attempts were made to measure the intracellular pH of *B. alkalophilus*.

An alternative method using the weak base methylamine was subsequently developed (Rottenberg *et al.*, 1972). Methylamine can also penetrate membranes in its neutral form and has successfully been used to determine the intracellular pH of *S. lactis* (Kashket and Wilson, 1973) and *B. alkalophilus* (Guffanti *et al.*, 1978). The intracellular pH of most alkalophilic bacteria appears to be between pH 7.0 and 9.0, although the external pH of the medium may be in excess of pH 10.0.

It has been demonstrated that respiring bacteria, such as *E. coli* and *S. faecalis*, extrude protons and as a consequence the intracellular pH becomes more alkaline (Harold *et al.*, 1970; Padan *et al.*, 1976). However, protons must be returned to the cytoplasm since the intracellular pH remains relatively constant. The balance is maintained by antiport systems which are coupled to an electrochemical proton gradient which extrudes cations in exchange for protons. Such cation–proton antiporters have been described in numerous bacteria including *E. coli* (West and Mitchell, 1974), *S. faecalis* (Harold and Papineau, 1972), *Halobacterium* (Lanyi and MacDonald, 1975) and *Azotobacter* (Bhattacharyya and Barnes, 1978). Three distinct antiport systems have been identified in *E. coli* (K^+–H^+, Ca^{2+}–H^+ and Na^+–H^+), all of which aid in the extrusion of cations from the cell. Brey and Rosen (1980) considered that the K^+–H^+ antiporter is responsible for intracellular pH regulation since it exhibits an alkaline pH optimum and catalyses electroneutral exchange.

Antiport systems play an important role in the physiology of alkalophilic bacteria since they must combat the effects of respiration and external pH, both of which tend to raise the intracellular pH. *B. alcalophilus* has an internal pH of 9.0–9.5 over an external pH range of 9.0–11.5 (Guffanti *et al.*, 1978). Both an Na^+-H^+ and a K^+-K^+ antiport system have been demonstrated in this species, but it has been suggested that the Na^+-H^+ antiporter is responsible for the acidification of the cytoplasm (Mandel *et al.*, 1980). Further evidence to support the role of cation–H^+ antiporters as the regulatory system was provided by Krulwich *et al.* (1979). A mutant of *B. alcalophilus* which only grew at neutral pH and could not grow above pH 9.0 showed no evidence of Na^+ efflux, suggesting that the antiporter has a role in intracellular pH regulation. Similar observations were made with acid-sensitive mutants of *S. faecalis* (Kobayashi and Unemoto, 1980).

Products, Processes and Applications for Alkalophilic Microorganisms

Enzymes for Industry

Proteases. Alkaline proteases are not only produced by alkalophiles; two of the most widely studied enzymes are subtilins BPN′ and Carlsberg produced by neutrophilic strains of *B. subtilis* (Matsubara *et al.*, 1965; Smith *et al.*, 1968). Both enzymes are relatively similar with a pH optimum of pH 9.0–11.0 and isoelectric points between 9.0 and 10.5. They are uninhibited by metal chelators (ethylenediaminetetraacetate (EDTA) or thiol reagents (*p*-chloromercuribenzoate), the N-terminal amino acid of both enzymes was identified as alamine and both were inhibited by the serine-directed agent disopropyl phosphorofluoridate (DFP)).

Horikoshi (1971a) reported the production of an extracellular alkaline protease from an alkalophile. *Bacillus* species No 221, isolated from soil, was found to produce large amounts of alkaline protease which differed from the *B. subtilis* enzymes. The enzyme had a pH optimum of pH 11.5, although 75% of its activity remained at pH 13.0. It was characteristically inhibited by DFP and 6 M urea, but not by EDTA or *p*-chloromercuribenzoate, and alanine was found to be the N-terminal amino acid of the protein. The molecular weight of the enzyme was 30 000 daltons which is slightly higher than those of other reported alkaline proteases. Calcium ions affected both activity and stability of the enzyme. The addition of 5 mM calcium reflected a 70% increase in activity at the optimum temperature (60 °C). Without the addition of Ca^{2+} 80% of the activity was lost after 10 minutes at 60 °C and all activity was lost at 65 °C.

Subsequently two further *Bacillus* species, AB42 and PB12, were reported which also produced alkaline protease (Aunstrup *et al.*, 1972). The enzymes' molecular weights were 20 000 daltons and 26 000 daltons respectively and both had an isoelectric point of 11.0. They exhibited a broad pH range of pH

9.0–12.0 and the temperature optimum of the strains was 60 °C for AB42 and 50 °C for PB12. Both enzymes were inhibited by phenyl methyl sulphonyl fluoride indicating that they are "serine" enzymes. The proteases from these strains exhibited higher pH activity than those of subtiliin, Carlsberg or other *Bacillus* proteases.

Microbial alkaline proteases have a common sequence of amino acids at the active site: Thr–Ser–Met–Ala (Sanger, 1963; Markland and Smith, 1967; Smith *et al.*, 1968; Mikes *et al.*, 1969). The active site sequence of protease No 221 has been determined as Asp–Gly–Thr–Ser–Met which resembles the above sequence and is identical with those of other microbial alkaline proteases (Watanabe and Horikoshi, 1977).

In addition to their protease activity both enzymes 221 and AB42 exhibit esterase activity towards amino acid ethyl esters. In comparison with alkaline proteases from other neutrophilic *B. subtilis* species the K_m and V_{max} values of enzyme 221 would indicate a different substrate binding site or mode of binding (Horikoshi and Akiba, 1982).

A number of neutrophilic *Streptomyces* species have also been shown to produce alkaline proteases (Mizusawa *et al.*, 1964, 1966; Morihara *et al.*, 1967; Wahlby and Engstrom, 1968). However, more recent studies have shown the production of an alkaline protease from an alkalophilic *Streptomyces* species (Nakanishi *et al.*, 1974). This strain grew well at pH 10.5 and produced a protease with a pH optimum of 12.3 but which was also active at pH 13.0 and in 0.2 N NaOH. The enzyme had a molecular weight of 50 000 daltons and was inhibited by DFP but not *p*-chloromercuribenzoate and exhibited no esterase activity.

The enzyme was active against various keratinous proteins such as wool, feathers, hair and silk under alkaline conditions (Nakanishi and Yamamoto, 1974). The protease was also active against various peptide linkages involved in the oxidized insulin B chain, and the specificity of the enzyme was very different from those of other alkaline proteases (Matsubara and Feder, 1971; Nakanishi and Yamamoto, 1974).

Alkaline proteases have been extensively used in detergents for many years. Originally enzyme detergents were utilized for the cleaning of heavily soiled working clothes from the fish industry, slaughter houses, bakeries and hospitals where much of the staining was a result of highly proteinaceous material. The domestic use of enzyme washing powders quickly followed. Many alkaline microbial proteases have been developed for production on an industrial scale, particularly by the Japanese fermentation industry. However, only a few are now produced on a commercial scale which fulfil the required demands, e.g. range of pH and temperature activity, stability under alkaline conditions and good activity in the presence of surfactants. Two such commercially available enzymes are Alcalase® produced by *B. licheni-*

formis and commonly used in household detergents and Esperase® (Novo) which is used in heavy duty formulations and liquid detergents where greater alkali tolerance is necessary. Esperase® (Novo) is produced by an alkalophilic *Bacillus* species and exhibits high activity at pH 12.0.

Aunstrup *et al.* (1972) investigated the use of the proteases AB42 and PB12 for laundering purposes. The residual activities of the enzymes at pH 10.0 after 30 minutes at 50 °C in the presence of detergents was examined. The proteases exhibited quite different profiles. Protease AB42 retained over 85% activity in the presence of sodium tripolyphosphate (STPP) and sodium perborate (SPB). Lower activity was retained in the presence of sodium dodecyl benzene sulphonate (DBS) (31%), tallow alcohol sulphate (TAS) (22%) and soap (45%). Protease PB12 retained much higher activities over a wide range of detergents. Residual activity in excess of 65% was maintained for SPB, DBS, TAS and soap; however, in the presence of STPP all activity was lost.

Dehairing is one of six defined stages in the preparation of leathers. Traditionally this stage requires extreme alkaline conditions and high concentrations of lime and sodium sulphide. This has often led to severe pollution caused by effluent from tanneries. Simple reduction of concentrations of lime and sulphides or the pH proved unreliable and wasteful of leather owing to uneven dehairing. However, the incorporation of enzymes in the process has proved successful in improving leather quality and in reducing pollution.

In order to function in this process, enzymes, particularly proteases, need to have an alkaline pH optimum. Alkaline proteases have been utilized and commercially prepared enzymes include Milezyme® 8X, Novo Unhairing Enzyme® No 1, and an immobilized enzyme Rapidepilase® No 7. These enzymes are produced by industrially improved neutrophilic *Bacillus* species. However, a number of proteases from the obligate alkalophilic *Bacillus* species have been demonstrated to perform the dehairing of hides as efficiently as other alkaline proteases (Horikoshi and Akiba, 1982).

Amylases.   Horikoshi (1971b) described the first alkaline α-amylase, which was produced by an alkalophilic *Bacillus* species A-40-2 and which had pH and temperature optima of pH 10.5 and 55 °C. Boyer and Ingle (1972) described the production of an alkalophilic α-amylase from a *Bacillus* alkalophile NRRL B-3881, with similar pH and temperature optima of pH 9.2 and 50 °C.

Both these amylases, which have a molecular weight of 70 000 daltons, are unique. Amylase A-40-2 retains 50% of its activity at pH values between 9.0 and 11.5 and is not inhibited by 10 mM EDTA at 30 °C. Inactivation by 8 M urea is reversible and 95% activity can be regained after dialysis. The enzyme hydrolyses 70% of starch to yield glucose and to a lesser extent maltose and maltotriose. Amylase B-3881 has similar properties (Boyer and Ingle, 1972). 50% activity is retained at a slightly lower pH range (pH 7.0–10.5) and the enzyme is only slightly inhibited (12%) by 20 mM EDTA at 32 °C. The stability of both enzymes to EDTA is greater than that of most other amylases, e.g. *B. amyloliquefasciens* (Boyer and Ingle, 1972), *B. subtilis* or *Aspergillus oryzae* (Toda and Narita, 1968). Hydrolysis of starch yields maltose and maltotriose and to a lesser extent glucose and maltotetrose, all with a β-configuration.

Fukumoto and Okada (1963) classified *Bacillus* amylases into two groups according to their ability to hydrolyse soluble starch. Amylases with the ability to hydrolyse 30–40% of the substrate were classified as liquifying amylases. Amylases with the ability to hydrolyse 50–60% starch were classified as saccharifying amylases. Both amylases A-40-2 and B-3881 fall into the latter group of saccharifying amylases.

The alkaline amylases of *Bacillus* species have been further classified on their pH activity profiles (Yamamoto *et al.*, 1972). Four groups were identified after identifying a further eight amylase-producing strains. Type I exhibited a profile with a single amylase peak at pH 10.5 corresponding to the optimum pH of the enzymes. The enzymes were most stable between pH 7.0 and 9.0–9.5 (50% of the activity remained after incubation at 50 °C for 15 minutes), but complete inactivation was observed at 55 °C after 15 minutes even in the presence of 10 mM Ca^{2+}. Such type I enzymes were also completely adsorbed on DEAE cellulose at pH 9.0 and would appear to have a high isoelectric point pH.

Type II amylases exhibited profiles with two peaks, one in the acid range pH 4.0–4.5 and the other in the alkaline range pH 9.0–10.0, again corresponding closely to the pH optima. Type II amylases were most stable at pH values between 7.0 and 9.0. However, 70–95% of the original activity was retained after incubation at 55 °C for 15 minutes. In this respect type II enzymes are similar to type IV enzymes.

Type III amylases, characterized by a three-peak profile, were also the most thermostable, retaining 50% of their activity at pH 4.5 or 9.0 at 65 °C for 40 minutes. The enzyme studied was stable over a wide pH range (pH 5.0–10.5) and exhibited a high degree of cyclodextrin glycosyltransferase activity.

Type IV amylases exhibited a pH activity profile with a peak at pH 4.0 and a shoulder at pH 7.0–10.0 which corresponded to approximately 30–40% relative activity. The pH optimum of the enzymes was 4.5, but they were most stable between pH 6.5 and 10.0. They retained up to 95% activity after incubation at 55 °C for 15 minutes.

Amylase B-3881 has also been shown to exhibit hydrolysis of cyclomaltoheptose (Boyer and Ingle, 1972). Hydrolysis of this β-cyclodextrin yielded maltotriose and to a lesser degree glucose. This indicated that the enzyme did not require a non-reducing end of the substrate for reaction. Such activity has only previously been described for *B. polymyxa* (Robyt and French, 1964) and *B. subtilis* var. *amylosacchariticus* (Keay, 1970).

α-amylases are becoming increasingly important as detergent additives. Detergents in the catering industry are required to remove food residues containing starch from potatoes, oatmeal or spaghetti. The inclusion of α-amylase into detergents assists in the removal of these residues. Ideally the enzyme needs to be thermostable, alkali stable and resistant to chlorine compounds used as bleaching agents. Termamyl® produced by *B. licheniformis* is thermostable and alkalotolerant but sensitive to bleaching agents (Barfoed, 1983). Alkalophilic bacteria, particularly *Bacillus* species, may provide a source of suitable enzymes.

Cyclodextrin glycosyltransferases (CGTs). CGT catalyses the conversion of starch to cyclodextrins (Schardinger dextrins). These compounds comprise six to eight glucose units linked by α-1,4 bonds forming a cyclic molecule, the basic structures of which are shown below:

α-cyclodextrin

γ-cyclodextrin

β-cyclodextrin

G, α-glucose;

—, α-1,4 linkage.

Cyclodextrins have become increasingly important in the food, cosmetic, pharmaceutical and plastics industries. They are used for the stabilization of volatile materials, such as flavours and spices, deodorizing medicines and foods, protection against oxidation and ultraviolet light degradation during processing and storage, modification of certain chemical and physical properties (e.g. the water solubility of medicines, the stability of deliquescent chemicals and for reducing bitterness in foods and medicines), and the emulsification and solidification of numerous hydrocarbons (Mifune and Shima, 1977).

CGT has been demonstrated to catalyse coupling and disproportionation reactions (Norberg and French, 1950; French et al., 1954; Kitahata et al., 1978).

CGT activity was first observed in culture supernatant from B. macerans (French, 1957). Subsequently the enzyme has been observed in other strains, e.g. B. macerans ATCC 8514 (DePinto and Campbell, 1968) (an unusual strain since the production of CGT is intracellular), B. macerans IAM 1243 (Kobayashi et al., 1978), B. macerans IFO 3490 (Kitahata et al., 1974), B. megaterium (Kitahata and Okada, 1974; Kitahata et al., 1974) and Klebsiella pneumoniae (Bender, 1977). These enzymes all differ in their characteristics and in their hydrolytic products formed from starch. B. megaterium is unique in producing predominantly β-cyclodextrins, all the other strains producing predominantly α-cyclodextrins.

B. macerans CGTs have been widely used for the production of cyclodextrins. However, they are not wholly suitable for industrial processes because they are geneally thermolabile and their yield is rarely in excess of 55%. The requirement for a CGT with significant thermostability and a high conversion yield was met by a CGT from an alkalophilic Bacillus strain No 38-2 (ATCC 21783) (Nakamura and Horikoshi, 1976a, b, c, d).

Bacillus species No 38-2 produced CGT at alkaline pH values between pH 9.5 and 10.3. The most effective medium composition was determined to consist of 1% w/v soluble starch, 5% w/v corn steep liquor, 0.1% K_2HPO_4, 0.02% $MgSO_4-7H_2O$ and 1% Na_2CO_3 (Nakamura and Horikoshi, 1976a). The crude enzyme exhibited a broad pH–activity profile for both cyclizing and dextrinizing activity with two pH optima at pH 4.5–5.0 and pH 7.0–8.5. Cyclodextrin yields from potato starch were in excess of 70% depending on the pH. The crude enzyme preparation was separated into three fractions: acid, neutral and alkaline CGTs all of which predominantly produced β-cyclodextrin (Nakamura and Horikoshi, 1976b,c,d).

The acid CGT had a pH optimum of pH 4.3–4.7, a temperature optimum of 45 °C and a molecular weight of 88 000 daltons and was stable at 65 °C for 30 minutes. Yields of β-cyclodextrins from potato starch, amylopectin and maltose were 73%, 65% and 7% respectively. Dextrinizing activity was demonstrated in the presence of various cosubstrates including, in decreasing order of effectiveness, glucose, maltose and maltotriose.

The neutral CGT produced higher yields of cyclodextrins than the acid CGT. Yields were determined to be 85–90% from amylose, 75–80% from potato starch, 65–70% from amylopectin and 10–15% from maltose (Nakamura and Horikoshi, 1976d). The enzyme was stable up to 60 °C but could be stabilized at 70 °C by the addition of 10 mM Ca^{2+}. The pH and temperature optima were pH 7.0 and 50 °C and the molecular weight was 85 000–88 000 daltons. The properties of both the acid and the neutral CGTs were similar, except for their pH optima.

As a result of the similar physiochemical properties of the neutral and alkaline CGTs, the separation and purification of the alkaline CGT has not been accomplished. In crude enzyme preparations the enzyme has a pH optimum of pH 8.0–9.0 and a molecular weight of 85 000–88 000 daltons (Horikoshi and Akiba, 1982).

Nakamura and Horikoshi (1977) used an ion-exchange resin, vinylpyridine copolymer (60–100 mesh beads), to immobilize CGT from *Bacillus* strain No 382. The crude enzyme was succinylated prior to mixing with the activated resin (Klotz, 1967). A number of minor changes in the properties of the immobilized enzyme were observed. The optimum pH range of the enzyme shifted from pH 4.5–5.0 to pH 5.5–6.0 and the activities of the acid and neutral fractions were decreased. The optimum temperature also shifted from 50 to 55 °C, but no change in yield of cyclodextrins was observed. However, the yields were greater at alkaline pH values. The yield of cyclodextrins from starch could be influenced by the addition of alkalophilic pullulanase produced by *Bacillus* sp. No 202-1 (Nakamura *et al.*, 1975). Increases in yields of 4–6% were reported (Nakamura and Horikoshi, 1977). The immobilized enzyme showed no loss of activity when used continuously over 2 weeks at a temperature of 55 °C and at pH 8.0.

Kato and Horikoshi (1984) investigated the immobilization of cyclomalto-dextrin glucanotransferase from *Bacillus* sp. No 38-2 on synthetic adsorption resin (DIAION HP-20). No change in pH optimum or thermal stability was observed following immobilization; however, the optimum temperature for enzyme activity changed from 50 to 55 °C. The immobilized enzyme converted starch to cyclodextrin without significant loss of activity under continuous operation for approximately 2 weeks at 60 °C and pH 8.0. This system had advantages over the vinylpyridine copolymer system since the enzyme needed no modification for immobilization on DIAION HP-20; reactivation following loss of activity could be achieved by charging the enzyme without any treatment and the optimum pH for enzyme activity was pH 9.0, the highest so far reported. Such a high pH would prevent retrogradation of starch and hence lead to longer operation of the system.

Pectinases.   These are produced by a wide range of microorganisms and are used in the fruit and vegetable processing industry (Deuel and Stutz, 1958; Fogarty and Ward, 1977). Pectinases can be divided into three groups on the basis of their action or substrate activity (Fogarty and Ward, 1977). Esterases remove methoxyl residues from pectin to form pectic acid. Hydrolases, e.g. polygalacturonase and polymethylgalacturonase, cleave α-1,4 glycosidic linkages in pectin and pectic acid. This may be random (endo) or sequential (exo). Lyases catalyse the transelimination cleavage of α-1,4 glycosidic linkages which may also be random or sequential.

The most widely distributed microbial pectinases are endopolygalacturonases with a pH optimum of approximately 4.0–5.0, and endopolygalacturonate lyases with a pH optimum between 7.0 and 10.0. The latter are usually activated by Ca^{2+} ions.

Alkalophilic pectinase is produced by *Bacillus* species No. P-4-N (Horikoshi, 1972) and *Bacillus* species RK9 (Kelly and Fogarty, 1978). Strain P-4-N grew at pH 7.0–10.0, but optimally at pH 10.0, producing endopolygalacturonase. The enzyme was inducible provided that certain medium components were available, including Na_2CO_3, Mn^{2+} and pectin. The purified enzyme, with a molecular weight of 60 000–70 000 daltons, exhibited optimal activity at pH 10.0 and 65 6C. It was inhibited by 10 mM EDTA but not by 8 M urea or 0.1 mM *p*-chloromercuribenzoate. Activity was increased tenfold by the addition of 1 mM $CaCl_2$, an unusual property for hydrolytic pectinases. The addition of Ca^{2+} increased thermal stability, but inactivation occurred at 80–90 °C. Dimers, trimers and tetramers were the main products from pectic acid hydrolysis, although prolonged incubation resulted in the production of monomers and dimers of galacturonic acid.

Bacillus species RK9 grew at pH 9.7 constitutively producing endopolygalacturonate lyase (Kelly and Fogarty, 1978). The purified lyase had an optimum pH and temperature of 10.0 and 60 °C. Activity was increased by a factor of 2.9 by addition of 0.4 mM $CaCl_2$, but similar divalent cations (Mg^{2+}, Si^{2+}, Co^{2+}) had little or no effect, unlike other *Bacillus* pectinases (Nagel and Wilson, 1970; Ward and Fogarty, 1972). Complete inactivation occurred with the addition of 1 mM EDTA; however, activity was restored following dialysis. The enzyme was found to be most active with acid-soluble pectic acid.

Xylanases.   Xylan-rich biomass is relatively abundant, particularly in crop residues such as rice, straw and corn cob. Xylanases which hydrolyse β-1,4 linkages in xylan polysaccharides are produced by many groups of microorganisms (Dekker and Richards, 1976); however, alkaline xylanase may be potentially useful owing to the increased solubility of xylan in alkaline solutions.

Xylanases are produced by the alkalophiles *Bacillus* species No C-59-2 (Horikoshi and Atsukawa, 1973b) and *Bacillus* species No C-11 (Ikura and Horikoshi, 1977). Both strains grow and produce xylanase at pH 10.0.

The enzyme from strain C-59-2 produced xylanase in the presence of carbonate ions. Replacement of Na_2CO_3 by NaCl or KCl resulted in the inhibition of enzyme production. The purified enzyme had pH and temperature optima of 6.0–8.0 and 60 °C and hydrolyzed 40% of the substrate over the pH range 6.0–9.0. *B. subtilis* xylanase hydrolyses approximately 38% of the substrate (Takahashi and Hashimoto, 1963). Product inhibition has been considered as a reason for low hydrolysis and it has been demonstrated that 15% xylose or glucose will inhibit the activity of *Trichoderma viride* xylanase. The main products of xylanase C-59-2 are xylobiose, xylotriose and xylotetrose.

The purified xylanase from strain C-11 had an optimum pH of 7.0, although 37% of the activity remained at pH 10.0. The main products of hydrolysis were xylose, xylobiose and xylotriose (Ikura and Horikoshi, 1977).

Xylanases from alkalophilic *Bacillus* species exhibit better activity at higher pH values and may have potential for treating some alkaline industrial wastes (Ikura and Horikoshi, 1977).

Cellulases. Horikoshi *et al.* (1984) isolated a *Bacillus* species No N-4 (ATCC 21833) which produced an alkaline cellulase. This strain grew at pH 8.0–11.0 and optimally produced cellulase at pH 10.3 in complex medium. The crude enzyme obtained by ethanol precipitation had a wide pH spectrum of activity (pH 5.0–11.0) and the partially purified enzyme had maximum activity at pH 10.0. The cellulase was relatively thermostable, retaining approximately 100% residual activity at 50 °C for 10 minutes at pH 9.0. The crude enzyme showed two pH optima at pH 6.7 and 10.0, suggesting the presence of two cellulases. Two cellulases were partially purified (enzymes E1 and E2), both with a pH optimum of pH 10.0. Enzyme E1 was stable up to 60 °C and enzyme E2 was stable up to 80 °C. Both enzymes catalysed the hydrolysis of cellotetrose to cellobiose as a main product.

Other significant alkalophilic enzymes. β-1,3 glucanases hydrolyse β-1,3 glucosidic links of β-1,3 glucan which occurs as a cell wall component, cytoplasmic reserve material and extracellular products of microorganisms and higher plants. The majority of β-1,3 glucanases have pH optima in the acid range. However, alkalophilic enzymes from alkalophilic *Bacillus* sp. have been identified (Horikoshi and Atsukawa, 1973a, 1975).

The production of an extracellular inducible β-lactamase from an alkalophilic *Bacillus* species has been reported (Sunaga *et al.*, 1976). The enzyme comprised two active fractions, which were produced optimally at pH 9.0; both fractions were optimally active at pH 6.0–7.0. The optimum temperature for activity of the crude enzyme was 50 °C.

Extracellular lipases have been investigated with respect to their application as detergent additives. Alkaline lipases have been reported; however, only one obligate alkalophile, an *Achromobacter* species, has been shown to produce the enzyme (Horikoshi and Akiba, 1982). Optimal activity of this enzyme was observed at pH 10.0. Two alkalotolerant *Pseudomonas* strains also produce alkaline lipases (Watanabe *et al.*, 1977) with an optimal pH value of 9.5. The *Pseudomonas* lipases which exhibited a high degree of thermostability remain stable up to 70°C.

Alkalophilic catalase is produced by an alkalophilic *Bacillus* species No KU-1 (Kurono and Horikoshi, 1973). The enzyme which is uniquely extracellular has pH and temperature optima of 10.0 and 15 °C.

Nakamura *et al.* (1975) reported an alkalophilic *Bacillus* species No 202-1 producing extracellular pullulanase at pH 10.0. Complete hydrolysis of amylopectin and glycogen was achieved by the enzyme in conjunction with β-amylase. β-amylase alone only achieved partial hydrolysis. The enzyme was optimally active at 55 °C and had a pH optimum of 8.5–9.0.

Alkaline DNase and RNase were also detected in culture fluids of *Bacillus* alkalophiles and have been purified and investigated (Horikoshi *et al.*, 1973; Atsukawa and Horikoshi, 1976).

Waste Treatment

Alkalophilic bacteria may have potential in the treatment of certain wastes. Horikoshi *et al.* (1972, 1984) isolated alkalophilic bacteria producing alkaline cellulases which are relatively thermostable. Most commercially available cellulases are produced by mesophilic fungi and have optimal activity at pH 4.0–5.0 at 40–50 °C. Alkaline cellulases are advantageous since celluloses and hemicelluloses are more soluble under highly alkaline conditions. Such cellulases have shown some potential in the treatment of human sewage where treatment may cause the pH to rise to pH 8.0 or 9.0 owing to the production of ammonia (Horikoshi and Akiba, 1982). An alkalophilic *Aeromonas* species No 212 (ATCC 31085) produces an alkaline cellulase, which is stable in the presence of metal ion and hydrogen sulphate and hydrolyses cellulose in raw sewage (Horikoshi and Akiba, 1982).

Bacillus species have been isolated from alkaline rayon waste which contains significant amounts of hemicellulase (Ikura and Horikoshi, 1977). *Bacillus* species C-11 utilized rayon waste and may have potential in the biological treatment of rayon waste.

Antibiotics

An alkalophilic strain of *Paecilomyces lilacinus* was isolated from soil at pH 10.5 and was found to produce a novel antibiotic designated 1907 (Sato *et al.*, 1980). The strain produced two compounds, 1907-II and 1907-VIII, under alkaline conditions (pH 9.0–10.5). 1907-II is a peptide antibiotic with a molecular weight of 1203 daltons. Both 1907-II and 1907-VIII exhibit broad antibacterial and antifungal activities.

Reduction of Indigo

The reduction of indigo is a traditional and ancient use of alkalophilic bacteria. Indigo has long been reduced in an alkaline environment containing Na_2CO_3. A *Bacillus* species S-8 was isolated from an indigo ball and was subsequently used to improve the process (Takahara and Tanabe, 1960). By adding a seed culture of *Bacillus* S-8 to the reduction mixture the processing time was decreased by 75%. Reduction was easier and the product was considered better than that obtained from the traditional fermentation method.

Acidophilic Microorganisms

Definition and Occurrence

Many microorganisms are able to grow at pH values as low as pH 4.0, although most of these are also able to grow at neutral pH. These organisms are regarded as acid tolerant. Organisms which are able to grow at pH values of 2.0–4.0 but are unable to grow at neutral pH are regarded as obligate acidophiles. Moderately acidic natural environments between pH 3.0 and 4.0 are relatively common and include acid lakes, pine forests, soils and acid bogs. More extreme environments with pH values of less than pH 3.0 are often associated with coal tips, drainage waters and mining effluents. The production of acid in these environments is a consequence of the oxidation of sulphides and pyritic materials producing sulphuric acid.

Natural acidic environments are more commonly found than alkaline environments since acidity develops in an aerobic oxidizing environment. Sulphuric acid is formed from the oxidation of sulphides such as H_2S and FeS_2. Sulphides are common in volcanic and geothermal areas and in swamps, bogs and the sea. While the environment remains anaerobic the sulphides remain stable. However, if aerobic conditions prevail rapid oxidation leads to the formation of sulphuric acid. Naturally occurring acidic pools with pH values in the range pH 1.0–3.0 have been studied in Yellowstone National Park, Wyoming, USA, and volcanic lakes in New Zealand. Geothermal regions present a dramatic development of acidic environments. Gases coming from deep within the earth are often rich in hydrogen sulphide. As the gas reaches the surface it is oxidized, first to elemental sulphur and then to sulphuric acid. Examples of these acidic soil areas (known as solfataras) are found in Yellowstone National Park, and one of the most famous is found in a volcanic crater along the Bay of Naples.

A number of obligately acidophilic microorganisms have been isolated and described and many of these are also thermophilic. They fall into four distinct groups: *Thiobacillus*, *Sulfolobus*, *Thermoplasma* and *Bacillus*.

The Major Groups of Acidophilic Microorganisms

Three of the four major acidophilic groups of microorganisms are also thermophilic. *B. acidocaldarius* and *Sulfolobus* have been discussed previously with the thermophiles.

Thiobacillus species.   *T. thiooxidans*, isolated and characterized by Waksman and Jaffe (1922), oxidizes elemental sulphur with the production of sulphuric acid. It grows in the pH range 0.9–4.5 with an optimum at pH 2.5 (Rao and Berger, 1971). It has been isolated from acid hot springs and surrounding solfataras where the temperature was below 55 °C.

T. ferrooxidans described by Temple and Colmer (1951) oxidizes elemental sulphur and also ferrous iron to ferric iron with the production of sulphuric acid (Dugan and Lundgren, 1965). The optimum pH range is dependent upon the substrate utilized (McGoren *et al.*, 1969). The optimum for elemental sulphur oxidation is pH 5.6, but metallic sulphide minerals are oxidized at a lower pH, e.g. calcopyrite at pH 2.0, bornite at pH 3.0 and pyrite at pH 2.0 (Landesman *et al.*, 1966). Growth on ferrous iron is optimal at pH 2.5.

Le Roux *et al.* (1977) reported the isolation of acidophiles closely resembling *Thiobacillus* species from steam vents and mud springs in Iceland. They were able to oxidize ferrous iron, pyrite and metal sulphides at pH 1.4–3.0 and at 55 °C. Physiological investigations were reported by Brierley and Le Roux (1977) and Brierley *et al.* (1978).

Thermoplasma acidophilum.   *T. acidophilum* was isolated from a coal refuse pile by Darland *et al.* (1970). It has been classified as a mycoplasma owing to its lack of cell wall. The organism is resistant to vancomycin which inhibits cell wall growth. This characteristic may be used to exert selection pressure when isolating *Thermoplasma* from samples containing *B. acidocaldarius* and *B. coagulans*. Unlike other mycoplasma it is not osmotically unstable and is tolerant to triple-distilled water (Belly and Brock, 1972). *T. acidophilum* requires yeast extract for growth; Smith *et al.* (1975) found a specific requirement for a short polypeptide of eight to 10 amino acids. They considered that it may (i) act as an ion scavenger for trace metals, (ii) be involved in ion transport, (iii) give protection at low pH or (iv) be a supply of essential amino acids in a form readily assimilated by the cell. It has a pH optimum of pH 1.0–2.0 and a growth temperature of 59 °C. The physiology, isolation and ecology of this organism are discussed by Brock (1978).

Microbial Physiology in Acidic Environments

Growth at low pH requires the maintenance of a cytoplasmic environment at a relatively neutral pH. Acidophiles may achieve this by pumping protons out of the cell or by maintaining a cell surface barrier which is impermeable to protons. Hsung and Haug (1975) used a variety of methods including the use of DMO to establish the internal pH of *T. acidophila*. An internal pH of 6.5 was maintained when the cells were suspended in medium at pH 2.0, 4.0 and 6.0. Krulwich *et al.* (1978) reported an internal pH of 5.85–6.13 for *B. acidocaldarius* at 50 °C when the external environmental pH was 2.0–4.5. The pH optimum for β-galactosidase from *B. acidocaldarius* was pH 6.0–6.5 correlating closely with the cytoplasmic pH. The internal pH of *Sulfolobus* growing at pH 3.0 and 70 °C was found, following lysis with sodium dodecyl sulphate, to be pH 6.3 (De Rosa *et al.*, 1975). *Thiobacillus* was reported to maintain a stable internal pH of 6.5 when grown between pH 1.0 and 8.0. Krulwich and Guffanti (1983) reviewed the literature relating to the proposed mechanisms for maintaining the transmembrane pH.

Products, Processes and Applications for Acidophilic Microorganisms

Metal Leaching Processes

The acid leaching of mineral sulphides is a natural phenomenon but it has been accelerated by the exposure of mineral-rich ores through the mining of copper, iron and coal. The drainage of dissolved metals and acids from mines, coal tips and heaps of low grade ore has been observed for many years, but it is only relatively recently that the role of microorganisms has become apparent. There are a number of reviews involving the use of microorganisms in waste metal recovery and metal leaching, including those of Zajic (1969), Brierley (1977), Brierley (1978), Murr *et al.* (1978), Kelly *et al.* (1979), Lundgren and Silver (1980), Lundgren and Malouf (1983) and Curtin (1983).

The development of industrial microbiological processes for the solubilization of metals has centred around large-scale dump and heap leaching of low grade copper-containing ores. In the USA 11–15% of total copper production comes from microbial leaching of low grade copper ore by *T. ferrooxidans*. Metals are dissolved from insoluble minerals either directly as part of the organisms' metabolism or indirectly by the products of their metabolism. The reactions involved are oxidations of sulphur or mineral sulphides.

Studies have mainly been carried out under field-simulated conditions in the laboratory. Scanning electron micrographs indicate bacterial attachment to sulphide minerals in nutrient-supplemented solutions (Bennet and Tributsch, 1978). The bacteria appear to dissolve a sulphide surface of the crystal directly. Attachment to the surface of pyrite (FeS_2) and chalcopyrite ($CuFeS_2$) results in the following reactions: for pyrite

$$4FeS_2 + 15O_2 + 2H_2O \longrightarrow 2Fe_2(SO_4)_3 + 2H_2SO_4$$
$$4FeSO_4 + O_2 + 2H_2SO_4 \longrightarrow 2Fe_2(SO_4)_3 + 2H_2O$$
$$2S + 3O_2 + 2H_2O \longrightarrow 2H_2SO_4$$

and for chalcopyrite

$$4CuFeS_2 + 17O_2 + 2H_2SO_4 \longrightarrow 4CuSO_4 + 2Fe_2(SO_4)_3 + 2H_2O$$

In addition to the direct attachment or influence of microorganisms on metal solubilization from ores, an indirect effect is the formation of ferric iron which is a very effective solubilizing agent for numerous minerals (Kelly, 1976):

$$2\ FeS_2 + 2Fe_2(SO_4)_3 \longrightarrow 6FeSO_4 + 4S^0$$
$$2\ CuFeS_2 + 2Fe_2(SO_4)_3 \longrightarrow CuSO_4 + 5FeSO_4 + 2S^0$$

The microbial oxidation of Fe^{2+} and S^0 provides an energy source for the leaching microorganisms. The recycling of ferrous–ferric iron and the continued solubilization of S^{2-}, yielding oxidizable sulphur, is important for an effective leaching process (Lundgren and Malouf, 1983).

Dump leaching for the extraction of copper involves the deposition of large quantities of low grade ores or waste rock on impermeable ground, usually in valleys. Such material generally contains less than 0.4% copper and is not crushed or graded. One of the largest is at Bingham Canyon, Utah, USA, which contains 4 billion tons of ore and waste rock. Leach solutions are introduced naturally on the top of the dumps by spraying, flooding or injection through vertical pipes. The fluid percolates through the dump and is collected in catch basins at the base. Copper can be recovered either by precipitation with metallic iron or by electrowinning. The leachate may contain copper, iron, aluminium, magnesium and minor amounts of uranium. After removal of the metals the leach solution is returned to the surface for re-application. Leaching cycles for dumps may be carried on for several years.

Heap leaching is smaller in scale and is primarily used to extract copper and uranium from crushed or uncrushed ores of a higher grade than those used in dump leaching. The ore is deposited in mounds on prepared drainage pads and the leach cycle is complete within months. In addition to copper and uranium, gold has also been extracted using heap leaching.

Underground *in situ* leaching is applicable to oxide and sulphide minerals of uranium and copper. It has the advantage of reducing surface disturbance over the deposit. It eliminates many of the problems of solid waste disposal and acid drainage from sulphide-bearing rocks which would otherwise be brought to the surface. One of its main disadvantages is the difficulty in predicting and controlling the path of leach solutions which may contaminate ground and surface water (Brierley, 1978).

The leaching of uranium requires ferric iron for the oxidation of reduced uranium, and the role of bacteria is primarily to generate ferric ions. Manchee (1979) reported the extraction of 90% of the uranium from low grade ore. Leaching was carried out at 55 °C and ferric iron was generated in a separate reactor at 30 °C. Derry *et al.* (1976) extracted 95% of the available uranium from a low grade ore containing 0.12% U_3O_8. A series of submerged beds of ore were leached continuously for 10 days at pH 1.2 and 50 °C. The uranium was recovered by ion exchange and the ferrous solution was regenerated by *T. ferrooxidans* at 30 °C. A successful two-stage process for the extraction of uranium is currently being operated on a production scale in Canada (Gow *et al.*, 1971). Since the role of bacteria in the leaching of uranium is almost exclusively the generation of ferric iron, the two processes can in effect be carried out separately and the bacteria do not need to be in contact with the ore. The advantage in separating the two stages is twofold: it enables the ferric leaching to be carried out at elevated temperatures and the bacteria do not have to oxidize the iron in the presence of uranium. This is significant since *T. ferrooxidans* is sensitive to low concentrations of uranium. It is also sensitive to low concentrations of silver, mercury, gold, thallium, rubidium and molybdate (Kelly *et al.*, 1979). Brierley (1974) reported the isolation of an organism resembling *Sulfolobus* which oxidized and tolerated high concentrations of molybdenum.

Many inorganic industrial wastes contain valuable metals in low concentration. Ebner (1978) reported studies on the leaching of waste left after zinc manufacture and basic slag derived from lead smelting.

Desulphurization of Coal

One of the main disadvantages of the direct utilization of coal is the emission of sulphur- and nitrogen-containing gases into the atmosphere. Sulphur compounds in coal cause problems with equipment wear and corrosion, and acid mine waters cause environmental pollution problems. Reduction of the sulphur content of the coal before or after combustion is essential to eliminate the problem of pollution (Kargi, 1984). The leaching of pyrite sulphur from coal has been examined using *T. ferrooxidans* in continuously stirred tanks at 29 °C (Torma, 1977; Kargi, 1982; Myerson and Kline, 1984).

S. acidocaldarius has been utilized for the removal of pyritic sulphur and organic sulphur from coal (Kargi and Robinson, 1982a,b). The major advantages of using *Sulfolobus* at 70–80 °C were (i) the ability to operate the system at high cell and coal concentrations, (ii) reduced cooling costs and (iii) an improvement in the rate of chemical oxidation of pyritic sulphur by ferric ions produced by microbial oxidation of pyrite present in the coal (Kargi, 1984).

The use of an airlift-external recycle fermenter for the removal of pyritic sulphur from coal using *S. acidocaldarius* resulted in the removal of 30% of the initial pyritic sulphur from a 5% coal slurry (Kargi and Cervoni, 1983).

Acid-tolerant Strains in the Food Industry

In the food industry a wide range of processes involve the use of acid-tolerant microorganisms.

Yoghurt, cheese, sour cream and buttermilk are produced by fermentation and acid production using species of *Lactobacillus*, *Stretococcus* and *Leuconostoc* (Porubcon and Sellars, 1979). Vinegar is commercially produced from the oxidation of ethanol to acetic acid by *Acetobacter* species.

Lactic acid bacteria such as *Leuconostoc plantarum*, *L. brevis* and *L. bulgaris* are added to the dough for rye bread manufacture. Their growth imparts a tangy or sour flavour to the rye bread. The curing of coffee beans may utilize an acid fermentation step by *L. mesenteroides*, *L. brevis* and *L. plantarum*. *L. delbrueckii* is involved in the acidification of the koji and the mash in soy sauce production. *Pediococcus soyae* increases the acid in the mash, stimulating the growth of yeasts.

Acidophilic streptomycetes play an important role in the decomposition processes in acid soils and ground litter. They have been demonstrated to have a wide range of hydrolytic activities including the hydrolysis of starch, inulin, fucoidin, laminarin, galactan, xylan and chitin (Kahn and Williams, 1975; Williams and Flowers, 1977).

Nutrient preservation in silage is largely due to the fermentation by *Lactobacilli* and other lactic-acid-producing bacteria.

Enzymes

Since acidophiles maintain a relatively neutral internal pH, intracellular enzymes are generally neutrophilic. Extracellular enzymes, however, are required to function in the external environment and a number of acid-stable extracellular enzymes have been studied.

Acid amylases produced by *B. acidocaldarius* with a pH optimum of 3.5 have been discussed. Jensen and Norman (1984) described the isolation of an acidophilic pullulanase from *B. acidopullulyticus*. Starches used in the manufacture of glucose syrup often contain considerable amylopectin, a branched polymer consisting of chains of α-1,4-linked D-glucose residues joined by α-1,6 glucosidic linkages. These α-1,6 linkages inhibit exo-acting amylases; however, pullulanase enzymes are able to degrade these branch points. The enzyme from *B. acidopullulyticus* had pH and temperature optima of pH 4.5 and 60 °C. The organism's optimum growth temperature was 20–37 °C.

Halophilic Microorganisms

Definition and Occurrence

Salt-tolerant bacteria, yeasts, fungi, algae and protozoa have been isolated. Marine microorganisms which are considred to be slightly halophilic grow best in medium containing 0.2–0.5 M salt. Microorganisms considered to be moderately halophilic and growing in 0.5–2.5 M salt include *Vibrio* species (Reali *et al.*, 1977; Richard and Lhuillier, 1977), *Micrococcus* species (Hiwatashi *et al.*, 1958), *Pseudomonas* species (Hiramatsu *et al.*, 1976; Ohno *et al.*, 1976) and *Bacillus* species (Kamekura and Onishi, 1974a). Extreme halophiles growing in medium containing 2.5–5.2 M salt include *Halobacteria* and *Halococci* (Kushner, 1968, 1978; Dundas, 1977; Colwell *et al.*, 1979; Larsen, 1983). Salt-tolerant microorganisms include *Staphylococcus aureus* and *S. epidermis* (Komaratat and Kates, 1975).

Salterns, natural salt lakes and the Dead Sea have proved to be a rich source of extreme halophiles. They are distinguished by their red pigmentation and are referred to as the "red halophiles". Salt obtained following the solar evaporation of water from salterns is often contaminated with *Halobacteriaceae* by up to 10^5-10^6 viable cells per gram of salt. Solar salt is used for the preservation of fish and hides and under favourable conditions contaminating bacteria in the salt begin to grow, causing spoilage. In the fishing industry spoilage is evident from the formation of a pink slime, and in the leather industry the growth of these bacteria causes staining, termed "red heat", which damages the leather (Kushner, 1968). The use of refrigeration has reduced spoilage since they are unable to grow below 10 °C.

Salt-tolerant microorganisms have also been isolated from salted foods, soy sauce and miso paste.

The Extreme Halophiles

The eighth edition of *Bergey's Manual of Determinative Bacteriology* (Buchanan and Gibbons, 1974) recognized only two groups of extreme halophiles, *Halobacterium* and *Halococcus*. Both have a red or orange pigmentation and belong to the archaebacteria (Woese *et al.*, 1978). *Halobacterium* cells are generally rod shaped but some strains are pleomorphic, forming transitions between rods and spheres; disk shaped, rectangular and even triangular forms have been observed. When freshly isolated the cells contain gas vacuoles which are often lost following repeated subculture in the laboratory. Vacuole formation has been considered to be a plasmid-mediated function. The *Halobacteria* contain two DNA components: the larger (64–90% of total DNA) has a $G+C$ content of 66–68%; the minor component has a $G+C$ content of 57–60%.

Halococcus cells are coccoid and occur singly, in pairs, in regular sarcina packets or in irregular clusters. The $G+C$ content is within the range 60.5–65.8%; a minor component (31% of total DNA) with a $G+C$ content of 59% has been reported in *Halococcus morrhuae* (Moore and McCarthy, 1969). The presence of satellite DNA appears characteristic of *Halococcus* and *Halobacterium*.

The medium for cultivation of Halobacteriaceae is characterized by a high NaCl concentration, generally 25% w/v. The requirement for NaCl is specific and is not completely replaced by other salts. K^+ and Mg^+ may partially replace Na^+ and Br^- and NO_3^- may partially replace Cl^-. Many isolates require a concentration of 1–5% w/v $MgCl_2$. Most strains utilize amino acids as a carbon and energy source. The optimum growth temperature is in the range 40–45 °C and growth does not occur below 10 °C or above 55 °C. Most strains are aerobic, but some grow anaerobically in the presence of nitrate. The culture conditions and media are described by Larsen (1983). A standard medium for the cultivation of Halobacteriaceae comprises yeast extract (1% w/v), NaCl (25% w/v) and $MgCl_2$ (5% w/v) and is prepared in tap water at pH 7.0.

Kushner (1966) cultivated Halobacteriaceae in a polythene tank containing 70 l of unsterilized medium. Mixing was carried out using a sparger with the air flow set at 0.1 vvm. Maximum growth of *H. cutirubrum* was obtained after 3–5 days at 37 °C giving a cell yield of 1.0–1.5 g dry w l^{-1}.

Gibbons (1969) used a 150 l stainless steel fermenter with 100 l of medium and obtained cell yields of 5–6 g wet wt l^{-1} after 65 hours at 37 °C.

In our laboratory we have grown *H. halobium* in a stainless steel fermenter with a working volume of 150 l. The medium comprised Casamino acids (7.5 g), yeast extract (10 g), trisodium citrate (3.0 g), KCl (2 g), $MgSO_4 \cdot 7H_2O$ (20 g), NaCl (250 g), $FeCl_2 \cdot 4H_2O$ (35.4 mg) and distilled water (1000 ml). The pH of the medium was adjusted to 6.2 after sterilization, but it increased steadily to pH 8.0 over the 5 day fermentation. Aeration was 1 vvm and foaming was controlled by the addition of 25% v/v silicone RD antifoam. The yield after 5 days was 10 g of wet cell paste per litre of medium.

The Physiology of Extreme Halophiles and Stability of their Proteins

The physiological adaptation to growth and high salt and solute concentrations has been reviewed by Dundas (1977), Bayley and Morton (1978), Kushner (1978) and Morishita and Masui (1980). The salt dependence of halophilic enzymes has been reviewed by Lanyi (1974, 1979).

The ability of halophiles to grow at high external solute concentrations is not achieved by the maintenance of a low solute concentration within the cytoplasm but in the adaptation of cell components to function in high solute concentrations. Although the extreme halophiles require a high concentration of NaCl in the growth medium, they contain intracellular concentrations of K^+, which as KCl would saturate the available water. K^+ may constitute 30–40% of the dry weight of *H. halobium* depending upon the growth phase of the culture (Gochnauer and Kushner, 1971). The gradient of intracellular to extracellular K^+ can be up to 1000:1 (Ginzburg *et al.*, 1971) and the requirement for K^+ (up to 1 mg ml^{-1} in the medium) may become growth limiting (Gochnauer and Kushner, 1969). Halophilic enzymes require salt concentrations in the range 1–4 M to maintain their activity and stability (Lanyi, 1974), levels which would inactivate many non-halophilic enzymes. Halophilic proteins are generally more acidic than non-halophilic homologous enzymes owing to the incorporation of increased glutamic and aspartic acids. Studies of ribosomal proteins indicate a decrease in the frequency of hydrophobic amino acids in the halophiles and an increase in the occurrence of less hydrophobic residues such as serine and threonine (Lanyi, 1974). The stabilization of halophilic proteins has been considered to be due to three main mechanisms: (i) the electrostatic shielding of the molecule by the increase in acidic amino acid residues (effective at 0.5–1 M salt (Lanyi, 1974)); (ii) the increase in new hydrophobic interactions at high salt concentrations, resulting in a more tightly packed and folded structure (Lanyi, 1974, 1979); and (iii) the maintenance of a hydrated protein surface through the utilization of carboxyl groups in glutamate and aspartate (Pundak et al., 1981).

Rao and Argos (1981) reported a statistical comparison of the amino acid composition of ferredoxins from halophiles and non-halophiles. They considered that the bulkiness of amino acids used by halophiles was considerably reduced and the overall hydrophobicity of halophilic and non-halophilic molecules was essentially the same. They suggested the principal mode of structural stabilization of halophilic proteins to be through competition with the cytoplasmic salt for water. This was effected through the utilization of external carboxyl groups of glutamic and aspartic acids. The reduction in bulkiness was considered to reduce inactivity in the presence of high molarity antichaotropic KCl and activity was preserved through the avoidance of additional negative charge at the active site surface.

Products, Processes and Applications for Halophilic Microorganisms

Enzymes

Proteases. The isolation of proteases from *Halobacterium* was reported by Norberg and Hofsten (1969) who discussed the problems of purifying enzymes which are denatured in normal buffers. Protease produced by a moderately halophilic *Bacillus* isolated from unrefined solar salt had an optimum enzyme activity in 0.5 M NaCl and 0.75 M KCl; no activity was observed in 3 M NaCl (Kamekura and Onishi, 1974a). A protease isolated from a marine *Pseudomonas* species and purified to homogeneity had a molecular weight of 120 000 daltons and optimum activity was observed at pH 8.0 in 18% NaCl (Van Qua et al., 1981). Sakata and Kakimoto (1980) demonstrated the necessity for Ca^{2+} ions for the production of extracellular protease from a marine *Pseudomonas* species. Four protease fractions were produced, all having maximum activity at pH 10.0–11.0. Two of the fractions required Ca^{2+} for activity.

Amylases.   Amylase produced by *H. halobium* had optimal activity at pH 6.4–6.6 in sodium β-glycerophosphate buffer containing 0.05% NaCl at 55 °C. Ca^{2+} was not required for stability or activity (Good and Hartman, 1970). An extracellular amylase produced by a moderately halophilic *Micrococcus* species isolated from unrefined solar salt required Ca^{2+} and a high concentration of NaCl and KCl for its stability and activity. The enzyme had maximum activity at pH 6.0–7.0 in 1.4–2.0 M NaCl or KCl at 50 °C (Onishi, 1972). The activity of both amylases was lost following dialysis against distilled water. Activity was restored to the *Halobacterium* enzyme after dialysis against 2 M NaCl. The enzyme from *Micrococcus* required Ca^{2+} ions to restore its activity.

Nucleases.   Kamekura and Onishi (1974b, 1978) reported an extracellular nuclease with a molecular weight of 99 000 daltons produced by *Micrococcus varians* (ATCC 21971) isolated from soy sauce mash. The amino acid composition of the enzyme indicated a predominance of acidic amino acids over basic amino acids. The enzyme was produced in a medium containing 1–4 M NaCl or KCl and had a maximum activity in the presence of 2.9 M NaCl or 2.1 M KCl at 40 °C. Dialysis in low salt buffer resulted in loss of activity; however, dialysis against 3–4 M NaCl restored up to 77% of the original activity. The enzyme exhibited both DNA and RNA exonuclease activity. An extracellular halophilic nuclease with a molecular weight of 138 000 daltons was isolated from a moderately halophilic *Bacillus* species. The organism was cultured aerobically in a medium containing 1–2 M NaCl. Maximum enzyme activity occurred in the presence of 1.4–3.2 M NaCl or 2.3–3.2 M KCl at pH 8.5 and at 50 °C for DNAase activity and 60 °C for RANase activity. Enzyme activity was enhanced by Mg^{2+} and Ca^{2+}. Activity was lost following dialysis against water or low salt buffer but was protected by the addition of 10 mM Ca^{2+} to the buffer. Dialysis against 3.5 M NaCl restored up to 68% of the original activity (Onishi *et al.*, 1983).

Malate dehydrogenase and problems in the purification of halophilic enzymes.   The purification of halophilic enzymes has proved to be difficult using conventional fractionation procedures owing to their requirement for high salt concentrations to maintain stability and activity. Chromatography using calcium hydroxyl apatite has proved particularly useful since it is not disturbed by high salt concentrations (Norberg and Hofsten, 1970; Hochstein and Dalton, 1973).

Meverech *et al.* (1977) and Meverech and Neumann (1977) developed a process for the purification of malate dehydrogenase from 55 g of *Halobacterium*. The cells were lysed by sonication and purified by (i) ammonium sulphate precipitation, (ii) hydrophobic fractionation on Sepharose 4B using a decreasing $(NH_4)_2$ concentration gradient, (iii) gel permeation chromatography on Sephadex G-100, (iv) chromatography on hydroxylapatite and (v) affinity chromatography on 8-(6-aminohexyl)amino-NAD$^+$-Sepharose at 4.26 M NaCl. Leicht and Pundak (1981) described a large-scale process for the purification of malate dehydrogenase and glutamate dehydrogenase from *Halobacterium* species. The process was based on salting out mediated chromatography and yielded a few hundred milligrams of enzyme from 2.1 kg of cell paste.

The immobilization of halophilic malate dehydrogenase on cyanogen-bromide-activated agarose resulted in a considerable increase in enzyme stability and an optimum activity at a lower salt concentration (0.6 M NaCl) than with the soluble enzyme (1.2 M NaCl) (Koch-Schmidt *et al.*, 1979).

Barophilic Microorganisms

Definition, Occurrence and Classification

Hydrostatic pressure has interested marine ecologists for many years and led to research into the growth and survival of microorganisms at pressures in excess of 1000 atm. Naturally occurring high pressures can be found at the bottom of the ocean and in deep oil or sulphur wells.

In the ocean, pressure increases at the rate of 1 atm every 10 m to a maximum of approximately 1160 atm in the Challenger Deep of the Pacific Ocean. The average pressure on the ocean floor has been calculated to be approximately 380 atm (Kinne, 1972). Great ocean depths are also often associated with low temperatures (0–4 °C) (Oppenheimer and ZoBell, 1952), although recently discovered deep-sea hydrothermal vents have indicated temperatures of 350 °C which are also associated with microbial activity (Jannasch, 1984; Jannasch and Taylor, 1984). Pressure in deep oil and sulphur wells increases by approximately 0.1 atm m^{-1} and temperature also increases with depth at 0.014 °C m^{-1}, although this is not always uniform. Thermophilic sulphate-reducing bacteria have been isolated from such wells at a depth of 3500 m (400 atm) and at temperatures between 60 and 105 °C (ZoBell, 1958).

Microorganisms that grow and survive at such extremes of pressure are termed barophilic but may be further classified as barotolerant-baroduric or obligately barophilic (ZoBell and Morita, 1957). The definitions are a little obscure; obligate barophiles only grow at elevated pressures, barotolerant species grow between 1 and 600 atm and barosensitive species cannot grow or survive at pressures in excess of 1 atm. *Halobacterium* species are barosensitive, containing gas vacuoles which collapse at pressures around 2 atm

(Walsby, 1971, 1972); this affects their buoyancy and subsequently affects their metabolism (Dinsdale and Walsby, 1972). Kriss (1962) isolated baroduric bacteria from garden soil indicating their ubiquity. The relationship of the growth and survival of *S. faecalis* and *E. coli* at elevated pressures has been studied (Marquis, 1976; Marquis and Matsumura, 1978). Some strains of *E. coli* have been shown to grow well at pressures in excess of 500 atm and may be termed baroduric (Yayanos, 1975). The existence of obligate barophiles has been in question, but their existence appears to be verified by the isolation of a bacterium designated MT41 (Yayanos *et al.*, 1981). This strain, isolated at a depth of 10 476 m from the Mariana Trench, did not grow at 380 atm (average deep-sea pressure), but at 690 atm and at 2 °C growth was optimal.

The Major Groups of Barophilic Microorganisms

Barophilic bacteria are a diverse group of microorganisms with an equally diverse response to pressure and it is difficult to discern any pattern of barotolerance owing to a lack of well-defined taxonomic data. Comprehensive lists of barophilic microorganisms have been presented by ZoBell and Johnson (1949) and Oppenheimer and ZoBell (1952). Strains were incubated for 8 days at 27 °C at various hydrostatic pressures and it was possible to distinguish species according to their barotolerance. The most sensitive were strains of *Achromobacter thalassius, M. euryhalis, Ps. azotogena, Ps. marinopersica* and *Ps. hypothermus*, which were all killed at 200 atm. Strains of *B. borborokoites, M. aquivivius, Ps. perfectomarinus* and *Vibrio phytoplanktis* grew well at 600 atm. Other highly baroduric species have been reported, such as *Ps. bathycetes*, which grows at 1000 atm at 3 °C (Schwarz and Colwell, 1975). Sulphate-reducing bacteria, presumed to be *Desulfovibrio* species isolated from deep wells, have been cultivated at 1000 atm at 104 °C, although difficulties in subculturing these strains were encountered (ZoBell, 1958).

In recent years the concept of the deep sea as a very low nutrient environment has been challenged. Studies by Hessler *et al.* (1972) demonstrated the existence of large amphipod species. Some years later deep-sea hydrothermal vents were located, which were also associated with numerous large marine animals (Ballard, 1977; Lonsdale, 1977; Corliss *et al.*, 1979). Recent reports have demonstrated that barophilic bacteria can be isolated from deep-sea animals such as amphipods and halothurians (Schwarz and Colwell, 1976; Yayanos *et al.*, 1979, 1981; Deming *et al.*, 1981; Deming and Colwell, 1982). Also, a rich microbial fauna has been observed around hydrothermal vents (Jannasch, 1984). Deep-sea bacteria were originally considered to have slow metabolic rates. At 1000 atm and 3 °C *Ps. bathycetes* had a lag phase of 4 months and a mean generation time of 33 days in exponential phase and reached stationary phase after 12 months (Schwarz and Colwell, 1975). Bacteria associated with deep-sea animals often have higher metabolic activity. A spirillum-like bacterium isolated from an amphipod species was shown to have a mean generation time of 4–13 hours at 500 atm at a temperature of 2–4 °C, in comparison with a mean generation time of 3–4 days at 1 atm (Yayanos *et al.*, 1979). Schwarz *et al.* (1976) had shown similar results with other deep-sea bacteria which grew rapidly at 750 atm at 3 °C. Many of these animal-associated bacteria remain unidentified. An obligate barophilic strain designated MT41, isolated from an amphipod from the Mariana Trench at a depth of 10 476 m, was unable to grow at 380 atm (Yayanos *et al.*, 1981). An optimum generation time of 25 hours was observed at 690 atm at 2 °C; at 1035 atm, a pressure close to that of its origin, the generation time was extended to 33 hours. Yayanos *et al.* (1981) suggested that *Ps. bathycetes* is atypical of both sedimentary and animal-associated bacteria of the Mariana Trench and its low activity state is an uncommon feature of the indigenous microflora.

Numerous bacterial species have been identified in association with the recently discovered hydrothermal vents. Many of the species chemosynthetically reduce CO_2 to organic carbon while geothermally reduced inorganic compounds provide an energy source. Species isolated around these vents include *Thiomicrospora* (Ruby and Jannasch, 1982), Calothrix-like organisms (Jannasch and Wirsen, 1981), *Hyphomicrobium* and *Hyphomonas* (Poindexter, 1981), *Methanococcus* (Jones *et al.*, 1983) and strains of manganese-oxidizing bacteria (Ehrlich, 1983). Studies of the effect of pressure on the biochemistry, physiology and molecular aspects of barophiles has involved microorganisms which normally grow at 1 atm. This is due mainly to an interest in how pressure inhibits bacterial growth and the lack of suitable obligate barophiles for study and comparison. A number of reviews have extensively discussed the effect of pressure on microorganisms (Marquis, 1976; Morita, 1976; Marquis and Matsumura, 1978) and ecological aspects have recently been reviewed by Jannasch (1984) and Jannasch and Taylor (1984). Pressure can generally be said to inhibit reactions and processes which result in an increase in volume and to stimulate those which result in a decrease in volume. Pressure affects both chemical equilibrium and reaction rates. It can induce the dissociation of specific polymeric aggregates, such as multimeric enzymes, and can also cause denaturation of biopolymers. This is most obvious when other environmental conditions, such as temperature, pH or ionic strength, are not optimal. Pressure-induced changes can also be observed in catabolic and anabolic pathways. Most reaction rates decrease at pressures of 300 atm, although some increase. Differences from the normal rate of reaction may affect the cell metabolism and explain the barosensitivity of microorganisms. Pressure is known to be lethal to many microorganisms (ZoBell and Johnson, 1949; Oppenheimer and ZoBell, 1952). Hydrostatic pressure has been proposed as a sterilizing agent for labile products (Hite *et al.*, 1914), although more recent reports noted problems with resistant microorganisms (Timson and Short, 1965).

Processes and Applications for Barophilic Microorganisms

Microbial-enhanced Oil Recovery (MEOR)

MEOR is one important area in which barophilic microorganisms may play a vital role. Although pressure is not a critical factor in open tar-sand deposits close to the surface, it is an important factor in the "richest interval" for oil deposits at depths of 2130 m (Hunt, 1979). At these depths pressures range from 200 to 450 atm or more in overpressurized strata. Future wells may go as deep as 3000 m to achieve maximum recovery, leading to a concomitant increase in pressure, comparable with that found in the deep oceans (Kinne, 1972). At such depths temperatures range from 60–90 °C, as found in Texan and Louisiana coastal fields, to the 100–123 °C found in Middle East fields (Hunt, 1979).

Microorganisms used in MEOR are required to grow at high temperature and pressure, in extremes of pH (at least pH 4.0–9.0), in NaCl concentrations between 0.5 and 1.5 M (sometimes in excess of 6.0 M), under anaerobic conditions and possibly in the presence of heavy metals (Trudinger and Bubela, 1967). Such conditions do not exclude bacteria since sulphate-reducing bacteria can be readily isolated under these conditions and their potential in MEOR has been investigated (Kuznetsova, 1960; Kuznetsova and Pantskhava, 1962). They have proved to be too destructive to be used in most reservoirs and cause many serious and costly corrosion problems (Davis, 1967; Hamilton, 1983). In general the growth of these bacteria is suppressed by the use of biocides.

Other bacterial species which may have the necessary requirements or be adapted for MEOR include *Arthrobacter, Bacillus, Clostridium, Myobacterium, Peptococcus* and *Pseudomonas* (Lazar, 1983; Long-Kuan *et al.*, 1983). Hitzman (1962) patented a process which involved the injection of spores into deep wells. Pressure has been demonstrated to act as a germinant and the spores are assumed to germinate and grow *in situ*. However, outgrowing cells have been found to be pressure sensitive, suggesting that problems may be encountered in deep wells (Gould and Sale, 1972). The main role for bacteria in MEOR is to produce *in situ* gas and/or chemicals which reduce the viscosity of the remaining oil. Microorganisms were successfully used in oil recovery in the USA some 30 years ago, but the prevailing oil price made their use uneconomic (Moses, 1983). At today's oil prices tertiary oil recovery has become more economical. Microbial biosurfactants have been demonstrated in numerous bacteria and their potential in oil recovery has been investigated (Gerson and Zajic, 1979). Ideally MEOR also requires the bacteria to utilize the crude oil for production of gas and surfactants which eliminate the requirement for injecting an additional substrate such as molasses into the well (Moses, 1983). Many marine bacteria utilize crude oil and may be valuable in MEOR (Colwell *et al.*, 1976; Conrad *et al.*, 1976; Seesman *et al.*, 1976). Bacterial species isolated from hydrothermal vents in the deep sea may also be potentially useful. The inability to isolate an ideal strain may result in studies of mixed populations for oil mobilization or the genetic construction of strains with suitable characteristics. Some 10 years ago the production of microbial biosurfactants on an industrial scale for injection into wells was not cost effective (Gerson and Zajic, 1979); however, at current oil prices and with the upsurge in biotechnological expertise this may be reconsidered, particularly since larger fermenters are now available.

Other Applications

Both barophilic manganese-oxidizing and MnO_2-reducing bacteria have been isolated from ferromanganese nodules from the deep sea (Ehrlich, 1974a, b, 1975). Manganese(II)-oxidizing bacteria identified as *Arthrobacter* spp. are believed to be involved in the deposition of the nodules at extremely slow rates at 4 °C and between 340 and 476 atm. MnO_2 reduction has also been demonstrated by unknown deep-sea Gram-negative bacteria; the effect is inducible at pressures of 476 atm. Previous studies with manganese-oxidizing and manganese-reducing bacteria have shown that dried deposits of such bacteria contain approximately 28% iron and manganese with an Fe:Mn ratio of 0.9 (Wolfe, 1960). Such mineral leaching activity has a potential use in the mining industry.

The enzymes from some marine barophilic microoragnisms have been studied (Okami, 1979, 1982).

Process Considerations

Culturing microorganisms under pressure is not a significant problem and fermeters have been constructed to operate at high pressures (Vance, 1984). The main problem with barophiles may be their sensitivity to low pressures. Jannasch and Wirsen (1977a) suggested that barophiles may be sensitive to decompression and Yayanos *et al.* (1979) reported that growth of a *Spirillum* sp. slowed on decompression but was not inactivated. Devices for the retrieval of undecompressed microbial populations from the deep sea have been described (Jannasch and Wirsen, 1977b; Jannasch *et al.*, 1982). However, it has been suggested that an increase in temperature from environmental temperatures may be more deleterious than a reduction in pressure (Yayanos *et al.*, 1981).

Psychrophilic Microorganisms

Definition, Occurrence and Organisms

Psychrophilic microorganisms were defined by Ingraham and Stokes (1959) as those which grew well at 0 °C within 2 weeks. Psychrotrophs were described by Eddy (1960) as microorganisms able to grow at 5 °C. Morita (1975) proposed that psychrophiles be defined as organisms having an optimal growth temperature of 15 °C or below, a maximum growth temperature of 20 °C and a minimum of 0 °C or below. A number of classification schemes have been proposed and various definitions prevail in different areas of the food and dairy industry. These are discussed by Berry and Magoon (1934), Hucker (1954), Ingraham and Stokes (1959), Ingram (1965), Morita (1966), Morita (1975) and Baross and Morita (1978).

Psychrophiles have been isolated from the northern Pacific Ocean 67 miles off the Oregon coast at depths of 350 m and 2800 m (Morita and Burton, 1963). Harder and Veldkamp (1967) reported the isolation of 69 strains from the North Sea able to grow at low temperatures. Ten of those isolates had a maximum growth temperature below 20 °C and were identified as species of *Pseudomonas, Vibrio* and *Spirillum. Arthrobacter* species were isolated from sediments below Arctic glaciers and in Arctic caves (Gounot, 1968a, b, 1973; Moiroud and Gounot, 1969). One strain subsequently named *Arthrobacter glacialis* had an optimum growth temperature of 13 °C but grew in the range from −5 to 18 °C. Matches and Liston (1973) isolated 288 strains of *Clostridia* from a marine sediment in the Puget Sound area, of which seven did not grow above 20 °C. The ecological aspects of microbial life at low temperature have been discussed by Morita (1975) and Baross and Morita (1977). Michels and Visser (1976) reported the isolation of psychrophilic and psychrotrophic aerobic spore formers from soil in The Netherlands.

Microbial Physiology of Psychrophiles

Morita and Albright (1965) demonstrated that low cell yields and slow growth rates, often reported for psychrophiles, are not valid for true psychrophiles. They demonstrated that *V. marinus* MP-1 grew well at low temperatures yielding 1.3×10^{12} viable cells ml^{-1} at 15 °C after 24 hours, and 9×10^9 viable cells ml^{-1} at 3 °C after 24 hours. *V. marinus* had generation times of 80.7 minutes and 226 minutes at 15 °C and 3 °C respectively.

The genetic basis for psychrophily has not been established although there are reports of conversion of a mesophile to psychrotrophy by transduction (Olsen and Metcalf, 1968) and ultraviolet light (Azuma *et al.*, 1962) and of the ultraviolet light mutation of a psychrotroph to a mesophile (Tai and Jackson, 1969). These reports are similar to those relating to the conversion of mesophiles to thermophily (McDonald and Matney, 1963; Lindsey and Creaser, 1975) and similarly it is unlikely that mutagenesis, transduction or transformation would change the total enzymic make-up of the cells to promote psychrophily, mesophily or thermophily.

Krajewska and Szer (1967) compared amino acid incorporation in cell-free systems of *E. coli* and a psychrotrophic *Pseudomonas*. At 30 °C and 45 °C the ribosomes and cell-free supernatant fractions of the two organisms were interchangeable in a poly-U-promoted phenylalanine incorporation system. At 2 and 9 °C, however, only the *Pseudomonas* ribosomes continued to be active with both supernatant fractions. They also reported greater fidelity in the incorporation of amino acids when compared with mesophiles and thermophiles, indicating that psychrophiles may possess greater genetic stability. The physiology of growth at low temperature has been discussed by Morita (1975), Inniss (1975) and Inniss and Ingraham (1978).

Food Spoilage

The refrigeration of foods at temperatures above freezing is one of the most common methods of food preservation. Even 30 years ago it was estimated that some 85% of all foods in the USA had, at some stage, been refrigerated prior to consumption (Anderson, 1953). Psychrophilic microorganisms consequently represent a group of major economic importance to the food industry since they are considered to be the chief cause of spoilage of meat, fish, poultry, eggs, dairy produce, vegetables and fruit. They are also known to cause spoilage of some heat-processed canned foods (Heinrich *et al.*, 1970).

Psychrophilic spoilage microorganisms are a diverse group including bacteria, yeast and moulds. *Pseudomonas* species are probably the most predominant food spoilage bacteria. Non-pigmented species usually cause spoilage of meats, whereas pigmented species, e.g. *Ps. fluorescens* and *Ps. ovalis*, are often responsible for spoilage of shell eggs (Lorenze *et al.*, 1952). Other *Pseudomonas* species have been implicated in the spoilage of milk and vegetables (Nickerson and Sinskey, 1972). Most spoilage organisms have optimum growth temperatures of between 20 and 25 °C but grow well at 0 °C.

Achromobacter species are often the predominant spoilage organisms of foods which are irradiated and then subsequently refrigerated (Idziak and Incze, 1968). The spoilage of meats by *Achromobacter* species generally does not produce the bad odours and flavours characteristic of most *Pseudomonas* species. For organoleptic changes to be noticeable the cell concentration of *Achromobacter* species must be at least tenfold higher than that of *Pseudomonas* species (Nickerson and Sinskey, 1972).

The Lactobacteriaceae, another common food spoilage group, are responsible for the spoilage of dairy products. They are relatively heat resistant, grow at low temperatures, grow both aerobically and anaerobically and are tolerant to high salt concentrations and to high and low pH. The group includes such genera as *Streptococcus*, *Leuconostoc*, *Pediococcus* and *Lactobacillus*. Many species cause spoilage of dairy products.

A number of moulds including *Mucor* species, *Cladosporium* sp., *Sporotrichum* sp. and *Thaminidium* species are also involved in food spoilage. Spoilage microbiology has been extensively studied and reviewed (Frazier, 1967; Shewan and Hobbs, 1967; Panes and Thomas, 1968; Nickerson and Sinskey, 1972).

Dairy Industry

Psychrophilic microorganisms, and their enzymes in particular, may have future applications in the dairy industry. Calf rennet, used for the coagulation of milk during the manufacture of cheese, has in recent years become relatively scarce. Both milk and cheese production have increased, but from a reduced stock of cattle, and consequently there has been a reduction in the number of calves available for slaughter and hence the amount of available rennet. Alternative sources of suitable enzymes have been investigated. Plant and bacterial coagulants appear to be too proteolytic and often result in adverse flavouring of the product. Certain other animal-derived coagulants, e.g. porcine pepsins, have given good results, although a number of fungal enzymes have been successfully produced on a commercial scale. Enzymes from *Mucor miehei* (Rennilase®, Hannilase®, Marzyme®) and *M. pusillus lindt* (Emporase®) have been used for milk coagulation in a number of countries (Burgess and Shaw, 1983). These enzymes have a temperature range of 28–37 °C although Emporase® is reported to have a range between 5 and 50 °C. Processing at mesophilic temperatures results in residual coagulation activity when the whey is used in other dairy products. Psychrophilic enzymes would have the advantage that any residual activity can be removed simply by pasteurization. This has already been demonstrated using commercially available thermolabile microbial rennets, e.g. Marzyme 11®, Rennilase 50 TL® and Modilase® (Burgess and Shaw, 1983).

Lactose, which is a major component of milk and whey, may be hydrolysed to glucose and galactose resulting in increased solubility, digestability and sweetness. This enables its use in a more diverse range of products. β-galactosidase from *Kluyveromyces* or *Aspergillus* is normally used for hydrolysis depending on the pH of the milk or whey; the former has a pH optimum of 6.0–7.0 and the latter an optimum of pH 4.0–5.0. Hydrolysis can be achieved either by a batch process or by a continuous process using immobilized enzymes. Neutral pH and incubation at 30–40 °C, however, are ideal conditions for the growth of mesophilic bacteria and it is necessary to keep the incubation time to a minimum (4 hours). This reduces microbial contamination but results in reduced lactose hydrolysis. In preference, the temperature is often reduced to 5–10 °C and the incubation time increased to 16–24 hours, resulting in 70–80% hydrolysis. The use of highly active psychrophilic enzymes may reduce the problems of contamination, particularly in the operation of a continuous system. A commercially available system utilizing a yeast β-galactosidase and operating at low temperature has been reported (Burgess and Shaw, 1983).

Psychrophilic enzymes may be potentially useful in the accelerated ripening of cheeses. A wide range of enzymes have already been used with some success in many types of cheese. The commercial neutral protease Neutrase® from *B. subtilis* when added to cheese has been shown to produce in 1 month at 18 °C a cheese with a flavour intensity equivalent to that normally obtained in 4 months (Burgess and Shaw, 1983). Psychrophilic enzymes, particularly neutral proteases, would be more suited at these temperatures and may lead to further improvements in the process.

Other Applications

Wellinger and Kaufmann (1982) reported that methane could be generated on small farms in Switzerland using a psychrophilic methane digester. The fermentation process using pig slurry held at ambient temperatures (approximately 18 °C) compared favourably with mesophilic fermentations.

The detoxification of land contaminated with pesticides may be possible by the direct application of suitable degradative psychrophilic microorganisms or psychrophiles expressing cloned degradative enzymes.

Acknowledgements

We wish to thank Dr. A. V. Quirk, Dr. J. R. Court and Professor T. Atkinson for useful discussions during the preparation of this manuscript.

We also wish to thank Miss V. M. Bowden for typing the manuscript and for her considerable help and patience during its preparation.

References

Akiba, T. and Horikoshi, K. (1976). Identification and growth characteristics of α-galactosidase producing microorganisms. *Agricultural and Biological Chemistry* **40**, 1845–1849.

Ambroz, A. (1913). *Denitrobacterium thermophilum* spec. nova. Ein Beitrag zur Biologie der thermophilen Bakterien. *Zentralblatt für Bakteriologie und Parasitenkunde* **II**, 3–16.

Amelunxen, R. E. (1966). Crystallization of thermostable glyceraldehyde-3-phosphate dehydrogenase from *Bacillus stearothermophilus*. *Biochemica et Biophysica Acta* **122**, 175–181.

Amelunxen, R. E. and Lins, M. (1968). Comparative thermostability of enzymes from *Bacillus stearothermophilus* and *Bacillus cereus*. *Archives of Biochemistry and Biophysics* **125**, 765–769.

Amelunxen, R. E. and Murdock, A. L. (1978). Life at high temperatures: molecular aspects. *In* "Microbial Life in Extreme Environments" (Ed. D. J. Kushner), pp. 217–278. Academic Press, London and New York.

Anderson, O. A. (1953). "Refrigeration in America". Princeton University Press, Princeton, New Jersey.

Ando, A., Yabuki, M., Fujii, T. and Fukui, S. (1981a). General characteristics of an alkalophilic bacterium, *Bacillus* A-007. *The Technical Bulletin of the Faculty of Horticulture, Chiba University, Japan* **29**, 17–28.

Ando, A., Yabuki, M. and Kusaka, I. (1981b). Na$^+$-driven Ca^{2+} transport in alkalophilic *Bacillus*. *Biochimica et Biophysica Acta* **640**, 179–184.

Atkinson, A. (1976). Thermostable enzymes. *Journal of Applied Chemistry and Biotechnology* **26**, 577–578.

Atkinson, A., Ellwood, D. C., Evans, C. G. T. and Yeo, R. G. (1975a). Production of alcohol by *Bacillus stearothermophilus*. *Biotechnology and Bioengineering* **17**, 1375–1377.

Atkinson, A., Evans, C. G. T. and Yeo, R. G. (1975b). Behaviour of *Bacillus stearothermophilus* grown in different media. *Journal of Applied Bacteriology* **38**, 301–304.

Atkinson, A., Bruton, C. J., Comer, M. J. and Sharp, R. J. (1979b). UK patent application. "Production of Enzymes".

Atkinson, T., Banks, G. T., Bruton, C. J., Comer, M. J., Jakes, R., Kamalagharan,

T., Whitaker, A. R. and Winter, G. P. (1979a). Large-scale isolation of enzymes from *Bacillus stearothermophilus*. *Journal of Applied Biochemistry* **1**, 247–258.

Atsukawa, A. and Horikoshi, K. (1976). Alkaline ribonuclease of alkalophilic bacteria. *In* "Ribosomes and RNA Metabolism", Vol. 2, pp. 89–95. The Slovak Academy of Sciences, Bratislava.

Aunstrup, K., Outtrup, H., Andresen, O. and Dambmann, C. (1972). Proteases from alkalophilic *Bacillus* species. *In* "Fermentation Technology Today, Proceedings of the 4th International Fermentation Symposium", pp. 299–305. Society of Fermentation Technology, Osaka, Japan.

Avgerinos, G. C., Fank, H. Y., Biocic, I. and Wang, D. I. E. (1981). A novel, single step microbial conversion of cellulosic biomass to ethanol. *In* "Advances in Biotechnology; Fuels, Chemicals, Foods and Waste Treatment" (Ed. M. Moo-Young), pp. 119–131. Pergamon, Oxford.

Azuma, Y., Newton, S. B. and Witter, L. D. (1962). Production of psychrophilic mutants from mesophilic bacteria by ultraviolet irradiation. *Journal of Dairy Science* **45**, 1529–1530.

Baker, H., Frank, O., Pasher, I., Black, B., Hutner, S. H. and Sobotka, H. (1960). Growth requirements of 94 strains of thermophilic *Bacilli*. *Canadian Journal of Microbiology* **6**, 557–563.

Ballard, R. D. (1977). Notes on a major oceanographic find. *Oceanus* **20**, 35–44.

Barfoed, H. C. (1983). Detergents. *In* "Industrial Enzymology. The Application of Enzymes in Industry" (Eds T. Godfrey and J. Reichelt), pp. 284–293. Macmillan, London.

Baross, J. A. and Deming, J. W. (1983). Growth of "black smoker" bacteria at temperatures of at least 250 °C. *Nature, London* **303**, 423–426.

Baross, J. A. and Morita, R. Y. (1978). Microbial life at low temperatures; ecological aspects. *In* "Microbial Life in Extreme Environments" (Ed. D. J. Kushner), pp. 9–71. Academic Press, London, New York and San Francisco, California.

Bayley, S. T. and Morton, R. A. (1978). Recent developments in the molecular biology of extremely halophilic bacteria. *CRC Critical Reviews of Microbiology* **6**, 151–205.

Bechina, E. M., Loginova, L. G. and Gernet, M. V. (1983). Selection of products of amylolytic enzymes among thermophilic microorganisms. *Applied Biochemistry and Microbiology* **18**, 514–521.

Bellamy, W. D. (1977). Cellulose and lignocellulose digestion by thermophilic Actinomycetes for single cell protein production. *Developments in Industrial Microbiology* **18**, 249–254.

Belly, R. T. and Brock, T. D. (1972). Cellular stability of a thermophilic, acidophilic Mycoplasma. *Journal of General Microbiology* **73**, 465–469.

Bender, H. (1977). Cyclodextrin glucanotransferase from *Klebsiella pneumoniae*. I. Formation, purification and properties of the enzyme from *Klebsiella pneumoniae* M5 al. *Archives of Microbiology* **111**, 271–282.

Bennet, J. C. and Tributsch, H. (1978). Bacterial leaching patterns on pyrite crystal surfaces. *Journal of Bacteriology* **134**, 310–317.

Berry, J. A. and Magdon, C. A. (1934). Growth of microorganisms at and below 0 °C. *Phytopathology* **24**, 780–796.

Bhattacharyya, P. and Barnes, E. M., Jr. (1978). Proton-coupled sodium uptake by membrane vesicles from *Azotobacter vinelandii*. *Journal of Biological Chemistry* **253**, 3848–3851.

Bingham, A. H. A. and Darbyshire, J. (1982). Isolation of two restriction enzymes from *Chloroflexus aurantiacus*. (Cau I, Cau II). *Gene* **18**, 87–91.

Bingham, A. H. A., Atkinson, T., Sciaky, D. and Roberts, R. J. (1978). A specific endonuclease from *B. caldolyticus*. *Nucleic Acid Research* **5**, 3457–3467.

Boyer, E. W. and Ingle, M. B. (1972). Extracellular alkaline amylase from a *Bacillus* species. *Journal of Bacteriology* **110**, 992–1000.

Breed, R. S., Murray, E. G. D. and Parker-Hitchens, A. (1948). "Bergey's Manual of Determinative Bacteriology", 6th edition. Bailliere, Tindall and Cox, London.

Breed, R. S., Murray, E. G. D. and Smith, N. R. (1957). "Bergey's Manual of Determinative Bacteriology", 7th edition. Bailliere, Tindall and Cox, London.

Brey, R. N. and Rosen, B. P. (1980). Cation/proton antiport systems in *Escherichia coli*. Properties of the potassium/proton antiporter. *Journal of Biological Chemistry* **255**, 39–44.

Brierley, C. L. (1974). Molybdenite-leaching: use of a high-temperature microbe. *Journal of the Less-Common Metals* **36**, 237–247.

Brierley, C. L. (1977). Thermophilic microorganisms in extraction of metals from ores. *Developments in Industrial Microbiology* **18**, 273–284.

Brierley, C. L. (1978). Bacterial leaching. *CRC Critical Reviews in Microbiology* **6**, 207–262.

Brierley, C. L. and Brierley, J. A. (1973). A chemoautotrophic and thermophilic microorganism isolated from an acid hot spring. *Canadian Journal of Microbiology* **19**, 183–188.

Brierley, J. A. and Le Roux, N. W. (1977). A facultative thermophilic *Thiobacillus* like bacterium: oxidation of iron and pyrite. *In* "Conference on Bacterial Leaching" (Ed. W. Schwartz), GMB Monograph No 4, pp. 55–66. Verlag Chemie, Weinheim and New York.

Brierley, J. A., Norris, P. R., Kelly, D. P. and Le Roux, N. W. (1978). Characteristics of a moderately thermophilic and acidophilic iron-oxidising *Thiobacillus*. *European Journal of Applied Microbiology and Biotechnology* **5**, 291–299.

Brock, T. D. (1978). "Thermophilic Microorganisms and Life at High Temperatures". Springer, New York.

Brock, T. D. and Boylen, L. K. (1973). Presence of thermophilic bacteria in laundry and domestic hot water heaters. *Applied Microbiology* **25**, 72–76.

Brock, T. D. and Freeze, H. (1969). *Thermus aquaticus* gen. n. and sp. n. A nonsporulating extreme thermophile. *Journal of Bacteriology* **98**, 289–297.

Brock, T. D. and Gustafson, J. (1976). Ferric iron reduction by sulfur and iron-oxidizing bacteria. *Applied and Environmental Microbiology* **32**, 567–571.

Bryant, F. and Ljungdahl, L. G. (1981). Characterisation of an alcohol dehydrogenase from *Thermoanaerobacter ethanolicus* active with ethanol and secondary alcohols. *Biochemical and Biophysical Research Communications* **100**, 793–799.

Buchanan, R. E. and Gibbons, N. E. (1974). *In* "Bergey's Manual of Determinative Bacteriology", 8th ed. Williams and Wilkins, Baltimore, Maryland.

Buonocore, V., Caporale, C. De Rosa, M. and Gambacorta, A. (1976). Stable inducible thermoacidophilic α-amylase from *Bacillus acidocaldarius*. *Journal of Bacteriology* **128**, 515–521.

Buonocore, V., Sgambati, O., De Rosa, M., Esposito, E. and Gambacorta, A. (1980). A constitutive β-galactosidase from the extreme thermoacidophile Archaebacterium *Caldariella acidophila*: properties of the enzyme in the free state and in immobilized whole cells. *Journal of Applied Biochemistry* **2**, 390–397.

Burgess, K. and Shaw, M. (1983). Dairy. *In* "Industrial Enzymology. The Application of Enzymes in Industry" (Eds T. Godfrey and J. Reichelt), pp. 260–283. Macmillan, London.

Cameron, E. J. and Esty, J. R. (1926). The examination of spoiled canned foods, 2. Classification of flat sour organisms from non-acid foods. *Journal of Infectious Diseases* **39**, 89–105.

Campbell, L. L. (1955). Purification and properties of an α-amylase from facultative thermophilic bacteria. *Archives of Biochemistry and Biophysics* **54**, 154–161.

Carreira, L. H., Wiegel, J. and Ljungdahl, L. G. (1984). Production of ethanol from biopolymers by anaerobic, thermophilic and extreme thermophilic bacteria: I. Regulation of carbohydrate utilisation in mutants of *Thermoanaerobacter ethanolicus*. *Biotechnology and Bioengineering Symposium* **13**, 183–193.

Castenholz, R. W. (1969). Thermophilic blue–green algae and the thermal environment. *Bacteriological Reviews* **33**, 476–504.

Castenholz, R. W. and Pierson, B. K. (1983). Isolation of members of the family Chloroflexaceae. *In* "The Prokaryotes" (Eds M. P. Starr, H. Stolp, H. G. Trüper, A. Balows and H. G. Schlegel), pp. 290–298. Springer, Berlin, Heidelberg and New York.

Chen, W. (1980). Glucose isomerase (a review). *Process Biochemistry* **15**, 36–41.

Chen, M. (1983). Adaptation of mesophilic anaerobic sewage fermentor populations to thermophilic temperatures. *Applied and Environmental Microbiology* **45**, 1271–1276.

Chislett, M. E. and Kushner, D. J. (1961). A strain of *Bacillus circulans* capable of growing under highly alkaline conditions. *Journal of General Microbiology* **24**, 187–190.

Colwell, R. R., Walker, J. D., Conrad, B. F. and Seesman, P. A. (1976). Microbiological studies of Atlantic Ocean water and sediment from potential off-shore drilling sites. *Developments in Industrial Microbiology* **17**, 269–282.

Colwell, R. R., Litchfield, C. D., Vreeland, R. H., Kiefer, L. A. and Gibbons, N. E. (1979). Taxonomic studies of red halophilic bacteria. *International Journal of Systematic Bacteriology* **29**, 379–399.

Comer, M. J., Bruton, C. J. and Atkinson, T. (1979). Purification and properties of glycerokinase from *Bacillus stearothermophilus*. *Journal of Applied Biochemisry* **1**, 259–270.

Conrad, B. F., Walker, J. D. and Colwell, R. R. (1976). Utilization of crude oil and mixed hydrocarbon substrate by Atlantic Ocean microorganisms. *Developments in Industrial Microbiology* **17**, 283–291.

Converse, J. C., Zeikus, J. G., Graves, R. E. and Evans, G. W. (1977). Anaerobic degradation of dairy manure under mesophilic and thermophilic temperatures. *Transactions of the American Society for Agricultural Engineering* **20**, 336–341.

Cooney, C. L. and Wise, D. L. (1975). Thermophilic anaerobic digestion of solid waste for fuel gas production. *Biotechnology and Bioengineering* **17**, 1119–1135.

Corliss, J. B., Dymond, J., Gordon, L. I., Edmond, J. M., von Herzen, R. P., Ballard, R. D., Green, K., Williams, D., Bainbridge, A., Crane, K. and van Andel, T. H. (1979). Submarine thermal springs on the Galapagos Rift. *Science* **203**, 1073–1083.

Cowan, D. A., Daniel, R. M., Martin, A. M. and Morgan, H. W. (1984). Some properties of a β-galactosidase from an extremely thermophilic bacterium. *Biotechnology and Bioengineering* **26**, 1141–1145.

Curtin, M. E. (1983). Microbial mining and metal recovery: corporations take the long and cautious path. *Biotechnology* **1**, 229–235.

Cysewski, G. R. and Wilke, C. R. (1978). Process design and economic studies of alternative fermentation methods for the production of ethanol. *Biotechnology and Bioengineering* **20**, 1421–1444.

Darland, G. and Brock, T. D. (1971). *Bacillus acidocaldarius sp.* nov., an acidophilic thermophilic spore-forming bacterium. *Journal of General Microbiology* **67**, 9–15.

Darland, G., Brock, T. D., Samsonoff, W. and Conti, S. F. (1970). A thermophilic, acidophilic *Mycoplasma* isolated from a coal refuse pile. *Science* **170**, 1416–1418.

Daron, H. H. (1970). Fatty acid composition of lipid extracts of a thermophilic *Bacillus* species. *Journal of Bacteriology* **101**, 145–151.

Davies, R. D. (1982). Corrosion resistant properties of metals used in laboratory equipment. *Laboratory Practice* (February), 103–109.

Davis, J. B. (1967). "Petroleum Microbiology". Elsevier, Amsterdam.

Davis, P. E., Cohen, D. L. and Whitaker, A. (1980). The production of α-amylase in batch and chemostat culture by *Bacillus stearothermophilus*. *Antonie van Leeuwenhoek; Journal of Microbiology and Serology* **46**, 391–398.

Decker, S. J. and Lang, D. R. (1978). Membrane bioenergetic parameters in uncoupler-resistant mutants of *Bacillus megaterium*. *Journal of Biological Chemistry* **253**, 6738–6743.

Dekker, R. F. H. and Richards, G. N. (1976). Hemicellulases: their occurrence, purification, properties, and mode of action. *Advances in Carbohydrate Chemistry and Biochemisty* **32**, 277–352.

Deming, J. W. and Colwell, R. R. (1982). Barophilic bacteria associated with digestive tracts of abyssal Holothurians. *Applied and Environmental Microbiology* **44**, 1222–1230.

Deming, J. W., Tabor, P. S. and Colwell, R. R. (1981). Barophilic growth of bacteria from intestinal tracts of deep-sea invertebrates. *Microbial Ecology* **7**, 85–94.

DePinto, J. A. and Campbell, L. L. (1968). Purification and properties of the amylase of *Bacillus macerans*. *Biochemistry* **7**, 114–120.

De Rosa, M., Gambacorta, A. and Bu'lock, J. D. (1975). Extremely thermophilic acidophilic bacteria convergent with *Sulfolobus acidocaldarius*. *Journal of General Microbiology* **86**, 156–164.

Derry, R., Le Roux, N. W., Garrett, K. H. and Smith, S. E. (1976). Bacterially assisted plant process for leaching uranium ores. *In* "Geology, Mining and Extractive Processing of Uranium" (Ed. M. L. Jones), pp. 56–62. Institute of Mining and Metallurgy, London.

Deuel, H. and Stutz, E. (1958). Pectic substances and pectic enzymes. *Advances in Enzymology* **20**, 341–382.

Devanathan, T., Akagi, J. M. and Hersh, R. T. (1969). Ferredoxin from two thermophilic *Clostridia*. *Journal of Biological Chemistry* **244**, 2846–2853.

Diers, I. V. (1977). U.S. Patent 4,042,460.

Dinsdale, M. T. and Walsby, A. E. (1972). The interrelations of cell turgor pressure, gas-vacuolation and buoyancy in a blue–green alga. *Journal of Experimental Botany* **23**, 561–570.

Donk, R. J. (1920). A highly resistant thermophilic organism. *Journal of Bacteriology* **5**, 373–374.

Dugan, P. R. and Lundgren, D. G. (1965). Energy supply for the chemolithotroph *Ferrobacillus ferrooxidans*. *Journal of Bacteriology* **89**, 825–834.

Dundas, I. E. D. (1977). Physiology of Halobacteriaceae. *Advances in Microbial Physiology* **15**, 85–120.

Ebner, H. G. (1978). Metal recovery and environmental protection by bacterial leaching of inorganic waste materials. *In* "Metallurgical Applications of Bacterial Leaching and Related Microbiological Phenomena" (Eds L. E. Murr, A. E. Torma and J. A. Brierley), pp. 195–206. Academic Press, New York, San Francisco, California, and London.

Eddy, B. P. (1960). The use and meaning of the term psychrophilic. *Journal of Applied Bacteriology* **23**, 189–192.

Ehrlich, H. L. (1974a). Response of some activities of ferromanganese module bacteria to hydrostatic pressure. *In* "Effect of the Ocean Environment on Microbial Activities" (Eds R. R. Colwell and R. Y. Morita), pp. 208–221. University Park Press, Baltimore, Maryland.

Ehrlich, H. L. (1974b). Induction of MnO_2-reductase activity under different hydrostatic pressures at 15 °C. *In* "Abstracts of the Annual Meeting of the American Society for Microbiology", p. 47.

Ehrlich, H. L. (1975). The formation of ores in the sedimentary environment of the deep-sea with microbial participation: the case for ferromanganese concretions. *Soil Science* **49**, 36–41.

Ehrlich, H. L. (1983). Manganese oxidizing bacteria from a hydrothermally active area on the Galapagos Rift. *Ecological Bulletin* **35**, 357–366.

Elsworth, R., Herbert, D., Sargeant, K. and Christie, A. A. (1973). Improvements in or relating to L-asparaginase production. UK Patent Specification No 1,314,530.

Emanuilova, E. I. and Toda, K. (1984). α-amylase production in batch and continuous cultures by *Bacillus caldolyticus*. *Applied Microbiology and Biotechnology* **19**, 301–305.

Endo, S. (1962). Studies on protease produced by thermophilic bacteria. *Fermentation Technology* **40**, 346–353.

Epstein, I. and Grossowicz, N. (1969). prototrophic thermophilic *Bacillus*: isolation, properties and kinetics of growth. *Journal of Bacteriology* **99**, 414–417.

Fogarty, W. M. and Ward, P. O. (1977). Pectinases and pectic polysaccharides. *Progress in Industrial Microbiology* **13**, 59–119.

Fogg, G. E. (1956). The comparative physiology and biochemistry of the blue green algae. *Bacteriological Reviews* **20**, 148–165.

Frazier, W. C. (1967). "Food Microbiology". McGraw-Hill, New York.

French, D. (1957). The Schardinger dextrins. *Advances in Carbohydrate Chemistry* **12**, 189–260.

French, D., Levine, M. L., Norberg, E., Nordin, P., Pazur, J. H. and Wild, G. M. (1954). Studies on Schardinger dextrins. VII. Co-substrate specificity in coupling reaction of *Macerans* amylase. *Journal of the American Chemical Society* **76**, 2387–2390.

Fukumoto, J. and Okada, S. (1963). Studies on bacterial amylase, amylase types of *Bacillus subtilis* species. *Journal of Fermentation Technology* **41**, 427–434.

Gee, J. M., Lund, B. M., Metcalf, G. and Peel, J. L. (1980). Properties of a new group of alkalophilic bacteria. *Journal of General Microbiology* **117**, 9–17.

Gerson, D. F. and Zajic, J. E. (1979). Microbial biosurfactants. *Process Biochemistry* **14**, 20–29.

Ghose, T. K. and Ghosh, P. (1979). Cellulase production and cellulose hydrolysis. *Process Biochemistry* **14**, 20–24.

Gibbons, N. E. (1969). Isolation, growth and requirements of halophilic bacteria. *Methods in Microbiology* **3B**, 169–183.

Gibson, T. (1934). An investigation of the *Bacillus pasteurii* group. II. Special physiology of the organisms. *Journal of Bacteriology* **28**, 313–322.

Ginzburg, M., Sachs, L. and Ginzburg, B. Z. (1971). Ion metabolism in a *Halobacterium*. II. Ion concentration in cells at different levels of metabolism. *Journal of Membrane Biology* **5**, 78–101.

Gochnauer, M. B. and Kushner, D. J. (1969). Growth and nutrition of extremely halophilic bacteria. *Canadian Journal of Microbiology* **15**, 1157–1165.

Gochnauer, M. B. and Kushner, D. J. (1971). Potassium binding, growth and survival of an extremely halophilic bacterium. *Canadian Journal of Microbiology* **17**, 17–23.

Golovacheva, R. S., Egorova, L. A. and Loginova, L. G. (1965). Ecology and systematics of aerobic obligate thermophilic bacteria isolated from thermal localities on Mount Yangan-Tau and Kunashir Isle of the Kuril Chain. *Microbiology (USSR)* **34**, 693–698.

Good, W. A. and Hartman, P. A. (1970). Properties of the amylase from *Halobacterium halobium*. *Journal of Bacteriology* **104**, 601–603.

Goodman, R. E. and Pederson, D. M. (1976). β-galactosidase from *Bacillus stearothermophilus*. *Canadian Journal of Microbiology* **22**, 817–825.

Gordon, R. E. and Smith, N. R. (1949). Aerobic spore forming bacteria capable of growth at high temperatures. *Journal of Bacteriology* **58**, 327–341.

Gould, G. W. and Sale, A. J. H. (1972). Role of pressure in the stabilization and destabilization of bacterial spores. *In* "Proceedings of the 26th Symposium of the Society for Experimental Biology" (Eds M. A. Sleigh and A. G. Macdonald), pp. 147–157. Academic Press, New York and London.

Gounot, A. M. (1968a). Etude microbiologique de boues glaciaires arctiques. *Comptes Rendus de l'Academie des Sciences, Paris, Série D* **266**, 1437–1438.

Gounot, A. M. (1968b). Etude microbiologique des limons de deux grottes arctique. *Comptes Rendus de l'Academie des Sciences, Paris, Série D* **266**, 1619–1620.

Gounot, A. M. (1973). Importance of the temperature factor in the study of cold soils microbiology. *Bulletin for Ecology Research Communications (Stockholm)* **17**, 172–173.

Gow, W. A., McCreedy, H. H., Ritcey, G. M., McNamara, V. M., Harrison, V. and Lucas, B. H. (1971). Bacteria based processes for the treatment of low grade uranium ores. *In* "Recovery of Uranium", pp. 195–211. International Atomic Energy Agency, Vienna.

Graf, E. G. and Thauer, R. K. (1981). Hydrogenase from *Methanobacterium thermoautotrophicum*, a nickel-containing enzyme. *FEBS Letters* **136**, 165–168.

Grant, W. D., Mills, A. A. and Schofield, A. K. (1979). An alkalophilic species of *Ectothiorhodospira* from a Kenyan soda lake. *Journal of General Microbiology* **110**, 137–142.

Grant, W. D. and Tindall, B. J. (1980). The isolation of alkalophilic bacteria. *In* "Microbial Growth and Survival in Extremes of Environment" (Eds G. W. Gould and J. E. L. Corry), pp. 27–38. Academic Press, London and New York.

Grootegoed, J. A., Lauwers, A. M. and Heinen, W. (1973). Separation and partial purification of extracellular amylase and protease from *B. caldolyticus*. *Archives of Microbiology* **90**, 223.

Grueninger, H., Sonnleitner, B. and Fiechter, A. (1984). Bacterial diversity in thermophilic aerobic sewage sludge. *Applied Microbiology and Biotechnology* **19**, 414–421.

Guffanti, A. A., Susman, P., Blanco, R. and Krulwich, T. A. (1978). The protonmotive force and α-aminoisobutyric acid transport in an obligately alkalophilic bacterium. *Journal of Biological Chemistry* **253**, 708–715.

Guffanti, A. A., Blanco, R. and Krulwich, T. A. (1979a). A requirement for ATP for β-galactoside transport by *Bacillus alcalophilus*. *Journal of Biological Chemistry* **254**, 1033–1037.

Guffanti, A. A., Monti, L. Blanco, R., Ozick, D. and Krulwich, T. A. (1979b). β-galactoside transport in an alkaline-tolerant strain of *Bacillus circulans*. *Journal of General Microbiology* **112**, 161–169.

Guffanti, A. A., Blanco, R., Beneson, R. A. and Krulwich, T. A. (1980). Bioenergetic properties of alkaline-tolerant and alkalophilic strains of *Bacillus firmus*. *Journal of General Microbiology* **119**, 79–86.

Guy, G. R. and Daniel, R. M. (1982). The purification and some properties of a stereospecific D-asparaginase from an extremely thermophilic bacterium *Thermus aquaticus*. *Biochemical Journal* **203**, 787–790.

Hale, E. M. (1977). Isolation of alkalophilic bacteria from an alkaline reservoir used in an industrial process. *In* "Proceedings of the 77th Annual Meeting of the American Society for Microbiology", Abstract 150.

Hamilton, W. A. (1983). Sulphate-reducing bacteria and the off-shore oil industry. *Trends in Biotechnology* **1**, 36–40.

Hammer, B. W. (1915). Bacteriological studies on the coagulation of evaporated milk. *Iowa Agricultural Experimental Station Research Bulletin* **19**, 119–131.

Harder, W. and Veldkamp, H. (1967). A continuous culture study of an obligately psychrophilic *Pseudomonas* species. *Archives of Microbiology* **59**, 123–130.

Hardie, L. A. and Euguster, H. P. (1970). The evolution of closed basin brines. *Mineralogical Society of America, Special Publications* **3**, 273–290.

Harkin, J. M., Crawford, D. L. and McCoy, E. (1974). Bacterial protein from pulps and paper mill sludge. *Technical Association of the Pulp and Paper Industry* **57**, 131–134.

Harold, F. M. and Papineau, D. (1972). Cation transport and electrogenesis by *Streptoccus faecalis*. II. Proton and sodium extrusion. *Journal of Membrane Biology* **8**, 45–62.

Harold, F. M., Pavlasova, E. and Baarda, J. R. (1970). A transmembrane pH gradient in *Streptococcus faecalis*: origin, and dissipation by proton conductors and *N,N'*-dicyclohexylcarbodiimide. *Biochimica et Biophysica Acta* **196**, 235–244.

Hartley, B. S. and Payton, M. A. (1983). Industrial prospects for the thermophiles and thermophilic enzymes. *Biochemical Society Symposium* **48**, 113–146.

Hase, T., Ohmy, N., Matsubara, H., Mullingen, R. N., Rao, K. K. and Hall, D. O. (1976). Amino acid sequences of a 4-iron 4-sulphur ferredoxin from *B. stearothermophilus*. *Biochemical Journal* **159**, 55–63.

Hashimoto, A. G. (1983). Conversion of straw–manure mixtures to methane at mesophilic and thermophilic temperatures. *Biotechnology and Bioengineering* **25**, 185–200.

Heinen, U. J. and Heinen, W. (1972). Characteristics and properties of a caldoactive bacterium producing extraccellular enzymes and two related strains. *Archiv für Mikrobiologie* **82**, 1–23.

Heinen, W. and Lauwers, A. M. (1975). Variability of the molecular size of extracellular amylase produced by intact cells and protoplasts of *B. caldolyticus*. *Archives of Microbiology* **106**, 201–207.

Heinrich, A., Curtet, R. and Asnar, A. (1970). Observation de bombements de semi-conserves au froid provoqués par des germes à la fois psychrophiles et halophiles. *Revue de Médecine Vétérinaire* **121**, 901–904.

Hessler, R. R., Isaacs, J. D. and Mills, E. L. (1972). Giant amphipod from the abyssal Pacific Ocean. *Science* **175**, 636–637.

Hiramatsu, T., Yokoyama, T., Ohno, Y., Yano, I., Masai, M. and Iwamoto, T. (1976). Preparation and chemical properties of the outer membrane of a moderately halophilic Gram-negative bacterium. *Canadian Journal of Microbiology* **22**, 731–740.

Hite, B. H., Giddings, N. J. and Weakly, C. E. (1914). The effect of pressure on certain microorganisms encountered in the preservation of fruits and vegetables. *West Virginia Agricultural Experimental Station Bulletin* **146**, 1–67.

Hitzman, D. O. (1962). Microbiological secondary recovery of petroleum. US Patent No 3,032,472.

Hiwatashi, T., Hara, M. and Tamada, A. (1958). Biological properties of halophilic bacteria No 101 and No 203. *Journal of Osaka City Medical Centre* **7**, 550–552.

Hochstein, L. I. and Dalton, B. P. (1973). Studies of a halophilic NADH dehydrogenase. I. Purification and properties of the enzyme. *Biochimica et Biophysica Acta* **307**, 216–228.

Hocking, J. D. and Harris, J. I. (1976). Glyceraldehyde 3-phosphate dehydrogenase from an extreme thermophile, *Thermus aquaticus*. *In* "Enzymes and Proteins from Thermophilic Microorganisms" (Ed. H. Zuber), pp. 107–120. Birkhaüser, Basel.

Hocking, J. D. and Harris, J. I. (1980). D-glyceraldehyde-3-phosphate dehydrogenase. *European Journal of Biochemistry* **108**, 567–579.

Horikoshi, K. (1971a). Production of alkaline enzymes by alkalophilic microorganisms. Part I. Alkaline proteases produced by *Bacillus* No. 221. *Agricultural and Biological Chemistry* **35**, 1407–1414.

Horikoshi, K. (1971b). Production of alkaline enzymes by alkalophilic microorganisms. Part II. Alkaline amylase produced *Bacillus* No. A-40-2. *Agricultural and Biological Chemistry* **35**, 1783–1791.

Horikoshi, K. (1972). Production of alkaline enzymes by alkalophilic microorganisms. Part III. Alkaline pectinase of *Bacillus* No P-4-N. *Agricultural and Biological Chemistry* **36**, 285–293.

Horikoshi, K. and Akiba, T. (1982). "Alkalophilic Microorganisms: a New Microbial World". Japan Scientific Societies Press and Springer, Tokyo, Berlin, Heidelberg and New York.

Horikoshi, K. and Atsukawa, Y. (1973a). β-1,3-glucanase produced by alkalophilic bacteria *Bacillus* No. K-12-5. *Agricultural and Biological Chemistry* **37**, 1449–1456.

Horikoshi, K. and Atsukawa, Y. (1973b). Xylanase produced by alkalophilic *Bacillus* No. C-59-2. *Agricultural and Biological Chemistry* **37**, 2097–2103.

Horikoshi, K. and Atsukawa, Y. (1975). Production of β-1,3-glucanase by *Bacillus* No. 221, an alkalophilic microorganism. *Biochimica et Biophysica Acta* **384**, 477–483.

Horikoshi, K., Ikeda, Y. and Nakao, M. (1972). US Patent No 3,844,890.

Horikoshi, K., Ando, T. and Yoshida, K. (1973). US Patent No 3,956,062.

Horikoshi, K., Nakao, M., Kurono, Y. and Sashihara, N. (1984). Cellulases of an alkalophilic *Bacillus* strain isolated from soil. *Canadian Journal of Microbiology* **30**, 774–779.

Horwath, R. O., Lally, J. A. and Rotheim, P. (1977). US Patent No 4,011,139.

Hsung, J. C. and Haug, H. (1975). Intracellular pH of *Thermoplasma acidophila*. *Biochimica et Biophysica Acta* **389**, 477–482.

Hucker, G. J. (1954). Low temperature organisms in frozen vegetables. *Food Technology* **8**, 79–108.

Hunt, J. M. (1979). "Petroleum Geochemistry and Geology". Freeman, San Francisco, California.

Idziak, E. S. and Incze, K. (1968). Radiation treatment of foods. 1. Radurization of fresh eviscerated poultry. *Applied Microbiology* **16**, 1061–1066.

Ikura, Y. and Horikoshi, K. (1977). Isolation and some properties of alkalophilic bacteria utilizing rayon waste. *Agricultural and Biological Chemistry* **41**, 1373–1377.

Imhoff, J. F. and Trüper, H. G. (1977). *Ectothirohodospira halochloris* sp. nov., a new extremely halophilic phototrophic bacterium containing bacteriochlorophyll *b*. *Archives of Microbiology* **114**, 115–121.

Ingraham, J. C. and Stokes, J. L. (1959). Psychrophilic bacteria. *Bacteriological Reviews* **23**, 97–108.

Ingram, M. (1965). Psychrophilic and psychrotrophic microorganisms. *Annales de l'Instut Pasteur de Lille* **16**, 111–118.

Inniss, W. E. (1975). Interaction of temperature and psychrophilic microorganisms. *Annual Review of Microbiology* **29**, 445–465.

Inniss, W. E. and Ingraham, J. L. (1978). Microbial life at low temperatures. Mechanisms and molecular aspects. *In* "Microbial Life in Extreme Environments" (Ed. D. J. Kushner), pp. 73–104. Academic Press, London, New York, and San Francisco, California.

Ito, K. A. (1981). Thermophilic organisms in food spoilage: flat-sour aerobes. *Journal of Food Protection* **44**, 157–163.

Jacobsen, G. (1918). On factors influencing efficient pasteurisation. *Abstracts of Bacteriology* **2**, 215.

Jannasch, H. W. (1984). Microbes in the oceanic environment. *In* "The Microbe 1984. Part II. Prokaryotes and Eukaryotes" (Eds D. P. Kelly and N. G. Carr), pp. 97–122. Cambridge University Press, Cambridge.

Jannasch, H. W. and Taylor, C. D. (1984). Deep-sea microbiology. *Annual Review of Microbiology* **38**, 487–514.

Jannasch, H. W. and Wirsen, C. O. (1977a). Microbial life in the deep sea. *Scientific American* **236**, 42–52.

Jannasch, H. W. and Wirsen, C. O. (1977b). Retrieval of concentrated and undecompressed microbial populations from the deep sea. *Applied and Environmental Microbiology* **33**, 642–646.

Jannasch, H. W. and Wirsen, C. O. (1981). Morphological survey of microbial mats near deep-sea thermal vents. *Applied and Environmnetal Microbiology* **42**, 528–538.

Jannasch, H. W., Wirsen, C. O. and Taylor, C. D. (1982). Deep-sea bacteria: isolation in the absence of decompression. *Science* **216**, 1315–1317.

Jensen, B. F. and Norman, B. E. (1984). *Bacillus acidopullulyticus* pullulanase: application and regulatory aspects for use in the food industry. *Process Biochemistry* **19**, 129–134.

Jones, J. B. and Beck, J. F. (1976). Applications of biochemical systems in organic chemistry, part 1. *Techniques of Chemistry* **10**, 107–402.

Jones, W. G., Leigh, J. A., Mayer, F., Woese, C. R. and Wolfe, R. S. (1983). *Methanococcus jannaschii* sp. nov., an extremely thermophilic methanogen from a submarine hydrothermal vent. *Archives of Microbiology* **136**, 254–261.

Kahn, M. R. and Williams, S. T. (1975). Studies on the ecology of Actinomycetes in soil. *Soil Biology and Biochemistry* **7,** 345–348.

Kallas, T. and Castenholz, R. W. (1982a). Internal pH and ATP–ADP pools in the cyanobacterium *Synechococcus* sp. during exposure to growth-inhibiting low pH. *Journal of Bacteriology* **149,** 229–236.

Kallas, T. and Castenholz, R. W. (1982b). Rapid transient growth at low pH in the cyanobacterium *Synechococcus* sp. *Journal of Bacteriology* **149,** 237–246.

Kamekura, M. and Onishi, H. (1974a). Protease formation by a moderately halophilic *Bacillus* strain. *Applied Microbiology* **27,** 809–810.

Kamekura, M. and Onishi, H. (1974b). Halophilic nuclease from a moderately halophilic *Micrococcus varians*. *Journal of Bacteriology* **119,** 339–344.

Kamekura, M. and Onishi, H. (1978). Properties of the halophilic nuclease of a moderate halophile, *Micrococcus varians* subsp. halophilus. *Journal of Bacteriology* **133,** 59–65.

Kargi, F. (1982). Microbiological coal desulfurization. *Enzyme and Microbial Technology* **4,** 13–17.

Kargi, F. (1984). Microbial desulfurization of coal. *Advances in Biotechnological Processes* **3,** 242–274.

Kargi, F. and Cervoni, T. D. (1983). An airlift-recycle fermenter for microbial desulfurization of coal. *Biotechnology Letters* **5,** 33–38.

Kargi, F. and Robinson, J. M. (1982a). Removal of sulphur compounds from coal by the thermophilic organism *S. acidocaldarius*. *Applied and Environmental Microbiology* **44,** 878–883.

Kargi, F. and Robinson, J. M. (1982b). Microbial desulfurization of coal by thermophilic organism, *S. acidocaldarius*. *Biotechnology and Bioengineering* **24,** 2115–2121.

Kashket, E. R. and Wilson, T. H. (1973). Proton-coupled accumulation of galactoside in *Streptococcus lactis* 7962. *Proceedings of the National Academy of Science of the United States of America* **79,** 2866–2869.

Kato, T. and Horikoshi, K. (1984). Immobilized cyclomaltodextrin glucanotransferase of an alkalophilic *Bacillus* sp. No. 38-2. *Biotechnology and Bioengineering* **26,** 595–598.

Keay, L. (1970). The action of *Bacillus subtilis* saccharifying amylase on starch and β-cyclodextrin. *Die Starke* **22,** 153–157.

Keay, L. and Wildi, B. S. (1970). Proteases of the genus *Bacillus*. *Biotechnology and Bioengineering* **12,** 179–212.

Keay, L., Moseley, M. H., Anderson, R. G., O'Connor, R. J. and Wildi, B. S. (1972). Production and isolation of microbial proteases. *Biotechnology and Bioengineering Symposium* **3,** 63–92.

Kelly, C. T. and Fogarty, W. M. (1978). Production and properties of polygalacturonate lyase by an alkalophilic microorganism *Bacillus* sp. RK9. *Canadian Journal of Microbiology* **24,** 1164–1172.

Kelly, D. P. (1976). Extraction of metals from ores by bacterial leaching: present status and future prospects. *In* "Microbial Energy Conversion" (Eds H. G. Schlegel and J. Barnea), pp. 329–338. Goltze, Gottingen.

Kelly, D. P., Norris, P. R. and Brierley, C. L. (1979). Microbiological methods for the extraction and recovery of metals. *In* "Microbial Technology: Current State, Future Prospects" (Eds A. T. Bull, D. C. Ellwood and C. Ratledge) SGM Symposium 29, pp. 264–325. Cambridge University Press, Cambridge.

Khoo, T. C., Cowan, D. A., Daniel, R. M. and Morgan, H. W. (1984). Interactions of calcium and other metal ions with caldolysin, the thermostable proteinase from *Thermus aquaticus* strain T351. *Biochemical Journal* **221**, 407–413.

Kinne, O. (1972). Pressure, general introduction. *In* "Marine Ecology" (Ed. O. Kinne), Vol. 1, Part 3, pp. 1323–1360. Wiley–Interscience, London and New York.

Kitahata, S. and Okada, S. (1974). Action of cyclodextrin glycosyltransferase from *Bacillus megaterium* strain No. 5 on starch. *Agricultural and Biological Chemistry* **38**, 2413–2417.

Kitahata, S., Tsuyama, N. and Okada, S. (1974). Purification and some properties of cyclodextrin glycosyltransferase from a strain of *Bacillus* species. *Agricultural and Biological Chemistry* **38**, 387–393.

Kitahata, S., Okada, S. and Fukui, T. (1978). Acceptor specificity of transglycosylation catalyzed by cyclodextrin glycosyltransferase. *Agricultural and Biological Chemistry* **42**, 2369–2374.

Klaushofer, H. and Hollaus, F. (1970). Zur Taxonomie der hochthermophilen, in Zuckerfabrikssaften verkommenden aeroben Sporenbildner. *Zeitschrift für die Zuckerindustrie* **20**, 465–470.

Klaushofer, H. and Parkkinen, E. (1965). Zur Frage der Bedeutung Aerober und Anaerober Thermophiler Sporenbildner als Infektionsursache in Rubenzuckerfabriken. I. *Clostridium thermohydrosulfuricum* eine neue Art eines Saccharoseabbauenden, Thermophilen, Schwefelwasserstoffbildenden *Clostridiums*. *Zeitschrift für die Zuckerindustrie* **15**, 445–449.

Klaushofer, H., Hollaus, F. and Pollach, G. (1971). Microbiology of beet sugar manufacture. *Process Biochemistry* **6**, 39–41.

Klingenberg, P., Zickler, F., Leuchtenberger, A. and Ruttloff, H. (1979). Gewinnung und Charakterisierung von Proteasen. *Zeitschrift für Allgemeine Mikrobiologie* **19**, 17–25.

Klotz, I. M. (1967). Saccinylation. *Methods in Enzymology* **11**, 576–580.

Kobayashi, Y. and Horikoshi, K. (1980). Identification and growth characteristics of alkalophilic *Corynebacterium* sp. which produces NAD(P)-dependent maltose dehydrogenase and glucose dehydrogenase. *Agricultural and Biological Chemistry* **44**, 41–47.

Kobayashi, H. and Unemoto, T. (1980). *Streptococcus faecalis* mutants defective in regulation of cytoplasmic pH. *Journal of Bacteriology* **143**, 1187–1193.

Kobayashi, S., Kainuma, K. and Suzuki, S. (1978). Purification and some properties of *Bacillus macerans* cycloamylose (cyclodextrin) glucanotransferase. *Carbohydrate Research* **61**, 229–238.

Kobayashi, Y., Ueyama, H. and Horikoshi, K. (1980). NAD-dependent maltose dehydrogenase and NAD-dependent D-glucose dehydrogenase of alkalophilic *Corynebacterium* sp. No. 150–1. *Agricultural and Biological Chemistry* **44**, 2837–2841.

Koch-Schmidt, A. C., Mosbach, K. and Werber, M. M. (1979). A comparative study on the stability of immobilized halophilic and non-halophilic malate dehydrogenases at various ionic strengths. *European Journal of Biochemistry* **100**, 213–218.

Koffler, H. (1957). Protoplasmic differences between mesophiles and thermophiles. *Bacteriological Reviews* **21**, 227–240.

Koffler, H. and Gale, G. O. (1957). The relative thermostabilities of cytoplasmic proteins from thermophilic bacteria. *Archives of Biochemistry and Biophysics* **67**, 249–251.

Komaratat, P. and Kates, M. (1975). The lipid composition of a halotolerant species of *Staphylococcus epidermis. Biochimica et Biophysica Acta* **398**, 464–484.

Krajewska, E. and Szer, W. (1967). Comparative studies of amino acid incorporation in a cell free system from psychrophilic *Pseudomonas* species 412. *European Journal of Biochemistry* **2**, 250–256.

Krasna, A. I. (1979). Hydrogenase: properties and applications. *Enzyme and Microbial Technology* **1**, 165–171.

Kriss, A. E. (1962). "Marine Microbiology (Deep-Sea)". Wiley–Interscience, London and New York.

Krulwich, T. A. and Guffanti, A. A. (1983). Physiology of acidophilic and alkalophilic bacteria. *Advances in Microbial Physiology* **24**, 173–214.

Krulwich, T. A., Davidson, L. F., Filip, S. J., Zuckerman, R. S. and Guffanti, A. A. (1978). The proton motive force and β-galactoside transport in *Bacillus acidocaldarius. Journal of Biological Chemistry* **253**, 4599–4603.

Krulwich, T. A., Mandel, K. G., Bornstein, R. F. and Guffanti, A. A. (1979). A nonalkalophilic mutant of *Bacillus alcalophilus* lacks the Na^+/H^+ antiporter. *Biochemical and Biophysical Research Communications* **91**, 58–62.

Kuhn, H., Friederich, U. and Fiechter, A. (1979). Defined minimal medium for a thermophilic *Bacillus* sp. developed by a chemostat pulse and shift technique. *European Journal of Applied Microbiology and Biotechnology* **6**, 341–349.

Kurono, Y. and Horikoshi, K. (1973). Alkaline catalase produced by *Bacillus* No. KU-1. *Agricultural and Biological Chemistry* **37**, 2565–2570.

Kushner, D. J. (1966). Mass culture of red halophilic bacteria. *Biotechnology and Bioengineering* **8**, 237–245.

Kushner, D. J. (1968). Halophilic bacteria. *Advances in Applied Microbiology* **10**, 73–99.

Kushner, D. J. (1978). Life in high salt and solute concentrations: halophilic bacteria. *In* "Microbial Life in Extreme Environments" (Ed. D. J. Kushner), pp. 317–368. Academic Press, London, New York, and San Francisco, California.

Kushner, D. J. and Lisson, T. A. (1959). Alkali resistance in *Bacillus cereus. Journal of General Microbiology* **21**, 96–108.

Kuznetsova, V. A. (1960). Occurrence of sulphate-reducing organisms in oil-bearing formations of the Kuibyshev region with reference to salt composition of layer waters. *Mikrobiologiya* **29**, 298–301.

Kuznetsova, V. A. and Pantskhava, E. S. (1962). Effect of freshening of stratal waters on development of halophilic sulphate-reducing bacteria. *Mikrobiologiya* **31**, 103–106.

Lamed, R. and Zeikus, J. G. (1980). Ethanol production by thermophilic bacteria: relationship between fermentation product yields of and catabolic enzyme activities in *Clostridium thermocellum* and *Thermoanaerobium brockii. Journal of Bacteriology* **144**, 569–578.

Lamed, R. J. and Zeikus, J. G. (1981). Novel NADP-linked alcohol–aldehyde/ketone oxidoreductase in thermophilic ethanologenic bacteria. *Biochemical Journal* **195**, 183–190.

Lamed, R. J., Keinan, E. and Zeikus, J. G. (1981). Potential applications of an alcohol aldehyde/ketone oxidoreductase from thermophilic bacteria. *Enzyme and Microbial Technology* **3**, 144–147.

Landesman, J., Duncan, D. W. and Walden, C. C. (1966). Oxidation of inorganic sulphur compounds by washed cell suspensions of *Thiobacillus ferrooxidans. Canadian Journal of Microbiology* **12**, 957–964.

Lanyi, J. K. (1974). Salt-dependent properties of proteins from extremely halophilic bacteria. *Bacteriological Reviews* **38**, 272–290.

Lanyi, J. K. (1979). Physicochemical aspects of salt-dependence in Halobacteria. *In* "Strategies of microbial life in extreme environments" (Ed. M. Shilo), pp. 93–107. Dahlem Konferenzen, Berlin.

Lanyi, J. K. and MacDonald, R. E. (1975). Existence of electrogenic hydrogen ion/sodium ion antiport in *Halobacterium halobium* cell envelope vesicles. *Biochemistry* **15**, 4608–4610.

Larsen, H. (1983). The family Halobacteriaceae. *In* "The Prokaryotes" (Eds M. P. Starr, H. Stolp, H. G. Trüper, A. Balows and H. G. Schlegel), Vol. 1, pp. 985–994. Springer, Berlin, Heidelberg and New York.

Lauwers, A. M. and Heinen, W. (1973). Correlation between the fatty acid composition and the activity of extracellular enzymes from *B. caldolyticus. Archives of Microbiology* **91**, 241–254.

Laxa, O. (1900). Bakteriologische Studien über die Produkte des mormalen Zuckerfabriksbetriebes. *Zentralblatt für Bakteriologie und Parasitenkunde, Abteilung 2* **6**, 286–295.

Lazar, I. (1983). Some characteristics of the bacterial inoculum used for oil release from reservoirs. *In* "Microbial Enhanced Oil Recovery" (Eds J. E. Zajic, D. G. Cooper, T. R. Jack and N. Kosaric), pp. 73–82. Penn Well Publishing, Tulsa, Oklahoma.

Leicht, W. and Pundak, S. (1981). Large-scale purification of halophilic enzymes by salting-out mediated chromatography. *Analytical Biochemistry* **114**, 186–192.

Le Roux, N. L., Wakerley, D. S. and Hunt, S. D. (1977). Thermophilic Thiobacillus-type bacteria from Icelandic thermal areas. *Journal of General Microbiology* **100**, 197–201.

Le Ruyet, P., Dubourguier, H. C. and Albagnac, G. (1984). Thermophilic fermentation of cellulose and xylan by methanogenic enrichment cultures: preliminary characterisation of main species. *Systematic Applied Microbiology* **5**, 247–253.

Levy, P. L., Pangburn, M. K., Burstein, Y., Ericsson, L. H., Neurath, H. and Walsh, K. A. (1975). Evidence of homologous relationship between thermolysin and neutral protease A of *Bacillus subtilis. Proceedings of the National Academy of Sciences of the United States of America* **72**, 4341–4345.

Lindsay, J. A. and Creaser, F. H. (1975). Enzyme thermostability is a transformable property between *Bacillus* sp. *Nature, London* **255**, 650–652.

Ljungdahl, L. (1979). Physiology of thermophilic bacteria. *Advances in Microbial Physiology* **19**, 149–243.

Ljungdahl, L. and Sherod, D. (1976). Proteins from thermophilic microorganisms. *In* "Extreme Environments: Mechanisms of Microbial Adaptation" (Ed. M. R. Heinrich), pp. 147–188. Academic Press, New York and London.

Ljungdhal, L., Bryant, F., Carreira, L., Saiki, T. and Wiegel, J. (1981). Some aspects of thermophilic and extreme thermophilic anaerobic microorganisms. *Basic Life Sciences* **18**, 397–419.

Loginova, L. G. and Egorova, L. A. (1975). An obligately thermophilic bacterium *Thermus ruber* from hot springs in Kamchatka. *Microbiologyia* **44**, 661–665.

Long, M. E. and Lee, C. K. (1979). US Patent 4,179,335.

Long-Kuan, J., Chang, P. W., Findley, J. E. and Yen, T. F. (1983). Selection of bacteria with favourable transport properties through porous rock for the application of microbial-enhanced oil recovery. *Applied and Environmental Microbiology* **46**, 1066–1072.

Lonsdale, P. F. (1977). Clustering of suspension-feeding macrobenthos near abyssal hydrothermal vents at oceanic spreading centers. *Deep Sea Research* **24**, 857–863.

Lorenze, F. W., Starr, P. B., Starr, M. P. and Ogasawars, F. X. (1952). The development of *Pseudomonas* spoilage in shell eggs. 1. Penetration through the shell. *Journal of Food Research* **17**, 351–356.

Lovitt, R. V., Longin, R. and Zeikus, J. G. (1984). Ethanol production by thermophilic bacteria: physiological comparison of solvent effects on parent and alcohol-tolerant strains of *Clostridium thermohydrosulfuricum*. *Applied and Environmental Microbiology* **48**, 171–177.

Lundgren, D. G. and Malouf, E. E. (1983). Microbial extraction and concentration of metals. *Advances in Biotechnological Processes* **1**, 224–251.

Lundgren, D. G. and Silver, M. (1980). Ore leaching by bacteria. *Annual Review of Microbiology* **34**, 263–283.

Manchee, R. J. (1979). Microbial mining. *Trends in Biochemical Sciences* **4**, 77–80.

Mandel, K. G., Guffanti, A. A. and Krulwich, T. A. (1980). Monovalent cation/proton antiporters in membrane vesicles from *Bacillus alcalophilus*. *Journal of Biological Chemistry* **255**, 7391–7396.

Markland, F. S. and Smith, E. L. (1967). Subtilisin BPN'. VII. Isolation of cyanogen bromide peptides and the complete amino acid sequence. *Journal of Biological Chemistry* **242**, 5198–5211.

Marquis, R. E. (1976). High-pressure microbial physiology. *Advances in Microbial Physiology* **14**, 159–241.

Marquis, R. E. and Matsumura, P. (1978). Microbial life under pressure. In "Microbial Life in Extreme Environments" (Ed. D. J. Kushner), pp. 105–158. Academic Press, London and New York.

Matches, J. R. and Liston, J. (1973). Methods and techniques for the isolation and testing of Clostridia from estuarine environments. In "Estuarine Microbial Ecology" (Eds L. H. Stevenson and R. R. Colwell), pp. 345–361. University of South Carolina Press, Columbia, South Carolina.

Matsubara, H. and Feder, J. (1971). Other bacterial, mould and yeast proteases. In "The Enzymes" 3rd ed., Vol. 3, pp. 721–795. Academic Press, New York and London.

Matsubara, H., Kasper, C. B., Brown, D. M. and Smith, E. L. (1965). Subitilisin BPN'. I. Physical properties and amino acid composition. *Journal of Biological Chemistry* **240**, 1125–1130.

Matsuzawa, H., Hamaoki, M. and Ohta, T. (1983). Production of thermophilic extracellular proteases (aqualysins I and II) by *Thermus aquaticus* YT-1, an extreme thermophile. *Agricultural and Biological Chemistry* **47**, 25–28.

Matteuzzi, D., Hollaus, F. and Biavati, B. (1978). Proposal of neotype for *Clostridium thermohydrosulfuricum* and the merging of *Clostridium tartarivorum* with *Clostridium thermosaccharolyticum*. *International Journal of Systematic Bacteriology* **28**, 528–531.

McBee, R. H. (1948). The culture and physiology of a thermophilic cellulose-fermenting bacterium. *Journal of Bacteriology* **56**, 653–663.

McBee, R. H. (1950). The anaerobic thermophilic cellulolytic bacteria. *Bacteriological Reviews* **14**, 51–63.

McClelland, M. (1981). Purification and characterisation of two new modification methylases: M Cla 1 from *Caryophanon latum* L and M Taq 1 from *Thermus aquaticus* YT-1. *Nucleic Acids Research* **9**, 6795–6804.

McDonald, W. C. and Matney, S. T. (1963). Genetic transfer of the ability to grow at 55 °C in *Bacillus subtilis*. *Journal of Bacteriology* **85**, 218–220.

McGoran, C. J. M., Duncan, D. W. and Walden, C. C. (1969). Growth of *Thiobacillus ferrooxidans* on various substrates. *Canadian Journal of Microbiology* **15**, 135–138.

McLachlan, J. and Gorham, P. R. (1962). Effects of pH and nitrogen sources on growth of *Microcystis aeruginosa* Kutz. *Canadian Journal of Microbiology* **8**, 1–11.

Mela, L. (1966). Intramitochondrial pH changes. *Federation Proceedings, Federation of American Societies for Experimental Biology* **25**, 414.

Mevarech, M. and Neumann, E. (1977). Malate dehydrogenase isolated from extremely halophilic bacteria of the Dead Sea. 2. Effect of salt on the catalytic activity and structure. *Biochemistry* **16**, 3786–3791.

Mevarech, M., Eisenberg, H. and Neumann, E. (1977). Malate dehydrogenase isolated from extremely halophilic bacteria of the Dead Sea. 1. Purification and molecular characterization. *Biochemistry* **16**, 3781–3785.

Michels, M. J. M. and Visser, F. M. W. (1976). Occurrence and thermoresistance of spores of psychrophilic and psychrotrophic aerobic sporeformers in soil and foods. *Journal of Applied Bacteriology* **41**, 1–11.

Mifume, A. and Shima, J. (1977). Cyclodextrins and their application. *Journal of Organic Synthesis, Japan* **35**, 116–130.

Mikes, O., Turkova, J., Toan, N. B. and Sorm, F. (1969). Serine-containing active center of alkaline protease of *Aspergillus flavus*. *Biochimica et Biophysica Acta* **178**, 112–117.

Militzer, W., Sonderegger, T. B., Tuttle, L. C. and Georgi, C. E. (1949). Thermal enzymes. *Archives of Biochemistry* **24**, 75–82.

Mizusawa, K., Ichishima, E. and Yoshida, F. (1964). Studies on the proteolytic enzymes of thermophilic *Streptomyces*. Part I. Purification and some properties. *Agricultural and Biological Chemistry* **28**, 884–895.

Mizusawa, K., Ichishima, E. and Yoshida, F. (1966). Studies on the proteolytic enzymes of thermophilic *Streptomyces*. Part II. Identification of the organism and some conditions of protease formation. *Agricultural and Biological Chemistry* **30**, 35–41.

Mizushima, S., Machida, Y. and Kitahara, K. (1964). Quantitative studies on glycolytic enzymes in *Lactobacillus plantarum*. I. Concentration of inorganic ions and coenzymes in fermenting cells. *Journal of Bacteriology* **86**, 1295–1300.

Moiroud, A. and Gounot, A. M. (1969). Sur une bactérie psychrophile obligatoire isolée de limons glaciaires. *Comptes Rendus de l'Academie des Sciences, Paris, Série D* **269**, 2150–2152.

Moore, R. L. and McCarthy, B. J. (1969). Characterisation of the deoxyribonucleic acid of various strains of halophilic bacteria. *Journal of Bacteriology* **99**, 248–254.

Morgan, H. W., Daniel, R. M., Cowan, D. A. and Hickey, C. W. (1981). Thermostable microorganism and proteolytic enzyme prepared therefrom, and process for the preparation of the microorganism and the proteolytic enzyme. European Patent Application No 80,302,743, 2; Patent No 0,024,182.

Morihara, K., Oka, T. and Tsuzuki, H. (1967). Multiple proteolytic enzymes of *Streptomyces fradiae*. Production, isolation, and preliminary characterization. *Biochimica et Biophysica Acta* **139**, 382–397.

Morihara, K., Tsuzuki, H. and Oka, T. (1968). Comparison of the specificities of various neutral proteinases from microorganisms. *Archives of Biochemistry and Biophysics* **123**, 527–588.

Morishita, H. and Masui, M. (1980). "Saline Environment: Physiological and Biochemical Adaptation in Halophilic Microorganisms." Business Centre for Academic Societies, Osaka, Japan.

Morita, R. Y. (1966). Marine psychrophilic bacteria. *Oceanography and Marine Biology Annual Review* **5**, 187–203.

Morita, R. Y. (1975). Psychrophilic bacteria. *Bacteriological Reviews* **39**, 144–167.

Morita, R. Y. (1976). Survival of bacteria in cold and moderate hydrostatic pressure environments with special reference to psychrophilic and barophilic bacteria. *In* "The Survival of Vegetative Microbes" (Eds T. R. G. Gray and J. R. Postgate), pp. 279–298. Cambridge University Press.

Morita, R. Y. and Albright, L. J. (1968). Moderate temperature effects on protein, ribonucleic acid and deoxyribonucleic acid synthesis by *Vibrio marinus*, an obligately psychrophilic marine bacterium. *Zeitschrift für Allgemeine Mikrobiologie* **8**, 269–273.

Morita, R. Y. and Burton, S. D. (1963). Influence of moderate temperature on the growth and malic dehydrogenase of a marine psychrophile. *Journal of Bacteriology* **86**, 1025–1029.

Moses, V. (1983). Microbes and oil recovery: an overview. *In* "Biotech '83", pp. 415–422. Online Publications, Northwood, UK.

Munster, A. P., Munster, M. J., Woodrow, J. R. and Sharp, R. J. (1984). A numerical taxonomic study of *Thermus* species isolated from Yellowstone National Park. *In* Abstracts of Presentations at the SGM Symposium on Microbes in Extreme Environments, Dundee.

Murr, L. E., Torma, A. E. and Brierley, J. A. (1978). "Metallurgical Applications of Bacterial Leaching and Related Microbiological Phenomena". Academic Press, New York, San Francisco, California, and London.

Myerson, A. S. and Kline, P. C. (1984). Continuous bacterial coal desulfurization employing *Thiobacillus ferrooxidans*. *Biotechnology and Bioengineering* **26**, 92–99.

Nagel, C. W. and Wilson, T. M. (1970). Pectic acid lyase of *Bacillus polymyxa*. *Applied Microbiology* **20**, 374–383.

Nakamura, N. and Horikoshi, K. (1976a). Characterisation and some cultural conditions of a cyclodextrin glycosyltransferase-producing alkalophilic *Bacillus* sp. *Agricultural and Biological Chemistry* **40**, 753–757.

Nakamuara, N. and Horikoshi, K. (1976b). Purification and properties of cyclodextrin glycosyltransferase of an alkalophilic *Bacillus* sp. *Agricultural and Biological Chemistry* **40**, 935–941.

Nakamura, N. and Horikoshi, K. (1976c). Characterization of acid-cyclodextrin glycosyltransferase of an alkalophilic *Bacillus* sp. *Agricultural and Biological Chemistry* **40**, 1647–1648.

Nakamura, N. and Horikoshi, K. (1976d). Purification and properties of neutral cyclodextrin glycosyltransferase of an alkalophilic *Bacillus* sp. *Agricultural and Biological Chemistry* **40**, 1785–1791.

Nakamura, N. and Horikoshi, K. (1977). Production of Schardinger β-dextrin by soluable and immobilized cyclodextrin glycosyltransferase of an alkalophilic *Bacillus* sp. *Biotechnology and Bioengineering* **19**, 87–99.

Nakamura, N., Watanabe, K. and Horikoshi, K. (1975). Purification and some properties of alkaline pullulanase from a strain of *Bacillus* No. 202-1, an alkalophilic microorganism. *Biochimica et Biophysica Acta* **397**, 188–193.

Nakanishi, T. (1975). Ph.D. thesis. Osaka City University, Osaka, Japan.

Nakanishi, T. and Yamamoto, T. (1974). Action and specificity of a *Streptomyces* alkalophilic proteinase. *Agricultural and Biological Chemistry* **38**, 2391–2397.

Nakanishi, T., Matsumura, Y., Minamiura, N. and Yamamamoto, T. (1974). Purification and some properties of an alkalophilic proteinase of a *Streptomyces* species. *Agricultural and Biological Chemistry* **38**, 37–44.

Ng, T. K. and Zeikus, J. G. (1981). Comparison of extracellular cellulase activities of *Clostridium thermocellum* LQR1 and *Trichoderma reesei* QM9414. *Applied and Environmental Microbiology* **42**, 231–240.

Ng, T. K., Weimer, P. J. and Zeikus, J. G. (1977). Cellulolytic and physiological properties of *Clostridium thermocellum*. *Archives of Microbiology* **114**, 1–7.

Ng, T. K., Ben-Bassat, A. and Zeikus, J. G. (1981). Ethanol production by thermophilic bacteria: fermentation of cellulosic substrates by co-cultures of *Clostridium thermocellum* and *Clostridium thermohydrosulfuricum*. *Applied and Environmental Microbiology* **41**, 1337–1343.

Nickerson, J. T. and Sinskey, A. J. (1972). "Microbiology of Foods and Food Processing". Elsevier, New York.

Norberg, E. and French, D. (1950). Studies on the Schardinger dextrins. III. Redistribution reaction of *Macerans* amylase. *Journal of the American Chemical Society* **72**, 1202–1205.

Norberg, P. and Hofsten, B. V. (1969). Proteolytic enzymes from extremely halophilic bacteria. *Journal of General Microbiology* **55**, 251–256.

Norberg, P. and Hofsten, B. (1970). Chromatography of a halophilic enzyme on hydroxylapatite in 3.4 M sodium chloride. *Biochimica et Biophysica Acta* **220**, 132.

Ogasahara, K., Imanishi, A. and Isemura, T. (1970). Studies on thermophilic α-amylase from *Bacillus stearothermophilus*. *Journal of Biochemistry* **67**, 65–70.

Ogawa, S., Shulman, R. G., Glynn, P., Yamane, T. and Navon, G. (1978). On the measurement of pH in *Escherichia coli* by ^{31}P nuclear magnetic resonance. *Biochimica et Biophysica Acta* **502**, 45–50.

Ohno, Y., Yano, I., Hiramatsu, T. and Masui, M. (1976). Lipids and fatty acids of a moderately halophilic bacterium, No. 101. *Biochimica et Biophysica Acta* **424**, 337–350.

Ohta, K., Kiyomiya, A., Koyama, N. and Nosoh, Y. (1975). The basis of the alkalophilic property of a species of *Bacillus*. *Journal of General Microbiology* **86**, 259–266.

Okami, Y. (1979). Antibiotics from marine microorganisms with reference to plasmid involvement. *Journal of Natural Products (Lloydia)* **42**, 583–595.

Okami, Y. (1982). Potential use of marine microorganisms for antibiotics and enzyme production. *Pure and Applied Chemistry* **54**, 1951–1962.

Olsen, R. H. and Metcalf, E. S. (1968). Conversion of mesophilic to psychrophilic bacteria. *Science* **162**, 1288–1289.

Onishi, H. (1972). Halophilic amylase from a moderately halophilic *Microccus*. *Journal of Bacteriology* **109**, 580–574.

Onishi, H., Mori, T., Takeuchi, S., Tani, K., Kobayashi, T. and Kamekura, M. (1983). Halophilic nuclease of a moderately halophilic *Bacillus* sp.: production, purification and characterization. *Applied and Environmental Microbiology* **45**, 24–30.

Oppenheimer, C. H. and ZoBell, C. E. (1952). The growth and viability of sixty-three species of marine bacteria as influenced by hydrostatic pressure. *Journal of Marine Research* **11**, 10–18.

Oshima, T. and Imahori, K. (1971). Description of *Thermus thermophilus* comb. nov. A non-sporulating thermophilic bacterium from a Japanese thermal spa. *International Journal of Systematic Bacteriology* **24**, 102–112.

Outtrup, H. (1976). US Patent 3,979,261.

Padan, E., Zilberstein, D. and Rottenberg, H. (1976). The proton electrochemical gradient in *Escherichia coli*. *European Journal of Biochemistry* **63**, 533–541.

Panes, J. J. and Thomas, S. B. (1968). Psychrotrophic coli-aerogenes in refrigerated milk: a review. *Journal of Applied Bacteriology* **31**, 420–425.

Pask-Hughes, R. and Williams, R. A. D. (1975). Extremely thermophilic Gram-negative bacteria from hot tap water. *Journal of General Microbiology* **88**, 321–328.

Payton, M. A. (1984). Production of ethanol by thermophilic bacteria. *Trends in Biotechnology* **2**, 153–158.

Perutz, M. F. and Raidt, H. (1975). Deletion of globin genes in haemoglobin-H disease demonstrates multiple globin structural loci. *Nature, London* **225**, 256–259.

Pfueller, S. I. and Elliott, W. K. (1969). The extracellular amylase of *Bacillus stearothermophilus*. *Journal of Biological Chemistry* **244**, 48–54.

Pierson, B. K. and Castenholz, R. W. (1974). A phototrophic gliding filamentous bacterium of hot springs, *Chloroflexus aurantiacus*, gen. and sp. nov. *Archives of Microbiology* **100**, 5–24.

Poindexter, J. S. (1981). Oligotrophy: feast and famine existence. *Advances in Microbial Ecology* **5**, 67–93.

Porubcan, R. S. and Sellars, R. L. (1979). Lactic starter culture concentrates. *Microbial Technology* **1**, 130–155.

Prickett, P. S. (1928). Thermophilic and thermoduric microorganisms with special reference to species isolated from milk. Description of spore forming types. *New York State Agricultural Experimental Station Technical Bulletin 147*.

Priest, F. G. (1977). Extracellular enzyme synthesis in the genus *Bacillus*. *Bacteriological Reviews* **41**, 711–753.

Pundak, S., Alani, H. and Eisenberg, H. (1981). Structure and activity of malate dehydrogenase from extreme halophilic bacteria of the Dead Sea. *European Journal of Biochemistry* **118**, 471–477.

Rao, G. S. and Berger, L. R. (1971). The requirement of low pH for growth of *Thiobacillus thioxidans*. *Archives of Microbiology* **79**, 338–344.

Rao, J. K. M. and Argos, P. (1981). Structural stability of halophilic proteins. *Biochemistry* **20**, 6536–6543.

Reali, D., Caroli, G., Filippi, S. and Simonetti, S. (1977). Isolamento ed identificazione di Vibrioni alofili possibili agenti di tossinfezioni alimentari. *Annali Sclavo* **19**, 455–463.

Richard, C. and Lhuillier, M. (1977). *Vibrio parahaemolyticus* et vibrions halophiles: leur importance en pathologie humaine et dans l'environnement marin. *Bulletin de l'Institut Pasteur, Paris* **75**, 345–368.

Roberts, R. J. (1980). Restriction and modification enzymes and their recognition sequences. *Nucleic Acids Research* **8**, r63–r80.

Robson, E. and Pain, R. H. (1971). Analysis of the code relating sequence to conformation in proteins: possible implications for the mechanism of formation of helical regions. *Journal of Molecular Biology* **58**, 237–259.

Robyt, J. and French, D. (1964). Purification and action pattern of an amylase from *Bacillus polymyxa*. *Archives of Biochemistry and Biophysics* **104**, 338–345.

Rorick, M. B., Spahr, S. L. and Bryant, M. P. (1980). Methane production from cattle waste in laboratory reactors at 40 °C and 60 °C after solid–liquid separation. *Journal of Dairy Science* **63**, 1953–1956.

Rottenburg, H., Grunwald, T. and Avron, M. (1972). Determination of ΔpH in chloroplasts. 1. Distribution of (^{14}C) methylamine. *European Journal of Biochemistry* **25**, 54–63.

Rowe, J. J., Goldberg, I. D. and Amelunxen, R. E. (1975). Development of defined and minimal media for the growth of *Bacillus stearothermophilus*. *Journal of Bacteriology* **124**, 279–284.

Ruby, E. G. and Jannasch, H. W. (1982). Physiological characteristics of *Thiomicrospora* sp. L-12 isolated from deep-sea hydrothermal vents. *Journal of Bacteriology* **149**, 161–165.

Rupela, O. P. and Tauro, P. (1973). Isolation and characterisation of *Thiobacillus* from alkali soils. *Soil Biology and Biochemistry* **5**, 891–897.

Saiki, T., Kimura, R. and Arima, K. (1972). Isolation and characterisation of extremely thermophilic bacteria from hot springs. *Agricultural and Biological Chemistry* **36**, 2357–2366.

Sakata, T. and Kakimoto, D. (1980). Study of proteases of marine bacteria: effect of cations on the enzyme production and activity. *In* "Saline Environment: Physiological and Biochemical Adaptation in Halophilic Microorganisms" (Eds H. Morishita and M. Masui), pp. 123–134. Business Centre for Academic Societies, Japan.

Sanger, F. (1963). Amino-acid sequences in the active centres of certain enzymes. *Proceedings of the Chemical Society, London* 76–83.

Sergeant, K., East, D. N., Whitaker, A. R. and Ellsworth, R. (1971). Production of *Bacillus stearothermophilus* NCA 1503 for glyceraldehyde-3-phosphate dehydrogenase. *Journal of General Microbiology* **65**, iii.

Sato, M., Beppu, T. and Arima, K. (1980). Properties and structure of a novel peptide antibiotic No. 1907. *Agricultural and Biological Chemistry* **44**, 3037–3040.

Sato, S., Hutchison, C. A. and Harris, J. I. (1977). A thermostable sequence-specific endonuclease from *T. aquaticus*. *Proceedings of the National Academy of Sciences of the United States of America* **74**, 542–546.

Sattar, S. A., Synek, E. J., Westwood, J. C. N. and Neals, P. (1972). Hazard inherent in microbiological tracers: reduction of risk by the use of *Bacillus stearothermophilus* spores in aerobiology. *Applied Microbiology* **23**, 1053–1059.

Schwarz, J. R. and Colwell, R. R. (1975). Macromolecular biosynthesis in *Pseudomonas bathycetes* at deep-sea pressure and temperature. *In* "Abstracts of the Annual Meeting of the American Society for Microbiology", No. 162.

Schwarz, J. R. and Colwell, R. R. (1976). Microbial activities under deep-ocean conditions. *Developments in Industrial Microbiology* **17**, 299–304.

Schwarz, J. R., Yayanos, A. A. and Colwell, R. R. (1976). Metabolic activities of the intestinal flora of a deep-sea invertebrate. *Applied and Environmental Microbiology* **31**, 46–48.

Seesman, P. A., Walker, J. D. and Colwell, R. R. (1976). Biodegradation of oil by marine microorganisms at potential off-shore drilling sites. *Developments in Industrial Microbiology* **17**, 293–297.

Sharp, R. J. and Woodrow, J. R. (1982). Taxonomy of *Bacillus* thermophiles. *In* "Abstracts of the 13th International Congress of Microbiology, Boston, Massachusetts".

Sharp, R. J., Bingham, A. H. A., Comer, M. J. and Atkinson, A. (1979). Partial characterisation of a bacteriocin (thermocin) from *Bacillus stearothermophilus* RS 93. *Journal of General Microbiology* **111**, 449–451.

Sharp, R. J., Bown, K. J. and Atkinson, A. (1980). Phenotypic and genotypic characterisation of some thermophilic species of *Bacillus*. *Journal of General Microbiology* **117**, 201–210.

Shaver, K. J. and Schiff, T. (1969). 47th General Meeting, International Association of Dental Research, Houston, Texas.

Shelef, G., Kimchie, S. and Grynberg, H. (1980). High rate thermophilic anaerobic digestion of agricultural wastes. *Biotechnology and Bioengineering Symposium* **10**, 341–351.

Shewan, J. M. and Hobbs, G. (1967). The bacteriology of fish spoilage and preservation. *Progress in Industrial Microbiology* **6**, 171–208.

Shilo, M. (1979). "Strategies of Microbial Life in Extreme Environments". Dahlem Konferenzen, Berlin, and Springer, New York.

Shinomiya, T. and Sato, S. (1980). A site specific endonuclease from *Thermus thermophilus* 111 *Th111I*. *Nucleic Acids Research* **8**, 43–56.

Shiosaka, M. (1976). US Patent 3,988,206.

Shivers, D. W. and Brock, T. D. (1973). Oxidation of elemental sulfur by *Sulfolobus acidocaldarius*. *Journal of Bacteriology* **114**, 706–710.

Sidler, W. and Zuber, H. (1972). Neutral proteases with different thermostabilities from a facultative strain of *Bacillus stearothermophilus* grown at 40 °C and at 50 °C. *FEBS Letters* **25**, 292–294.

Sidler, W. and Zuber, H. (1977). The production of extracellular thermostable neutral proteinase and α-amylase by *Bacillus stearothermophilus*. *European Journal of Applied Microbiology* **4**, 255–266.

Sidler, W. and Zuber, H. (1980). Isolation procedures for thermostable neutral proteinases produced by *Bacillus stearothermophilus*. *European Journal of Applied Microbiology and Biotechnology* **10**, 197–209.

Singleton, R. (1976). A comparison of the amino acid compositions from thermophilic and non-thermophilic origins. *In* "Extreme Environments: Mechanisms of Microbial Adaption" (Ed. M. R. Heinrich), pp. 189–200. Academic Press, New York and London.

Singleton, R. and Amelunxen, R. E. (1973). Proteins from thermophilic microorganisms. *Bacteriological Reviews* **37**, 320–342.

Smith, E. L., DeLange, R. J., Evans, W. H., Landon, M. and Karkland, F. S. (1968). Subtilisin Carlsberg. V. The complete sequence; comparison with subtilisin BPN: evolutionary relationship. *Journal of Biological Chemistry* **243**, 2184–2191.

Smith, M. D. and Olson, C. L. (1975). Differential amperometric determination of alcohol in blood or urine using alcohol dehydrogenase. *Analytical Chemistry* **47**, 1074–1099.

Smith, P. F., Langworthy, T. A. and Smith, M. R. (1975). Polypeptide nature of growth requirement in yeast extract for *Thermoplasma acidophilum*. *Journal of Bacteriology* **124**, 884–892.

Sonnleitner, B. (1984). Biotechnology of thermophilic bacteria, growth, products and applications. *Advances in Biochemical Engineering* **30**, 70–138.

Sonnleitner, B., Cometta, S. and Fiechter, A. (1982). Equipment and growth inhibition of thermophilic bacteria. *Biotechnology and Bioengineering* **24**, 2597–2599.

Sonnleitner, B., Fiechter, A. and Giovannini, F. (1984). Growth of *Thermoanaero-*

bium brockii in batch and continuous culture at supraoptimal temperatures. *Applied Microbiology and Biotechnology* **19**, 326–334.

Souza, K. A. and Deal, P. H. (1977). Characterization of a novel extremely alkalophilic bacterium. *Journal of General Microbiology* **101**, 103–109.

Souza, K. A., Deal, P. H., Mack, H. M. and Turnbill, C. E. (1974). Growth and reproduction of microorganisms under extremely alkaline conditions. *Applied Microbiology* **28**, 1066–1068.

Spencer, R. R. (1979). Enhancement of methane production in the anaerobic digestion of sewage sludges. *Biotechnology and Bioengineering Symposium* **8**, 257–268.

Sunaga, T., Akiba, T. and Horikoshi, K. (1976). Production of penicillinase by an alkalophilic *Bacillus*. *Agricultural and Biological Chemistry* **40**, 1363–1367.

Suzuki, Y., Kishigami, T. and Abe, S. (1976). Production of extracellular α-glucosidase by a thermophilic *Bacillus* species. *Applied and Environmental Microbiology* **31**, 807–812.

Swartz, R. D. and Keller, F. A. (1982a). Isolation of a strain of *Clostridium thermoaceticum* capable of growth and acetic acid production at pH 4.5. *Applied and Environmental Microbiology* **43**, 117–123.

Swartz, R. D. and Keller, F. A. (1982b). Acetic acid production by *Clostridium thermoaceticum* in pH controlled batch fermentations at acidic pH. *Applied and Environmental Microbiology* **43**, 1385–1392.

Tai, P. C. and Jackson, H. (1969). Mesophilic mutants of an obligate psychrophile, *Micrococcus cryophilus*. *Canadian Journal of Microbiology* **15**, 1145–1150.

Tajima, M., Urabe, I., Yutani, K. and Okada, H. (1976). Role of calcium ions in the thermostability of thermolysin and *Bacillus subtilis* var. *amylosacchariticus* neutral protease. *European Journal of Biochemistry* **64**, 243.

Takahara, Y. and Tanabe, O. (1960). Studies on the reduction of indigo in industrial fermentation vat (VII). *Journal of Fermentation Technology* **38**, 329–331.

Takahara, Y. and Tanabe, O. (1962). Studies on the reduction of indigo in industrial fermentation vat (XIX). Taxonomic characteristics of strain No. S-8. *Journal of Fermentation Technology* **40**, 181–186.

Takahara, Y., Takasaki, Y. and Tanabe, O. (1961). Studies on the reduction of indigo in industrial fermentation vat (XVIII). On the growth factor of the strain No. S-8(4). *Journal of Fermentation Technology* **39**, 183–187.

Takahashi, M. and Hashimoto, Y. (1963). Studies on bacterial xylanase (iii). Crystallization and some enzymatic properties of bacterial xylanase. *Journal of Fermentation Technology* **41**, 181–186.

Tanaka, M., Haniu, M., Matsueda, G., Yasunobu, K. T., Himes, R. H., Akagi, J. M., Barnes, E. M. and Devanathan, T. (1971). The primary structure of the *Clostridium tartarivorum* ferredoxin, a heat stable ferredoxin. *Journal of Biological Chemistry* **246**, 3958–3960.

Temple, K. L. and Colmer, A. R. (1951). The autotrophic oxidation of iron by a new bacterium *Thiobacillus ferrooxidans*. *Journal of Bacteriology* **62**, 605–611.

Te-Pang, H. (1890). "Manufacture of Soda, with Special Reference to the Ammonia Process: a Practical Treatise", American Chemical Society Monograph No 65. Hafner, New York.

Thomas, J. A., Cole, R. E. and Langworthy, T. A. (1976). Intracellular pH measurements with a spectroscopic probe generated *in situ*. *Federation Proceedings, Federation of American Societies for Experimental Biology* **35**, 1455.

Tiller, A. K. (1982). Aspects of microbial corrosion. *In* "Corrosion Processes" (Ed. R. N. Parkins). Applied Science, London.

Timson, W. J. and Short, A. J. (1965). Resistance of microorganisms to hydrostatic pressure. *Biotechnology and Bioengineering* **7**, 139–159.

Tindall, B. J., Mills, A. A. and Grant, W. D. (1980). An alkalophilic red halophilic bacterium with a low magnesium requirement from a Kenyan soda lake. *Journal of General Microbiology* **116**, 257–260.

Toda, H. and Narita, K. (1968). Correlation of the sulfhydryl group with the essential calcium in *Nacillus subtilis* saccharifying α-amylase. *Journal of Biochemistry* **63**, 302–307.

Torma, A. E. (1977). The role of *T. ferrooxidans* in hydrometallurgical processes. *Advances in Biochemical Engineering* **6**, 1–37.

Trudinger, P. A. and Bubela, B. (1967). Microorganisms and natural environment. *Mineralium Deposita* **2**, 147–157.

Uchino, F. (1982). A thermophilic and unusually acidophilic amylase produced by a thermophilic acidophilic *Bacillus* species. *Agricultural and Biological Chemistry* **46**, 7–13.

Uchino, F. and Doi, S. (1967). Acido-thermophilic bacteria from thermal waters. *Agricultural and Biological Chemistry* **31**, 817–822.

Uchino, F. and Nakane, T. (1981). A thermostable xylanase from a thermophilic acidophilic *Bacillus* species. *Agricultural and Biological Chemistry* **45**, 1121–1127.

Ulrich, J. T., McFeters, G. A. and Temple, K. L. (1972). Induction and characterisation of β-galactosidase in an extreme thermophile. *Journal of Bacteriology* **110**, 691–698.

Vance, I. (1984). Bacterial growth at elevated pressures and temperatures. SGM Symposium on Microbes in Extreme Environments, Dundee.

Van Qua, D., Simidu, U. and Taga, N. (1981). Purification and some properties of halophilic protease produced by a moderately halophilic marine *Pseudomonas* species. *Canadaian Journal of Microbiology* **27**, 505–510.

Varel, V. H. (1984). Characteristics of some fermentative bacteria from a thermophilic methane producing fermenter. *Microbial Ecology* **10**, 15–24.

Varel, V. H., Isaacson, H. R. and Bryant, M. P. (1977). Thermophilic methane production from cattle waste. *Applied and Environmental Microbiology* **33**, 298–307.

Vedder, A. (1934). *Bacillus alcalophilus* n. sp. benevens enkle ervaringen met sterk alcalische voedingsbodems. *Antonie van Leeuwenhoek; Journal of Microbiology and Serology* **1**, 141–147.

Waddell, W. J. and Butler, T. C. (1959). Calculation of intracellular pH from the distribution of 5,5-dimethyl-2,4-oxazolidinedoine (DMO). Application to skeletal muscle of the dog. *Journal of Clinical Investigation* **38**, 720–729.

Wahlby, S. and Engstrom, L. (1968). Studies on *Streptomyces griseus* protease. II. The amino acid sequence around the reactive serine residue of DFP-sensitive components with esterase activity. *Biochimica et Biophysica Acta* **151**, 402–408.

Wakerley, D. S. (1979). Microbial corrosion in UK industry; a preliminary survey of the problem. *Chemistry and Industry*, 656–658.

Waksman, S. A. and Jaffe, J. S. (1922). Microorganisms concerned in the soil. II. *Thiobacillus thiooxidans*, a new sulphur-oxidising organism isolated from soil. *Journal of Bacteriology* **7**, 239–256.

Walker, J. E., Wonacott, A. J. and Harris, J. I. (1980). Heat stability of a tetrameric enzyme D-glyceraldehyde-3-phosphate dehydrogenase. *European Journal of Biochemistry* **108**, 581–586.

Walker, P. D. and Wolf, J. (1971). The taxonomy of *Bacillus stearothermophilus*. *In* "Spore Research" (Eds. A. N. Barker, G. W. Gould and J. Wolff), pp. 247–262. Academic Press, London.

Walsby, A. E. (1971). The pressure relationships of gas vacuoles. *Proceedings of the Royal Society of London, Series B* **178**, 301–326.

Walsby, A. E. (1972). Gas-filled structures providing buoyancy in photosynthetic organisms. *In* "The Effects of Pressure on Living Organisms" (Eds M. A. Sleigh and A. G. MacDonald), pp. 233–250. Academic Press, New York and London.

Ward, O. P. and Fogarty, W. M. (1972). Polygalacturonate lyase of a *Bacillus* species associated with increase in permeability of Sitka spruce (*Picea sitchensis*). *Journal of General Microbiology* **73**, 439–446.

Watanabe, K. and Horikoshi, K. (1977). Amino acid sequence of an active center peptide containing serine isolated from alkaline protease of *Bacillus* No. 221. *Agricultural and Biological Chemistry* **41**, 715–716.

Watanabe, N., Ota, Y., Minoda, Y. and Tamada, K. (1977). Isolation and identification of alkaline lipase producing microorganisms, culture conditions and some properties of crude enzymes. *Agricultural and Biological Chemistry* **41**, 1353–1358.

Weast, R. C. and Selby, S. M. (Eds.) (1966). "CRC Handbook of Chemistry and Physics." The Chemical Rubber Company, Cleveland, Ohio.

Weimer, P. J. and Zeikus, J. G. (1977). Fermentation of cellulose and cellobiose by *Clostridium thermocellum* in the absence and the presence of *Methanobacterium thermoautotrophicum*. *Applied and Environmental Microbiology* **33**, 289–297.

Weimer, P. J., Wagner, L. W., Knowlton, S. and Ng, T. K. (1984). Thermophilic anaerobic bacteria which ferment hemicellulose: characterisation of organisms and identification of plasmids. *Archives of Microbiology* **138**, 31–36.

Weiss, R. L. (1973). Attachment of bacteria to sulphur in extreme environments. *Journal of General Microbiology* **77**, 502–507.

Welker, N. E. (1976). Microbial endurance and resistance to heat stress. *In* "The Survival of Vegetative Microbes" (Eds T. R. G. Gray and J. R. Postgate), pp. 241–277. Cambridge University Press, Cambridge.

Wellinger, A. and Kaufmann, R. (1982). Psychrophilic methane generation from pig manure. *Process Biochemistry* **17**, 26–30.

West, I. C. and Mitchell, P. (1974). Proton/sodium ion antiport in *Escherichia coli*. *Biochemistry Journal* **144**, 87–90.

Wiegel, J. (1980). Formation of ethanol by bacteria. A pledge for the use of extreme thermophilic anaerobic bacteria in industrial ethanol fermentation processes. *Experientia* **36**, 1434–1446.

Wiegel, J. and Ljungdahl, L. G. (1981). *Thermoanaerobacter ethanolicus* gen. nov., spec. nov., a new, extreme thermophilic, anaerobic bacterium. *Archives of Microbiology* **128**, 342–348.

Wiegel, J., Carreira, L. H., Mothershed, C. P. and Puls, J. (1984). Production of ethanol from biopolymers by anaerobic, thermophilic and extreme thermophilic bacteria. II. *Thermoanaerobacter ethanolicus* JW200 and its mutants in batch cultures and resting cell experiments. *Biotechnology and Bioengineering Symposium* **13**, 193–207.

Wilke, C. R. (1975). Cellulase as a chemical and energy resource. *Biotechnology and Bioengineering Symposium* **5**.

Williams, R. A. D. (1975). Caldoactive and thermophilic bacteria and their thermostable proteins. *Science Progress* **62**, 373–393.

Williams, S. T. and Flowers, T. H. (1977). The influence of pH on starch hydrolysis by neutrophilic and acidophilic streptomycetes. *Microbios* **20**, 99–106.

Woese, C. R., Magrum, L. J. and Fox, G. E. (1978). Archaebacteria. *Journal of Molecular Evolution* **11**, 245–252.

Wolf, J. and Barker, A. N. (1968). The genus *Bacillus*: aids to the identification of its species. *In* "Identification Methods for Microbiologists" (Eds M. Gibbs and D. A. Shatpon), Part B, pp. 91–109. Academic Press, London.

Wolf, J. and Sharp, R. J. (1981). Taxonomic and related aspects of thermophiles within the genus *Bacillus*. *In* "The Aerobic Endospore-forming Bacteria" (Eds R. C. W. Berkeley and M. Goodfellow), pp. 251–296. Academic Press, London and New York.

Wolfe, R. S. (1960). Microbial concentration of iron and manganese in water with low concentrations of these elements. *Journal of the American Water Works Association* **52**, 1335.

Yamamamoto, M., Tanaka, Y. and Horikoshi, K. (1972). Alkaline amylase of alkalophilic bacteria. *Agricultural and Biological Chemistry* **36**, 1819–1823.

Yayanos, A. A. (1975). Stimulatory effect of hydrostatic pressure on cell division in cultures of *Escherichia coli*. *Biochimica et Biophysica Acta* **392**, 271–275.

Yayanos, A. A., Dietz, A. S. and van Boxtel, R. (1979). Isolation of a deep-sea barophilic bacterium and some of its growth characteristics. *Science* **205**, 808–809.

Yayanos, A. A., Dietz, A. S. and van Boxtel, R. (1981). Obligately barophilic bacterium from the Mariana Trench. *Proceedings of the National Academy of Sciences of the United States of America* **78**, 5212–5215.

Yutani, K. (1976). Role of calcium ions in the thermostability of amylase produced by *Bacillus stearothermophilus*. *In* "Enzymes and Proteins from Thermophilic Microorganisms" (Ed. H. Zuber), pp. 91–103. Birkauser, Basel and Stuttgart.

Zajic, J. E. (1969). "Microbial Biogeochemistry". Academic Press, New York and London.

Zeikus, J. G. (1979). Thermophilic bacteria: ecology, physiology and technology. *Enzyme and Microbial Technology* **1**, 243–252.

Zeikus, J. G. and Wolfe, R. S. (1972). *Methanobacterium thermoautotrophicus* sp. n., an anaerobic, autotrophic, extreme thermophile. *Journal of Bacteriology* **109**, 707–713.

Zeikus, J. G., Hegge, P. W. and Anderson, M. A. (1979). *Thermoanaerobium brockii* gen. nov. and sp. nov., a new chemoorganotrophic caldoactive, anaerobic bacterium. *Archives of Microbiology* **122**, 41–48.

Zeikus, J. G., Ben-Bassat, A., Ng, T. K. and Lamed, R. J. (1981). *In* "Trends in the Biology of Fermentations for Fuels and Chemicals" (Ed. A. Hollaender), pp. 441–461. Plenum, New York.

Zertuche, L. and Zall, R. R. (1982). A study of producing ethanol from cellulose using *Clostridium thermocellum*. *Biotechnology and Bioengineering* **24**, 57–68.

Zillig, W., Stetter, K. O., Wunderl, S., Schulz, W., Priess, H. and Scholz, I. (1980). The Sulfolobus–"Caldariella" group: taxonomy on the basis of the structure of DNA-dependent RNA polymerases. *Archives of Microbiology* **125**, 259–269.

ZoBell, C. E. (1958). Ecology of sulphate reducing bacteria. *Producers Monthly* **22**, 12–29.

ZoBell, C. E. and Johnson, F. H. (1949). The influence of hydrostatic pressure on the growth and viability of terrestrial and marine bacteria. *Journal of Bacteriology* **57**, 179–189.

ZoBell, C. E. and Morita, R. Y. (1957). Barophilic bacteria in some deep-sea sediments. *Journal of Bacteriology* **73,** 563–568.

Zuber, H. (1978). Comparative studies of thermophilic and mesophilic enzymes: objectives, problems, results. *In* "Biochemistry of Thermophily" (Ed. S. M. Friedman), pp. 267–287. Academic Press, New York, San Francisco, California, and London.

10

Genetic Applications of Alkalophilic Microorganisms

K. HORIKOSHI

Department of Applied Microbiology, The RIKEN Institute, Wako-shi, Saitama 351, Japan

Introduction

Since 1969 we have been working on alkalophilic microorganisms in various fields. The points of interest concerning these microorganisms are as follows.

(1) No bacterial growth is observed at a neutral pH value.
(2) Unless the intracellular pH is increased to between 7 and 8 and the extracellular pH to between 9 and 11, no active transport of nutrients is exhibited.
(3) Almost all alkalophilic *Bacillus* strains require Na^+ ions for their growth, germination and sporulation.
(4) During cell growth they can change the pH value of the medium so that it becomes the optimum pH for their growth.
(5) The protein-synthesizing machinery exhibits no significant difference for alkalophilic and neutrophilic bacteria.

An analysis of cell walls and membranes showed several differences in their compositions. Both extracellular and intracellular enzymes which have several unique enzymatic properties have been isolated and characterized. One of these, cyclodextrin glycosyl transferase, has led to the establishment of a cyclodextrin industry (Horikoshi and Akiba, 1982).

However, no genetic investigation of alkalophilic microorganisms had been carried out before 1980 although mutants had been isolated after ultraviolet (UV) irradiation or nitrosoguanidine (NTG) treatments. The classical NTG mutation of alkalophilic *Bacillus* strains could increase the production of alkaline proteases, but no modern genetic technique nor any recombinant DNA technology had been applied to alkalophilic bacteria.

Alkalophilic microorganisms which can grow in extreme environments have many unique DNAs with unknown properties. They have genes coding for alkaline enzymes, and the physiological properties of host cells can be changed by the introduction of new DNA from alkalophilic microorganisms. It will be possible to clarify the formation of some extracellular multienzyme systems of carbohydrases. Furthermore, cloning of DNA is certain to be a valuable tool to study antiporter and synporter systems in the membrane of alkalophilic *Bacillus* strains.

Our main interest has focused on changing the physiological properties of host cells by the introduction of new DNA from alkalophilic microorganisms. *Escherichia coli*, which has been extensively studied in the field of genetic engineering, does not excrete gene products into the culture broth because it has a three-layered cell surface. Proteins passing the inner membrane are trapped in the periplasmic space, and nothing of this molecular size can be excreted through the outer membrane. This property is one of the major bottlenecks in the industrial production processes involving *E. coli* because maximum production yield is limited, and furthermore it is difficult to extract and purify the products from the cells. If intracellular proteins could be excreted into the culture broth, these problems would be solved automatically. *Bacillus subtilis*, which can excrete proteins into the culture broth, is a candidate for this purpose. However, the plasmid introduced in the host cells is not stable enough for industrial production, and the products are sometimes hydrolysed by their proteases during the cultivation process. Another host is yeast, but the technology itself is not fully developed yet. It is therefore clear that, if the outer membrane of *E. coli* can be modified to enable proteins to be excreted into the culture broth, this will be the best host from the industrial point of view.

Recently we succeeded in changing the outer membrane of *E. coli* by introducing a 2.4×10^3 bp DNA fragment of the alkalophilic *Bacillus* No. 170 strain. This DNA fragment containing the penicillinase gene was able to make the outer membrane leaky and the proteins including penicillinase trapped in the periplasmic space were excreted. In addition to this DNA fragment, another DNA fragment containing a xylanase gene has also been shown to make the outer membrane of *E. coli* leaky, although no critical studies have been undertaken to explain how these DNA fragments can change the outer membrane. In alkalophilic bacteria, no host–vector system has been established. Several plasmids can replicate in alkalophilic *Bacillus* strains and plasmid pC194 can transform alkalophilic *Bacillus* strain No. A-59 and yield chloramphenicol-resistant transformants with a high frequency. This fundamental research will be important in the development of new biotechnology using alkalophilic microorganisms.

This chapter will describe details of some genetic engineering and host–

vector systems of alkalophilic bacteria. Also, I would like to show how the isolation of new DNA has opened up a new field of biotechnology.

Cloning of Penicillinase-excretion Gene

An alkalophilic *Bacillus* strain No. 170 has been isolated which produced penicillinase in an alkaline medium (Sunaga *et al.*, 1976). Cloning of the penicillinase was investigated to study the excretion mechanisms involved. These experiments produced the interesting observation that most of the plasmid-borne penicillinase in *E. coli* was found in the culture medium during cultivation (Kato *et al.*, 1983; Kudo *et al.*, 1983).

Construction of pEAP1 and pEAP2

The alkalophilic *Bacillus* strain No. 170, which produced penicillinase in alkaline culture conditions, was isolated from soil. The best growth and maximum enzyme production were observed at pH 9.0. The optimal pH and temperature of the penicillinase were 6.0–7.0 and 50 °C respectively. The stable pH range of the enzyme was comparatively broad, ranging from about pH 7.0 to 10.0. *E. coli* K-12 HB101 and *E. coli* C600 were used in all experiments. The plasmids pBR322 and pMB9 were used as vectors. Bacterial chromosomal DNA was purified according to the method of Saito and Miura (1963). The chromosomal DNA of alkalophilic *Bacillus* No. 170 which was grown aerobically to the early stationary phase at 37 °C in an alkaline medium and plasmid pMB9 DNA were digested with *Hind*III or *Eco*RI restriction endonucleases respectively and ligated with T4 DNA ligase. The recombinant DNAs were introduced to *E. coli* C600, and transformants having *Eco*RI or *Hind*III fragments were selected on LB agar plates containing 20 μg of ampicillin and/or 50 μg of tetracycline. A plasmid pEAP1 was obtained from the transformant (Tc^r, Ap^r, penicillinase$^+$) which contained a 4.5×10^3 bp *Eco*RI fragment. Another plasmid, isolated from the transformant (Ap^r, penicillinase$^+$), contained a 2.4×10^3 bp *Hind*III fragment of alkalophilic *Bacillus* strain No. 170 DNA. The cleavage maps of pEAP1 and pEAP2 are shown in Fig. 1. The 2.4×10^3 bp *Hind*III fragment which contained a penicillinase gene was located in the middle of the 4.5×10^3 bp *Eco*RI fragment.

Excretion of Enzymes

E. coli carrying plasmids were aerobically grown in LB broth for 20 hours at

Table 1. *Bacterial strains and plasmids*

Bacterial strains and plasmids	Genotypes and properties
Alkalophilic *Bacillus*	
No 17 (ATCC 31006)	Production of CGTase[a]
No 38-2 (ATCC21783)	Production of CGTase[a]
No C-125 (FERM7344)	Production of xylanase
No 170 (FERM3221)	Production of penicillinase
No H331	Production of protease
No N-4 (ATCC21833)	Production of alkaline cellulase
Alkalophilic *Aeromonas* sp.	
No 212 (ATCC31085)	Production of semi alkaline cellulase and xylanase
E. coli HB101	*pro, leu*B, B1, *lac*Y, *hsd*R, *hsd*M, *ara*14, *gal*kz, *xyl*5, *mtl*1, *sup*E44, F⁻, *endo*I⁻, *rec*A⁻, *str*ʳ
E. coli C600	*leu*B, *thr, trp*B, *thi, lac*Y, *sup*E44, *hsd*R, *hsd*M
B. subtilis	
CU741	*leu*C7, *trp*C2
1012	*leu*A8, *met*B5
Plasmids	
pMB9	Tcʳ
pBR322	Tcʳ, Ampʳ
pUB110	Kmʳ
pGR71	Kmʳ, (Cmʳ)
pEAP1	Tcʳ, penicillinase⁺, pMB9 + 4.5 × 10³ bp *Eco*RI fragment of alkalophilic *Bacillus* No. 170
pEAP2	penicillinase⁺, pMB9 + 2.4 × 10³ bp *Hin*dIII fragment of alkalophilic *Bacillus* No. 170
pSAP21	pEAP1 + pUB110
pAX1	Ampʳ, xylanase⁺, pBR322 + 6 × 10³ bp *Hin*dIII fragment of alkalophilic *Aeromonas* No. 212
pCX311	Ampʳ, xylanase⁺, pBR322 + 4.6 × 10³ bp *Hin*dIII fragment of alkalophilic *Bacillus* No. C-125
pNK1	Ampʳ, cellulase⁺, pBR322 + 2.0 × 10³ bp *Hin*dIII fragment of alkalophilic *Bacillus* No. N-4
pNK2	Ampʳ, cellulase⁺, pBR322 + 2.8 × 10³ bp *Hin*dIII fragment of alkalophilic *Bacillus* No. N-4
pXP102	penicillinase⁺, xylanase⁺, pEAP2 + 4 × 10³ bp *Bgl*II fragment of pAX1
pAB3311	Cryptic 16 × 10³ bp, alkalophilic *Bacillus* No. H331
pAB3312	Cryptic 6 × 10³ bp, alkalophilic *Bacillus* No. H331
pAB13	Cryptic 7.9 × 10³ bp, alkalophilic *Bacillus* No. 13

[a] Cyclodextrin glycosyltransferase.

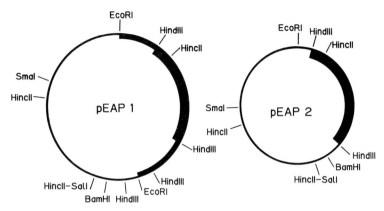

Fig. 1. Physical maps of plasmids pEAP1 and pEAP2. The thin lines represent the pMB9 DNA. The thick lines represent the penicillinase gene from alkalophilic *Bacillus* sp. strain No 170.

37 °C and enzymatic activities were determined. As shown in Table 2, most of the penicillinase produced by *E. coli* HB101 (pEAP1 or pEAP2) and *E. coli* C600 (pEAP2) was detected in the culture medium. Less than 15% of the total activity was observed in the periplasmic and cellular fractions. However, almost all the β-lactamase produced by *E. coli* HB101 (pBR322) was trapped in the periplasmic space. The distributions of β-galactosidase and alkaline phosphatase activities were investigated. As shown in Table 3,

Table 2. *Distribution of penicillinase in* E. coli

Strain	Extracellular activity	Periplasmic activity	Cellular activity	Total activity
Alkalophilic *Bacillus* No. 170	29.6	—	0.0	29.6
E. coli HB101 (pEAP1)	6.5	0.3	0.4	7.2
E. coli HB101 (pEAP2)	16.7	0.3	1.7	18.7
E. coli C600 (pEAP2)	16.5	0.1	3.0	19.6
E. coli HB101 (pBR322)	4.0	21.7	6.4	32.1

The values represent units of penicillinase activity per millilitre of the culture. *E. coli* strains were aerobically grown in the LB broth for 20 hours at 37 °C. Alkalophilic *Bacillus* No. 170 was cultured in the alkaline medium with continuous shaking for 13 hours at 37 °C, since the best production of penicillinase was observed in these conditions.

Table 3. *Distribution of enzymes in* E. coli *HB101 carrying plasmids*

Plasmids		Extracellular	Periplasmic	Cellular	Total
None	Protein	0.04 (4%)	0.07 (6%)	0.97 (90%)	1.08 (100%)
	APase	0.02 (2%)	1.05 (79%)	0.26 (20%)	1.33 (100%)
	β-gal	0.03 (2%)	0.00 (0%)	1.20 (98%)	1.23 (100%)
pMB9	Protein	0.01 (1%)	0.03 (4%)	0.75 (95%)	0.79 (100%)
	APase	0.01 (2%)	0.32 (67%)	0.15 (31%)	0.48 (100%)
	β-gal	0.00 (0%)	0.00 (0%)	0.80 (100%)	0.80 (100%)
pEAP2	Protein	0.18 (21%)	0.12 (14%)	0.55 (65%)	0.85 (100%)
	PCase	10.70 (83%)	0.40 (3%)	1.80 (14%)	12.90 (100%)
	APase	0.29 (58%)	0.15 (30%)	0.06 (12%)	0.50 (100%)
	β-gal	0.09 (10%)	0.03 (3%)	0.78 (87%)	0.90 (100%)

E. *coli* strains were aerobically grown in the LB broth for 20 hours at 37 °C. Enzymatic activities of alkaline phosphatase (APase) and β-galactosidase (β-gal) are expressed as absorbance at 420 nm. Penicillinase (PCase) activity is expressed as units per millilitre of the broth. Protein concentration was expressed as milligrams in 1 ml of the broth.

essentially neither protein nor enzymatic activities were detected in the culture broths of *E. coli* HB101 or *E. coli* HB101 (pMB9). About two-thirds of the alkaline phosphatase activity was observed in the periplasmic space and almost all the β-galactosidase was in the cellular fraction. In contrast, in *E. coli* HB101 (pEAP2), it is striking that 21% of the total protein, 58% of the alkaline phosphatase and 83% of the penicillinase were found in the culture broth. However, about 87% of the β-galactosidase was detected in the cellular fraction and not in the periplasmic fraction. These results suggest that the outer membrane of *E. coli* was changed by the introduction of pEAP2 into the cells because a periplasmic enzyme, alkaline phosphatase, was released from the periplasmic space in *E. coli* HB101 (pEAP2). Also a typical cellular enzyme, β-galactosidase, was not excreted into the periplasm. This gives rise to two possibilities.

(1) Penicillinase affects the outer membrane directly or indirectly and is excreted into the medium.
(2) The unknown gene products from the 2.4×10^3 bp fragment affect the outer membrane and penicillinase is excreted.

A DNA sequencing experiment will give us more information.

Fine Mapping and DNA Sequence

Fine mapping of the fragment is shown in Fig. 2. An open reading frame of 774 bp for the penicillinase was detected in a *Dra*I fragment. The *Dra*I

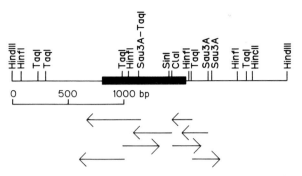

Fig. 2. Fine restriction map of pEAP2. The thick line represents the structural gene of the penicillinase.

fragment connected with an *Eco*RI linker inserted in the *Eco*RI site of pMB9 could be expressed in *E. coli* HB101; however, no significant amount of penicillinase was detected in the culture broth. Although no crucial experiment has been conducted to indicate which gene products in the 2.4×10^3 bp fragment directly cause the outer membrane to become leaky, penicillinase itself would not be a principal candidate. Further work on this point will be published elsewhere.

Stabilities in B. subtilis *and* E. coli

To compare the gene expression and stability of the penicillinase gene, a shuttle plasmid pSAP21 was constructed from pUB110 and pEAP1 (Kato *et al.*, 1984). Most of the penicillinase activity was detected in the culture broth of *B. subtilis* carrying pSAP21 or *E. coli* HB101 carrying pSAP21. The plasmid-encoded penicillinase was quite stable in the culture broth of *E. coli* (pSAP21). *B. subtilis* CU741 (pSAP21) secreted all the penicillinase into the culture broth, but the enzyme activity was about one-third of that of *E. coli* HB101 (pSAP21). Furthermore, the enzyme produced decreased rapidly and after 40 hours of cultivation only 20% of the maximum activity remained, perhaps owing to hydrolysis by proteases of *B. subtilis*.

Application of pEAP2 as an Excretion Vector

From the industrial point of view, the extracellular production of proteins is one of the most important technologies used to isolate gene products from

```
                     GTTTGAGAATTATGGAATGGAAGATTAAGTTTTAAAGCGTACA

       AAATTTTGTACGCTTTTTTGTTAATTACATAAAAGTATGCAAATGAAGATGGAACAACAT

       TTGAGATGAATTGTCTAATATAGGTAATAACTATTTAGCTTGAAAGAAAGGGTTGATAAC

            10        20        30        40        50        60
       ATGAAAAAGAATACGTTGTTAAAAGTAGGATTATGTGTAAGTTTACTAGGAACAACTCAA
       MetLysLysAsnThrLeuLeuLysValGlyLeuCysValSerLeuLeuGlyThrThrGln

            70        80        90       100       110       120
       TTTGTTAGCACGATTTCTTCTGTACAAGCATCACAAAAGGTAGAGCAAATAGTAATCAAA
       PheValSerThrIleSerSerValGlnAlaSerGlnLysValGluGlnIleValIleLys

           130       140       150       160       170       180
       AATGAGACGGGAACCATTTCAATATCTCAGTTAAACAAGAATGTATGGGTTCATACGGAG
       AsnGluThrGlyThrIleSerIleSerGlnLeuAsnLysAsnValTrpValHisThrGlu

           190       200       210       220       230       240
       TTAGGTTATTTTAATGGAGAAGCAGTTCCTTCGAACGGTCTAGTTCTTAATACTTCTAAA
       LeuGlyTyrPheAsnGlyGluAlaValProSerAsnGlyLeuValLeuAsnThrSerLys

           250       260       270       280       290       300
       GGGCTAGTACTTGTTGATTCTTCTTGGGATAACAAATTAACGAAGGAACTAATAGAAATG
       GlyLeuValLeuValAspSerSerTrpAspAsnLysLeuThrLysGluLeuIleGluMet

           310       320       330       340       350       360
       GTAGAAAAGAAATTTCAGAAGCGCGTAACAGATGTCATTATTACACATGCGCACGCTGAT
       ValGluLysLysPheGlnLysArgValThrAspValIleIleThrHisAlaHisAlaAsp

           370       380       390       400       410       420
       CGAATTGGCGGAATAACAGCGTTGAAAGAAAGAGGCATTAAAGCGCATAGTACAGCATTA
       ArgIleGlyGlyIleThrAlaLeuLysGluArgGlyIleLysAlaHisSerThrAlaLeu

           430       440       450       460       470       480
       ACCGCAGAACTAGCAAAGAAAAGTGGATATGAAGAGCCACTTGGAGATTTACAAACAGTT
       ThrAlaGluLeuAlaLysLysSerGlyTyrGluGluProLeuGlyAspLeuGlnThrVal

           490       500       510       520       530       540
       ACGAATTTGAAGTTTGGCGAATACAAAAGTAGAAACGTTCTATCCACGGGAAAGGAGCAT
       ThrAsnLeuLysPheGlyGluTyrLysSerArgAsnValLeuSerThrGlyLysGluHis

           550       560       570       580       590       600
       ACAGAAGATAATATTGTTGTTTGGTTGCCACAATATCAAATTTTAGCTGGAGGCTGTTTA
       ThrGluAspAsnIleValValTrpLeuProGlnTyrGlnIleLeuAlaGlyGlyCysLeu

           610       620       630       640       650       660
       GTAAAATCTGCGGAAGCTAAAAATTTAGGAAATGTTGCGGATGCGTACGTAAATGAATGG
       ValLysSerAlaGluAlaLysAsnLeuGlyAsnValAlaAspAlaTyrValAsnGluTrp

           670       680       690       700       710       720
       TCCACATCGATTGAGAATATGCTGAAGCGATATAGAAATATAAATTTGGTAGTACCTGGT
       SerThrSerIleGluAsnMetLeuLysArgTyrArgAsnIleAsnLeuValValProGly

           730       740       750       760       770
       CACGGGAAAGTAGGAGACAAGGGATTACTTTTTACATACATTGGATTTATTAAAATAAGAA
       HisGlyLysValGlyAspLysGlyLeuLeuLeuHisThrLeuAspLeuLeuLys***

       ATTGTAGAAATACAAAAGAGAGGAGAAATAATTTTCTCCTCTCTTTCTTTTCAACTA
```

Fig. 3. Complete nucleotide sequence of the penicillinase gene of the alkalophilic *Bacillus* sp. strain No 170. The 18 amino acids underlined constitute the signal peptide.

recombinant cells. A xylanase gene isolated from *Aeromonas* sp. strain 212 was cloned by using pBR322 (see next section). The plasmid-encoded xylanase was mainly detected in the periplasmic space and was not detected in the culture broth. The xylanase gene (4×10^3 bp of *Bgl*II fragment) was inserted in the *Bam*HI site of pEAP2 and introduced into *E. coli* HB101. A plasmid pXP102 (11.7×10^3 bp) was obtained from a transformant (penicillinase$^+$, xylanase$^+$). *E. coli* HB101 carrying pXP102 was aerobically cultured in the LB broth for 20 hours at 37 °C. As shown in Table 4, enzymatic activities, which had been detected only in the periplasmic space, were observed as extracellular enzymes in the culture broth of *E. coli* HB101 carrying pXP102.

Thus, the major advantages of our process are as follows.

(1) It eliminates the process of extraction of enzymes or proteins from the cells.
(2) The gene products excreted into the culture broth are generally stable during the cultivation process.

Table 4. *Distribution of enzymes in* E. coli *HB101 carrying plasmids*

Plasmids	Enzymes	Activity[a] in the following fractions		
		Extracellular (%)	Periplasmic (%)	Cellular (%)
None	β-gal	0.00 (0)	0.00 (0)	0.38 (100)
	APase	0.00 (0)	0.33 (77)	1.10 (23)
	β-lactamase	0.00 (0)	0.00 (0)	0.00 (0)
pBR322	β-gal	0.00 (0)	0.01 (3)	0.32 (97)
	APase	0.06 (11)	0.38 (72)	0.09 (17)
	β-lactamase	2.30 (13)	21.20 (86)	0.60 (1)
pAX1	Xylanase	0.10 (0)	0.64 (39)	1.00 (60)
	APase	0.12 (1)	0.07 (83)	0.04 (16)
	β-lactamase	1.70 (9)	15.20 (82)	0.40 (9)
	β-gal	0.01 (3)	0.00 (0)	0.33 (97)
pXP102	Xylanase	0.30 (85)	0.03 (8)	0.01 (5)
	APase	0.12 (48)	0.07 (42)	0.04 (10)
	β-lactamase	2.10 (90)	0.10 (4)	0.15 (6)
	β-gal	0.03 (9)	0.01 (2)	0.30 (89)

E. coli strains were aerobically grown in LB medium for 12 hours at 37 °C.

[a] Enzymatic activities of β-galactosidase (β-gal) and alkaline phosphatase (APase) are expressed as absorbance at 420 nm. Enzymatic activities of β-lactamase and xylanase are expressed as units per millilitre of broth.

Extracellular Production of Xylanase from *E. coli*

The alkalophilic *Bacillus* sp. strain C-125 (Honda *et al.*, 1985b) produced two types of xylanases (xylanase N and A) in the culture medium. Xylanase N had an optimum pH at 7.0 and xylanase A showed a very broad pH–activity curve (pH 5–11). We cloned the gene of xylanase A in *E. coli* by using pBR322 (Honda *et al.*, 1985a).

Construction of pCX311

The *E. coli* K-12 strain HB101 and the plasmid pBR322 were used throughout these experiments. DNAs were digested with *Hind*III at 37 °C for 1 hour (plasmid pBR322 DNA) or for 4 hours (alkalophilic *Bacillus* sp. strain C-125 DNA). After the digestion, 1 µg of plasmid and 3 µg of bacterial chromosomal DNA were mixed and ligated with T4 ligase overnight at room temperature. This ligated DNA mixture was used for the transformation of *E. coli* HB101.

The xylanase activity was detected directly on the plates, because a clear zone was formed around a colony producing xylanase on an LB agar plate containing 0.5% xylan. Three transformants out of 8000 (Amp^r,Tc^s) transformants produced clear zones around colonies. In addition, a plasmid pCX311 was obtained from the transformants (Ap^r, xylanase$^+$) which contained two *Hind*III fragments (2.6×10^3 and 2.0×10^3 bp). The restriction map of the plasmid pCX311 is shown in Fig. 4. Two *Hind*III fragments were inserted in an *Hind*III site of pBR322. To analyse which fragment is

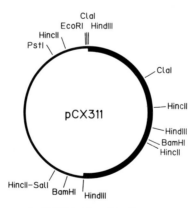

Fig. 4.   Restriction map of pCX311. The thin line represents pBR322 and the thick line is the xylanase gene cloned from alkalophilic *Bacillus* sp. strain C-125.

necessary for the expression of xylanase, two $Hind$III fragments $(2.6 \times 10^3$ and 2.0×10^3 bp) were inserted separately into the $Hind$III site of pBR322. These two plasmids (pCX312 and pCX313) were introduced into $E.\ coli$ HB101 and xylanase activity was assayed. Neither pCX312 nor pCX313 exhibited xylanase activity in $E.\ coli$. These results indicate that two $Hind$III fragments are essential for the expression of xylanase gene.

By using the genomic hybridization method, we found that radioactively labelled pCX311 hybridized to the 2.6×10^3 and 2.0×10^3 bp $Hind$III DNA fragments from alkalophilic $Bacillus$ sp. strain C-125 on nitrocellulose filters.

Extracellular Xylanase from E. coli

$E.\ coli$ HB101 carrying pCX311 was aerobically grown in LB broth for 15 hours at 37 °C. The xylanase activities in extracellular, periplasmic and intracellular fractions were assayed. Of the total activity (0.57 U), 82% (0.47 U), 14% (0.08 U) and 3.5% (0.02 U) was observed in the extracellular, periplasmic and intracellular fractions respectively.

Xylanase produced by $E.\ coli$ HB101 carrying pCX311 in the culture fluid was purified in the same manner as the xylanases (A and N) of the alkalophilic $Bacillus$ sp. strain C-125 (Honda $et\ al.$, 1985a), and antiserum

Table 5. $Distribution\ of\ enzymes\ in$ E. coli $HB101\ carrying\ plasmids$

Plasmids	Enzymes	Activity[a] in the following fractions		
		Extracellular (%)	Periplasmic (%)	Cellular (%)
None	β-gal	0.00 (0)	0.00 (0)	0.24 (100)
	APase	0.00 (0)	0.33 (77)	1.10 (23)
	β-lactamase	0.00 (0)	0.00 (0)	0.00 (0)
pBR322	β-gal	0.00 (0)	0.00 (0)	0.19 (100)
	APase	0.06 (11)	0.38 (72)	0.09 (17)
	β-lactamase	2.30 (13)	15.30 (86)	0.20 (1)
pCX311	β-gal	0.03 (16)	0.00 (0)	0.16 (84)
	APase	0.12 (52)	0.07 (30)	0.04 (17)
	β-lactamase	3.40 (67)	0.90 (18)	0.80 (16)
	Xylanase	0.44 (76)	0.06 (10)	0.08 (14)

$E.\ coli$ strains were aerobically grown in LB medium for 12 hours at 37 °C.

[a] Enzymatic activities of β-galactosidase (β-gal) and alkaline phosphatase (APase) are expressed as absorbance at 420 nm. Enzymatic activities of β-lactamase and xylanase are expressed as units per millilitre of broth.

against this xylanase was prepared. The results of Ouchterlony double-diffusion tests between three purified enzymes showed that the pCX311-borne xylanase immunologically crossed with xylanase A. Other properties were as follows.

(1) The molecular weight of pCX311-encoded enzymes by SDS–polyacrylamide gel electrophoresis was about 43 000, which is the same as that of xylanase A.
(2) The effect of pH on xylanase activity: no significant difference was observed between the xylanase of *E. coli* carrying pCX311 and xylanase A of alkalophilic *Bacillus* sp. strain C-125, both of which have sufficient enzyme activity at pH 12 (Fig. 5). Both xylanases were stable at 4 °C for 24 hours in the pH range 5–12 but were inactivated at pH 4 and pH 12.5.
(3) The effect of termperature: both xylanases were stable up to 50 °C and most active at 70 °C owing to the stabilization of the enzyme by the substrate.

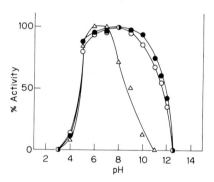

Fig. 5. Effect of pH on xylanase activity: ○——○, xylanase of *E. coli* HB101 carrying pCX311; ●——●, xylanase A of alkalophilic *Bacillus* sp. strain C-125; △——△, xylanase N of alkalophilic *Bacillus* sp. strain C-125.

Some xylanases produced by microorganisms have multienzyme systems. Studies on the homology of DNA fragments and immunological experiments suggest that the xylanases in alkalophilic *Bacillus* sp. strain C-125 are derived from different genes. It is noteworthy that not only xylanase but also β-lactamase were excreted into the culture broth as the pEAP plasmids exhibited in *E. coli*. No homology was observed between the active fragments of pEAP plasmids and pCX311 plasmid. It is therefore highly possible that different excretion processes are caused by the introduction of the plasmids.

Other Xylanases Cloned

A gene coding of semi-alkaline xylanase of alkalophilic *Aeromonas* strain No 212 (ATCC31085) was cloned in *E. coli* HB101 using pBR322 (Kudo *et al.*, 1985b). A plasmid pAX1 was isolated from a transformant producing xylanase only in the cells, and the xylanase gene was found in 4×10^3 bp of a *Bgl*II fragment. The plasmid-encoded xylanase had a molecular weight of 135 000, an optimum pH of about 8 and an optimum temperature of about 50 °C, which are the same as those of xylanase L, one of three xylanases produced by alkalophilic *Aeromonas* strain No 212, except for the molecular weight. This xylanase gene was inserted in a *Bam*HI site of pEAP2 and the plasmid-encoded xylanase was excreted from *E. coli* HB101 (see previous section).

Cloning of Alkaline Cellulase

The alkalophilic *Bacillus* sp. strain N-4 produced several cellulases (Horikoshi *et al.*, 1984). The enzymes had strong activity towards carboxymethyl cellulose (CMC) and very weak activity towards Avicel. We have started to clone the CMCase genes to analyse their genetic information on the multicomponents of the cellulase (Sashihara *et al.*, 1984).

Construction of pNK1 and pNK2

The chromosomal DNA alkalophilic *Bacillus* sp. strain No. N-4 which was cultured in medium II for 12 hours at 37 °C and plasmid pBR322 DNA were digested with *Hin*dIII restriction endonuclease and ligated with T4 DNA ligase. The recombinant DNAs were introduced to *E. coli* HB101, and transformants having *Hin*dIII fragments were selected on LB agar plates containing 20 µg of ampicillin. CMCase activity was detected on the plates when a shallow crater was formed around the colony producing CMCase. A plasmid pNK1 was obtained from the transformant (Ap^r,CMCase$^+$) which contained a 2.1×10^3 bp *Hin*dIII fragment. Another recombinant plasmid pNK2 was also isolated from a transformant (Ap^r, CMCase$^+$) containing a 2.8×10^3 bp *Hin*dIII fragment. Restriction maps of pNK1 and pNK2 are shown in Fig. 6. The homology between the plasmids and chromosomal DNA was analysed by the Southern hybridization method. The pNK1 and pNK2 hybridized to the 2.0×10^3 and 2.8×10^3 bp fragments from alkalophilic *Bacillus* No. N-4, and there was partial homology between pNK1 and

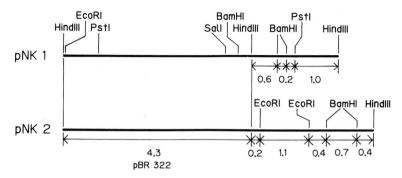

Fig. 6. Restriction maps of plasmids pNK1 and pNK2.

pNK2 fragments. However, no sequences complementary to pNK1 and pNK2 were detected in *E. coli* DNA fragments.

Expression and Localization of the CMCase in E. coli

The *E. coli* HB101 carrying pNK1 or pNK2 plasmids were aerobically grown in brain heart infusion broth (BH broth) for 24 hours at 37 °C. As shown in Table 6, most of the enzymatic activity was found in the periplasmic space. The synthesis of CMCase in *E. coli* was constitutive and no effect of CMC supplement was observed.

Table 6. *Distribution of CMCases in* E. coli *HB101[a] carrying plasmids*

Plasmid	CMCase activity[b] in the following fractions		
	Extracellular (%)	Periplasmic (%)	Cellular (%)
pNK1	50 (4.4)	850 (73.9)	250 (21.7)
pNK2	90 (13.9)	240 (36.9)	320 (49.2)
pBR322	0	0	0

[a] *E. coli* strains were aerobically grown in the BH infusion (Difco) for 24 hours at 37 °C.
[b] CMCase activity (at pH 10.0) is expressed as milliunits per millilitre of the broth.

Properties of Plasmid-encoded CMCase

The periplasmic CMCases of *E. coli* HB101 carrying pNK1 and pNK2 plasmids were studied using an antiserum prepared against the alkalophilic *Bacillus* strain No. N-4 crude CMCase fraction. The extract of *E. coli* HB101 carrying pNK plasmids gave lines of precipitation that fused with that for CMCase from alkalophilic *Bacillus* strain No. N-4. No reaction was observed with an extract of *E. coli* HB101. The plasmid-encoded CMCases were also analysed by immunological blotting. A 58×10^3 dalton protein was made in *E. coli* HB101 carrying pNK1 and a 52×10^3 dalton protein by pNK2. Their molecular weights were confirmed by gel filtration (Sephacryl S-200). Other enzymatic properties were as follows.

(a) The effects of pH on the stability and the activity of the enzyme were studied. The stability of the enzyme was investigated in buffer solutions of various pH values. The mixture was incubated at 60 °C for 10 minutes. The enzymes were stable from pH 5 to 10 without substrate. The optimum pH values for enzyme action were between 5.5 and 10.0.

(b) To investigate thermal stability, the enzyme was heated at various temperatures for 10 minutes and the residual activities were measured at pH 5.5 and 10.0. The enzyme was stable up to 75 °C.

Table 7. *Effect of pH on the activity of CMCase*

pH[a]	CMCase activity[b] (%)		
	pNK1	pNK2	Crude enzyme from N-4
4.4	18.4	30.3	27.8
5.0	81.0	82.4	98.1
7.5	96.7	79.7	86.4
8.0	106.8	89.1	103.3
9.2	102.3	87.9	92.3
10.1	100.0	100.0	100.0
10.9	107.7	100.0	119.9
11.7	36.8	62.1	54.7
12.8	7.4	10.6	14.7

[a] Acetate buffers were used for the assays at pH 4.4 and 5.0. Tris-HCl buffers were used for the assays at pH 7.5 and 8.0. Glycine buffers were used for the assays at pH 9.2, 10.1, 10.9, 11.7 and 12.8.

[b] The enzymes used were 10 mU and the enzyme activity at pH 10.1 is expressed as 100%.

Many cellulases produced by microorganisms are multienzyme systems. Attempts have been made to analyse the multicomponents of these enzymes. Two possibilities have been considered: (1) processing or modification during production occurs with proteases or other enzyme systems; (2) the DNA contains genes for multienzymes. On the DNA level, the pNK1 fragment was different from pNK2 in the restriction map, although there was partial homology between these fragments. These results strongly suggest that the cellulase gene is present in duplicate on another *Hin*dIII fragment. One interesting result is that the plasmid-encoded cellulases have very broad pH–activity curves, as was observed in the enzymes of alkalophilic *Bacillus* strain No. N-4. The plasmid-encoded enzymes cannot be a mixture of two or three enzymes, because the 2.8×10^3 or 2.0×10^3 bp DNA fragments are too small to encode multienzymes. As a result, we concluded that two of the cellulases produced by alkalophilic *Bacillus* strain No. N-4 are synthesized by genetically different genes, but that there is partial homology between these genes.

Isolation of Developmentally Regulated Promoter

Almost all extracellular enzymes begin to be synthesized at the end of the logarithmic phase and are fully produced at the stationary phase. In the case of alkaline cyclodextrin glycosyltransferase production, the alkalophilic *Bacillus* strain No. 38–2 must be grown for at least 48–72 hours. Why are these extracellular enzymes made at particular times? To answer this question, an alkalophilic *Bacillus* DNA bank cloned in the expression-probe plasmid pGR71 has been screened for the presence of developmentally regulated genetic elements (Kudo *et al.*, 1984). Competent *B. subtilis* 1012 cells were transformed with the alkalophilic *Bacillus* strain No. 38-2 DNA bank carried in the expression-probe plasmid pGR71. After 90 minutes of incubation to allow expression of the pGR71 kanamycin resistance, the cells were plated onto LB plates containing kanamycin ($20 \, \mu g \, ml^{-1}$). To analyse the chloramphenicol acetyltransferase (CAT) activity profile, *B. subtilis* carrying the pGR71 derivatives were cultivated in the 2xSSG sporulation medium containing $10 \, \mu g \, ml^{-1}$ kanamycin. The culture was sampled at various stages of growth and sporulation, and the cell extracts were prepared by sonication for CAT enzyme activity. As shown in Fig. 7, the CAT activity profile of pGR71-5 indicated that CAT expression from the plasmid was induced after the cessation of vegetative growth and that CAT specific activity increased until late in sporulation. The expression of CAT activity of pGR71-5 was strongly suppressed by the addition of glucose in the 2xSSG medium. This expression was altered in *spo0E* and *spo0H* mutants which had very low levels of the CAT activity. We determined the nucleotide sequence of the entire 508 bp fragment and located the site of regulated transcription initiation by high resolution S1 nuclease mapping of the *in vivo* transcript. We deduced that the promoter sequences upstream from this start site were 5′C–G–A–A–T–C–A–T–G–A3′ at -10 and 5′A–G–G–A–A–T–C3′ at -35. This transcript was detected in neither *spo0E* nor *spo0H* mutants, indicating that these gene products control the developmentally regulated CAT expression at the level of transcription.

In conclusion, the *Hin*dIII fragment of pGR71-5 was developmentally

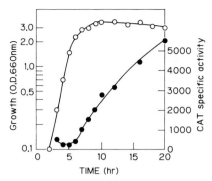

Fig. 7. The CAT activity as a function of growth and sporulation in *B. subtilis* 1012 carrying pGR71-5 plasmid: ○, cell growth; ●, CAT specific activity (nanomoles of 5,5′-dithiobis-(2-nitrobenzoic acid) reduced per minute per milligram of total soluble protein).

regulated in the sporulation medium of *B. subtilis* and was controlled by the *spo0E* and *spo0H* loci. We believe that it is significant that the promoter obtained from alkalophilic *Bacillus* resembled the promoters recognized by E-sigma37 of *B. subtilis* and that this promoter was under glucose and *spo0* control in a heterologous system.

New Host–Vector Systems

Alkalophilic *Bacillus* strains have several unique properties (Horikoshi and Akiba, 1982). The intracellular pH values of the cells are around 7–8, although pH values of the culture media are 9.5–11.0. It is quite possible that the pH difference is caused by the cell envelope. Analysis of the cell envelope is much too complicated for us to be sure of this. If suitable host–vector systems of alkalophilic *Bacillus* strains are established, more information on alkalophilicity will be obtained. Several physiological properties of these bacteria have been studied, but no information on plasmids in alkalophilic microorganisms has been reported. Plasmids of alkalophilic bacteria are required to make new host–vector systems for industrial applications.

Plasmid screening was performed using a rapid-isolation method with about 200 *Bacillus* strains from our stock culture collection of alkalophilic bacteria (Usami *et al.*, 1983). Plasmids were found in two strains of alkalophilic *Bacillus* No. H331 and No. 13.

In *Bacillus* strain No. H331 two plasmids, pAB3311 (16×10^3 bp) and pAB3312 (6×10^3 bp), were observed in an electron micrograph. The molecular size of pAB13 in *Bacillus* strain No. 13 was estimated to be

7.9×10^3 bp from the molecular size of the fragments produced by restriction endonucleases and from the electron micrograph. The numbers of the digestion sites in pAB13 for *Hinc*II, *Eco*RI, *Hind*III, *Pst*I and *Sal*I were 4, 2, 2, 2 and 1 respectively. Sizes of fragments after treatment with restriction endonucleases were as follows (Fig. 8): 4150 and 3750 bp for *Eco*RI; 4350 and 3550 bp for *Hind*III; 6000 and 1900 bp for *Pst*I; 5450, 1700, 480 and 270 bp for *Hinc*II.

Although these plasmids are still cryptic, the estimated molecular sizes, stabilities and restriction maps indicate that pAB13 is most suitable for use as a vector. Consequently, we have been trying to insert a suitable marker into pAB13. Many antibiotics are very unstable in alkaline media except chloramphenicol and decompose after 24 hours of incubation at 37 °C. Therefore other markers such as nutrient markers are expected to be used. Further work is in progress.

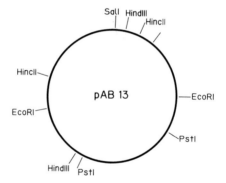

Fig. 8.   Restriction map of pAB13.

Acknowledgement

The author express his thanks to Mr Peter Ingham of the University of Kanazawa for his critical reading of the manuscript.

References

Honda, H., Kudo, T. and Horikoshi, K. (1985a). Molecular cloning and expression of xylanase gene of alkalophilic *Bacillus* sp. strain C-125 in *Escherichia coli*. *Journal of Bacteriology* **161**, 784–785.
Honda, H., Kudo, T. and Horikoshi, K. (1985b). Two types of xylanases of alkalophilic *Bacillus* sp. No. C-125. *Canadian Journal of Microbiology* **31**, 538–542.

Horikoshi, K. and Akiba, T. (1982). "Alkalophilic microorganisms". Springer-Verlag, Heidelberg.

Horikoshi, K., Nakao, M., Kurono, Y. and Sashihara, N. (1984). Cellulases of an alkalophilic *Bacillus* strain isolated from soil. *Canadian Journal of Microbiology* **30**, 774–779.

Kato, C., Kudo, T., Watanabe, K. and Horikoshi, K. (1983). Extracellular production of *Bacillus* penicillinase by *Escherichia coli* carrying pEAP2. *European Journal of Applied Microbiology and Biotechnology* **18**, 339–343.

Kato, C., Kudo, T. and Korikoshi, K. (1984). Gene expression and production of *Bacillus* No. 170 penicillinase in *Escherichia coli* and *Bacillus subtilis*. *Agricultural and Biological Chemistry* **48**, 397–401.

Kudo, T., Kato, C. and Horikoshi, K. (1983). Excretion of the penicillinase of an alkalophilic *Bacillus* sp. through the *Escherichia coli* outer membrane. *Journal of Bacteriology* **156**, 949–951.

Kudo, T., Yoshitake, J., Kato, C., Usami, R. and Horikoshi, K. (1985a). Cloning of developmentally regulated element from alkalophilic *Bacillus* DNA. *Journal of Bacteriology* **161**, 158–163.

Kudo, T., Ohkoshi, A. and Horikoshi, K. (1985b). Molecular Cloning and Expression of Xylanase Gene of Alkalophilic *Aeromonas* sp. No. 212 in *Escherichia coli*. *Journal of General Microbiology* **131**, 2825–2830.

Saito, H. and Miura, K. (1963). Preparation of transforming DNA by phenol treatment. *Biochimica Biophysica Acta* **72**, 619–629.

Sashihara, N., Kudo, T. and Horikoshi, K. (1984). Molecular cloning and expression of cellulase genes of alkalophilic *Bacillus* sp. strain N-4 in *Escherichia coli*. *Journal of Bacteriology* **158**, 503–506.

Sunaga, T., Akiba, T. and Horikoshi, K. (1976). Production of penicillinase by an alkalophilic *Bacillus*. *Agricultural and Biological Chemistry* **40**, 1363–1367.

Usami, R., Honda, H., Kudo, T. and Horikoshi, K. (1983). Plasmids in alkalophilic *Bacillus* sp. *Agricultural and Biological Chemistry* **47**, 2101–2102.

Index

Special Publications of the Society for General Microbiology

Publications Officer: Colin Ratledge, 62 London Road, Reading, UK

1. Coryneform Bacteria,
   eds I. J. Bousfield and A. G. Callely

2. Adhesion of Microorganisms to Surfaces,
   eds D. C. Ellwood, J. Melling and P. Rutter

3. Microbial Polysaccharides and Polysaccharases,
   eds R. C. W. Berkeley, G. W. Gooday and D. C. Ellwood

4. The Aerobic Endospore-forming Bacteria:
   Classification and Identification,
   eds R. C. W. Berkeley and M. Goodfellow

5. Mixed Culture Fermentations,
   eds M. E. Bushell and J. H. Slater

6. Bioactive Microbial Products: Search and Discovery,
   eds J. D. Bu'Lock, L. J. Nisbet and D. J. Winstanley

7. Sediment Microbiology,
   eds D. B. Nedwell and C. M. Brown

8. Sourcebook of Experiments for the Teaching of Microbiology,
   eds S. B. Primrose and A. C. Wardlaw

9. Microbial Diseases of Fish,
   ed. R. J. Roberts

10. Bioactive Microbial Products, Volume 2,
    eds L. J. Nisbet and D. J. Winstanley

11. Aspects of Microbial Metabolism and Ecology,
    ed. G. A. Codd

12. Vectors in Virus Biology,
    eds M. A. Mayo and K. A. Harrap

13. The Virulence of *Escherichia coli*,
    ed. M. Sussman

14. Microbial Gas Metabolism,
    eds R. K. Poole and C. S. Dow

15. Computer-Assisted Bacterial Systematics,
    eds M. Goodfellow, D. Jones and F. G. Priest

16. Bacteria in Their Natural Environments,
    eds M. Fletcher and G. D. Floodgate

17. Microbes in Extreme Environments,
    eds R. A. Herbert and G. A. Codd

In preparation
18. Bioactive Microbial Products, Volume 3,
    eds J. D. Stowell, P. J. Bailey and D. J. Winstanley